GOLD DISTRICTS OF CALIFORNIA

By William B. Clark

Geologist, California Division of Mines & Geology,
Sacramento, California

BULLETIN 193
California Division of Mines and Geology
Sacramento, California 95814

The Fricot Nugget. This 201-ounce (troy) cluster of gold crystals is on display in the Division of Mines and Geology mineral exhibit in San Francisco's Ferry Building. Melted down as gold, it would be worth some seven or eight thousand dollars, though its value as a historical object and museum piece is much more. The nugget is shown here at slightly less than half its actual size. *Photo by Mary Hill.*

STATE OF CALIFORNIA
Ronald Reagan, Governor

THE RESOURCES AGENCY
Norman B. Livermore, Jr., Secretary

DEPARTMENT OF CONSERVATION
James G. Stearns, Director

DIVISION OF MINES AND GEOLOGY
Ian Campbell, State Geologist

BULLETIN 193

Manuscript submitted for publication 1963
Some revisions through 1969
SIXTH PRINTING 1992

FOREWORD

This bulletin is an overall guide to the gold deposits in California. Although a vast number of publications have been written on gold and gold mining in California, there is no single report or treatise on all of the known gold-bearing districts in the state. A number of very excellent reports have been written on the gold deposits of certain districts or certain types of deposits within the state, mostly in the Sierra Nevada. Some of these reports are classics now. Among them are J. D. Whitney's 1875 survey of the auriferous gravels of the Sierra Nevada, Lindgren's 1911 professional paper on the Tertiary channels of the Sierra Nevada and the geologic folios of the U. S. Geological Survey by Lindgren (1890s), Turner (1890s), Diller (1900s), and Ransome (1900s). Also of considerable importance are reports on the Mother Lode belt by Knopf and Logan, and reports on the Grass Valley, Alleghany, and Randsburg districts by Johnston, Ferguson and Gannett, and Hulin, respectively.

In this bulletin the principal features of each gold-bearing district are described. The longer district descriptions contain sections on the location and extent, history, geology and character of the ore deposits, a list of mines, and a bibliography. Production figures are given whenever possible. Unfortunately, there is scant information available on many important gold mines in the state.

The first mention of gold in California was in *Las Sergus de Esplandian*, a romance published in Spain in 1510, in which "California" was believed to have been a great island north of Mexico where gold and precious stones were abundant. Richard Hakluyt expressed a similar opinion in his *The Principall Navigations Voiages and Discoveries of the English Nation*, published in London in 1589. Hakluyt, in his account of Sir Francis Drake's voyage and 1579 visit to California, stated, "There is no part of the earth here to be taken up wherein there is not a reasonable quantitie of gold and silver". Gold was mined in southern California in the latter part of the 18th and early part of the 19th Centuries under Spanish and Mexican rule, but little has been written on these operations. Soon after the beginning of the gold rush in 1848, many publications were written on various phases of gold mining.

The reports of John Trask, the first State Geologist in 1853–56, described a few important mines. From 1867 to 1876, the U. S. Commissioner of Mineral Statistics prepared reports of mine production and gold-mining activity. The California Mining Bureau, now the California Division of Mines and Geology, was established in 1880, and since then has published a fairly continuous record of gold-mining operations; the later ones appeared in the California Journal of Mines and Geology and the County Report series. The Division has published also a number of bulletins on certain phases of gold mining and reports on various districts and regions. The most popular recent publications are Bulletin 141, *Geologic Guidebook along Highway 49—Sierran Gold Belt*, The Mother Lode Country, in 1948; and *The Elephant as They Saw It*, an historical treatise, in 1949.

While collecting data for this bulletin, the author became greatly impressed with the vast amount of valuable information that has been amassed by the technical staff of the California Division of Mines and Geology and its predecessor, the California State Mining Bureau. Much of this work was done when this agency had a very small staff and limited funds and when many of the mining districts were accessible only by primitive roads or trails. He would like to pay tribute to a number of former staff members who helped make this bulletin possible. These men worked for this organization during much or all of their professional careers. They were Charles V. Averill, Walter W. Bradley, Fletcher Hamilton, Olaf P. Jenkins, C. McK Laizure, Clarence A. Logan, J. C. O'Brien, Reid J. Sampson, W. H. Storms, W. Burling Tucker, Clarence A. Waring, and Charles G. Yale. In 1969 only Jenkins, O'Brien, and Logan were living.

A number of other geologists and mining engineers prepared reports on gold mining or gold districts that were published by this agency. These were E. S. Boalich, Stephen Bowers, Ross E. Browne, Henry DeGroot, J. E. Doolittle, R. L. Dunn, A. S. Eakle, H. W. Fairbanks, W. A. Goodyear, C. S. Haley, John Hays Hammond, Paul Henshaw, J. B. Hobson, Emile Huguenin, C. D. Hulin, Charles Janin, Errol MacBoyle, F. J. H. Merrill, E. B. Preston, and W. B. Winston.

Tribute also is paid to the U. S. Geological Survey, especially to the three men of that organization who did a vast amount of pioneer work in the Mother Lode region of the Sierra Nevada: Waldemar Lindgren, F. L. Ransome, and H. W. Turner. Others of the Geological Survey who have contributed to knowledge of California's gold deposits have been John Albers, Josiah S. Diller, H. G. Ferguson, D. F. Hewitt, J. M. Hill, W. D. Johnston, Jr., Adolph Knopf, S. C. Creasy, R. W. Gannett, W. Yeend, and L. Noble.

During the preparation of this bulletin, the author visited nearly all the districts. At some, only a general reconnaissance was made, but at others all of the important mines were visited. Little detailed geologic mapping was done, but efforts were made in a number of districts to determine the nature and extent of the mineralized zones and vein systems. The publications and files of the California Division of Mines and Geology and the United States Geological Survey, including the folios of the Geologic Atlas of the United States, were important sources of information. The U. S. Geological Survey Bulletin 507, *The Mining Districts of the Western United States* (Hill, 1912), was a special source of data. Information on the earlier history of individual districts is found in the reports of the Commissioner of Mining Statistics of the U. S. Treasury Department (Browne, 1868, and Raymond, 1869–76). Other publications that were consulted included the Mining and Scientific Press, Engineering and Mining Journal, U. S. Bureau of Mines reports and records, and private reports. Some county records were examined. The author was assisted by the following persons who reviewed chapters on certain districts: John Albers, C. A. Bennett, O. E. Bowen, Clarence Carlson, F. F. Davis, Willard Fuller, Earl Hart, Paul Morton, B. W. Troxel, F. H. Weber, and John Wells. Credit is also given to the large number of mine owners and operators, mining engineers, and miners with whom the author became acquainted. The maps and other drawings were drafted by Hugo H. Hawkins, of the Division of Mines and Geology drafting section.

CONTENTS

CONTENTS—Continued

CONTENTS—Continued

CONTENTS—Continued

ABSTRACT

California has been the source of more than 106 million troy ounces * of gold, the most productive state in the Union. However, production has greatly declined in recent years because of high costs and depletion of easily accessible deposits.

Although gold was mined in California in the late 18th and early 19th Centuries, the gold rush did not begin until after Marshall's discovery at Sutter's Mill in 1848. Thousands of gold seekers soon arrived, and in a few years much of the state was permanently settled. Gold production attained an all-time high of $81 million in 1852 but then declined because of the exhaustion of the rich surface placers. At the last government-set price of $35 per ounce, the 1852 amount would have been about $138 million.

Hydraulic mines became the largest sources of gold until curtailed by court order in 1884. Lode mines and dredges were the principal sources after that date. During the depression years of the 1930s, gold output in the state was nearly as high as it had been during the gold rush. Gold mining was curtailed during World War II and has not recovered since.

A number of spectacular nuggets and masses of pure gold were recovered in California during the early days. The most famous were the 195-pound mass of gold from Carson Hill and the 54-pound Willard nugget from Magalia. Small high-grade ore shoots or pockets have been found in many districts, but the richest and most numerous have been in the Alleghany district of Sierra County.

Although gold is found in many areas in California, the most productive districts are in the northern and central portions of the Sierra Nevada. The primary deposits usually consist of gold-quartz veins in metamorphic rocks and are associated with the intrusion of the Sierra Nevada batholith. The most productive lode-gold districts in the Sierra Nevada have been the Grass Valley, Nevada City, Alleghany, and Sierra City districts, those of the Mother Lode belt, and several in the so-called East and West Gold Belts. Several districts are in the southern end of the range. The Sierra Nevada placer deposits are divisible into the older or Tertiary deposits, which were mined by hydraulicking and drifting, and the younger or Quaternary stream deposits, which have been mined by dredging. The principal Tertiary deposits are in the La Porte, Poker Flat, Magalia, Cherokee, North Bloomfield, North Columbia, Dutch Flat, Damascus, Forest Hill, Iowa Hill, Mokelumne Hill, and Columbia districts. The largest dredging fields were at Hammonton, Folsom, Oroville, Camanche, La Grange, and Snelling.

In the Klamath Mountain, the second most-productive province, the largest sources of gold have been the streams of the Klamath-Trinity River system. The older terrace deposits along the sides of the present stream channels also have yielded much gold and were mined by hydraulicking. The most productive source of lode-gold has been the French Gulch district of Shasta and Trinity Counties. Other important lode-mining centers were the Harrison Gulch, Liberty, Callahan, Sawyers Bar, Weaverville, and Old Diggings districts.

The Basin Ranges and Mojave Desert provinces of eastern and southern California have yielded substantial amounts of gold. The gold occurs either in epithermal deposits in brecciated silicified zones of Tertiary volcanic rocks or in mesothermal quartz veins of older metamorphic and granitic rocks. Gold also has been recovered from dry

---

* More than 8.8 million troy pounds, 7.2 million avoirdupois pounds, 3630 tons. Conversion factors: one troy ounce = about 1.1 av. ounce, but one troy pound = about .8 av. pound.

placers in several districts. The Bodie district has been the most important gold source in the Basin Ranges, while the Mojave, Rand, Stedman, and Cargo Muchacho districts contain the most productive mines in the Mojave Desert.

Moderate amounts of gold have been mined in the Transverse and Peninsular Ranges in southern California, the principal sources having been the Frazier Mountain, Saugus, Acton, Pinacate, Julian-Banner, and Cuyamaca districts. The mineral also has been recovered from the Modoc Plateau province in northeastern California, the main source having been the Hayden Hill district. Small amounts of gold have been produced in a number of places in the Coast Ranges.

Photo 1. Early Gold-Mining Scene. One miner operates a horse-powered arrastra, a second pans ore and a third works with a rocker.

GOLD DISTRICTS OF CALIFORNIA

By WILLIAM B. CLARK

INTRODUCTION

Between 1848 and 1967, California was the source of more than 106 million troy ounces of gold. This total was far greater than that for any other state in the Union and represented about 35 percent of the total United States production.

California's gold mining has been important in the history and development of the western United States. The influence it has had on the development and perfection of mining and metallurgical processes also has been significant. Although world gold production has

Photo 2. Winnowing Gold Near Chinese Camp.

gradually increased in recent years, chiefly because of increased output in the Union of South Africa and the Soviet Union, United States production, particularly that in California, has diminished. This diminishing trend is attributable to increased costs for labor and supplies combined with, until recently, a fixed price for domestically mined gold ($35 per fine ounce), the expense of reconditioning mines shut down during World War II, and depletion of many gold deposits. Another factor in California is the increased real estate value of many gold-bearing properties. In addition, a number of gold mines and gold-bearing deposits have been inundated by reservoirs.

The word "district" as used in this publication denotes an area or zone of gold mineralization. The location and extent of these districts are determined by the occurrence of deposits that have yielded gold in commercial amounts. Often the limits of the individual districts are not well defined, because the boundaries between rocks that have yielded commercial ore and those that have not are indefinite. Except in portions of the desert regions, the limits of the named mining districts in California often are uncertain. Commonly, what has been referred to as an organized mining district actually has been nothing more than a center of mining operations with an appropriate geographic name. The names used in this report are either those of the corresponding organized districts or geographic names long used to designate centers of mining operations. If a district has had several names, the most common name is used in this report. A list of alternate names appears in an index at the back of the book.

The size and productivity of the gold districts of California vary widely. Some are scores of square miles in extent and others cover only a few square miles. However, size often is no indication of the richness of the deposits or of the total value of output. Some lode-gold districts, such as Grass Valley, Alleghany, and Randsburg, contain many rich veins in a small area. Some placer-gold districts contain channel deposits of several different ages, and others contain deposits only of one age. Some districts are mostly lode, some mostly placer, and others have both lode and placer deposits. In the Sierra Nevada and Klamath Mountains, the gold mineralization is extensive. However, studies show that the bulk of the gold production has come from distinct districts within these major regions.

The organized mining districts were important during the days of the frontier. These were organized by the miners themselves to establish law and order. The miners would meet to draft bylaws defining the size of claims and territorial jurisdiction of the district. Commonly these laws included procedures for the punishment of claim jumpers, sluice robbers, and murderers. A recorder was appointed to keep records. The customs and laws were derived chiefly from European mining districts. The importance of the organized mining districts diminshed after the Federal Mining Acts of 1866 and 1872. The official records of some of the old districts still exist and are on file in county recorders' offices.

HISTORY OF GOLD MINING IN CALIFORNIA

California's gold-mining history is a brilliant lure, and many books, pamphlets, periodicals and articles have been published on the subject. The old mining districts and settlements, including "ghost" towns, are visited by increasing numbers of tourists each year. In a few districts the old camps have been reconstructed. Several old gold mining towns, such as Columbia, Johnsville, Coloma, Shasta, and Bodie, are California state parks or recreation areas. In recent years more people have become aware of the importance of California's gold rush in the history and development of the western United States, and steps have been made to preserve historical structures and equipment closely associated with gold mining.

Unfortunately, little visible evidence remains of many of California's important gold-quartz mines other than caved shafts and tunnels and heavily overgrown dumps. The surface plants of the large underground lode mines at Grass Valley and along the Mother Lode belt, which for years accounted for a major part of California's gold output, have been almost completely dismantled. More evidence remains of the large-scale placer-mining operations. The old hydraulic mine pits and the extensive tailing piles in the dredging fields still exist; some are used as commercial sources of sand and gravel. A number of the old ditches, flumes, and reservoirs that once supplied water to the hydraulic mines now are parts of hydroelectric and irrigation systems.

Photo 3. Dry-Washing Gold.

Table 1. Gold Production in California, 1848–1968.

Year	Fine Ounces	Value	Year	Fine Ounces	Value
1848	11,866	$ 245,301	1907	809,214	$16,727,928
1849	491,072	10,151,360	1908	907,590	18,761,559
1850	1,996,586	41,273,106	1909	979,007	20,237,870
1851	3,673,512	75,938,232	1910	953,734	19,715,440
1852	3,932,631	81,294,700	1911	954,870	19,738,908
1853	3,270,803	67,613,487	1912	953,640	19,713,478
1854	3,358,867	69,433,931	1913	987,187	20,406,958
1855	2,684,106	55,485,395	1914	999,113	20,653,496
1856	2,782,018	57,509,411	1915	1,085,646	22,442,296
1857	2,110,513	43,628,172	1916	1,035,745	21,410,741
1858	2,253,846	46,591,140	1917	971,733	20,087,504
1859	2,217,829	45,846,599	1918	799,588	16,528,953
1860	2,133,104	44,095,163	1919	807,667	16,695,955
1861	2,026,187	41,884,995	1920	692,297	14,311,043
1862	1,879,595	38,854,668	1921	759,721	15,704,822
1863	1,136,897	23,501,736	1922	709,678	14,670,346
1864	1,164,455	24,071,423	1923	647,210	13,379,013
1865	867,405	17,930,858	1924	636,140	13,150,175
1866	828,367	17,123,867	1925	632,035	13,065,330
1867	883,591	18,265,452	1926	576,798	11,923,481
1868	849,265	17,555,867	1927	564,586	11,671,018
1869	881,830	18,229,044	1928	521,740	10,785,315
1870	844,537	17,458,133	1929	412,479	8,526,703
1871	845,493	17,477,885	1930	457,200	9,451,162
1872	748,951	15,482,194	1931	523,135	10,814,162
1873	726,554	15,019,210	1932	569,167	11,765,726
1874	835,186	17,264,836	1933	613,579	15,683,075
1875	816,377	16,876,009	1934	719,064	25,131,284
1876	755,169	15,610,723	1935	890,430	31,165,050
1877	798,249	16,501,268	1936	1,077,442	37,710,470
1878	911,343	18,839,141	1937	1,174,578	41,110,230
1879	949,439	19,626,654	1938	1,311,129	45,889,515
1880	968,986	20,030,761	1939	1,435,264	50,234,240
1881	929,920	19,223,155	1940	1,455,671	50,948,585
1882	829,458	17,146,416	1941	1,408,793	49,307,755
1883	1,176,329	24,316,873	1942	847,997	29,679,895
1884	657,900	13,600,000	1943	148,328	5,191,480
1885	612,478	12,661,044	1944	117,373	4,108,055
1886	711,911	14,716,506	1945	147,938	5,177,830
1887	657,349	13,588,614	1946	356,824	12,488,840
1888	616,000	12,750,000	1947	431,415	15,099,525
1889	542,425	11,212,913	1948	428,473	14,751,555
1890	595,486	12,309,793	1949	417,231	14,603,085
1891	615,759	12,728,869	1950	412,118	14,424,130
1892	608,166	12,571,900	1951	339,732	11,890,620
1893	606,564	12,538,780	1952	258,176	9,036,160
1894	670,636	13,863,282	1953	234,591	8,210,685
1895	741,798	15,334,317	1954	237,888	8,326,010
1896	831,158	17,181,562	1955	251,737	8,810,795
1897	767,779	15,871,401	1956	193,816	6,783,560
1898	769,476	15,906,478	1957	170,885	5,980,975
1899	741,881	15,336,031	1958	185,400	6,489,000
1900	767,390	15,863,355	1959	146,141	5,114,935
1901	821,845	16,989,044	1960	123,713	4,329,955
1902	818,037	16,910,320	1961	97,648	3,417,680
1903	788,544	16,300,653	1962	106,272	3,719,520
1904	901,484	18,633,676	1963	86,867	3,040,345
1905	914,217	18,898,545	1964	71,028	2,485,980
1906	906,182	18,732,452	1965	62,885	2,220,975

Table 1. Gold Production in California, 1848–1968—Continued

Year	Fine Ounces	Value	Year	Fine Ounces	Value
1966	64,764	$2,266,740	1968	15,682	$616,000
1967	40,570	1,420,000	1969 (est.)	7,950	335,000

Totals: 106,276,163 ounces valued at $2,428,330,901 through 1968.

Note: The price of gold was $20.67 a fine ounce until 1933, when it was increased to $25.56. The figure rose to $34.95 the following year and again, to $35, in 1935. On March 15, 1968, the U.S. Treasury suspended purchases, leaving miners free to sell their gold on the open market; domestic prices have since risen. Dollar amounts above for production since that date are based on the New York selling price.

Table 2. Significant Dates in the History of Gold Mining in California.

1775–80 The first known discovery of gold in California was made in the Potholes district, Imperial County. Mining extended into the Cargo Muchacho and Picacho districts.

1828 A small placer gold deposit was found at San Ysidro, San Diego County.

1835 The placer deposits in San Francisquito Canyon, Los Angeles County, were discovered.

1842 Gold was discovered in Placerita Canyon, Los Angeles County. Some sources give the date of this discovery as 1841.

1848 Gold was discovered at Sutter's Mill at Coloma on the American River by James Marshall. Although the exact date has been the subject of some discussion, it is officially designated as January 24. The first printed notice of the discovery was in the March 15 issue of "The Californian" in San Francisco. Shortly after Marshall's discovery General John Bidwell discovered gold in the Feather River and Major Pearson B. Reading found gold in the Trinity River. The gold rush was soon in full sway as thousands of gold seekers poured into California.

1849 Quartz mining began at the Mariposa mine, Mariposa County. A stamp mill, probably the first in the state, was installed.

1850 Gold-bearing quartz was found at Gold Hill at Grass Valley. This led to the development of the great underground mines in that district and a major industry that continued for more than 100 years.

1851 Gold was discovered in Greenhorn Creek, Kern County. This discovery led to the rush to the upper Kern River region.

1852 California's annual gold production reached an all-time high of $81 million.

1852 Hydraulic mining began at American Hill just north of Nevada City, Nevada County, and at Yankee Jims, Placer County.

1852 The first extensive underground mining of buried river channels commenced in the Forest Hill district, Placer County.

1853 The placers at Columbia, Tuolumne County, began to yield vast amounts of gold. This continued until the early 1860s. At that time Columbia was one of the largest cities in the state.

1853 The Fraser River rush in British Columbia caused a partial exodus of miners from the state.

1854 A 195-pound mass of gold, the largest known to have been discovered in California, was found at Carson Hill, Calaveras County.

Continued on p. 7

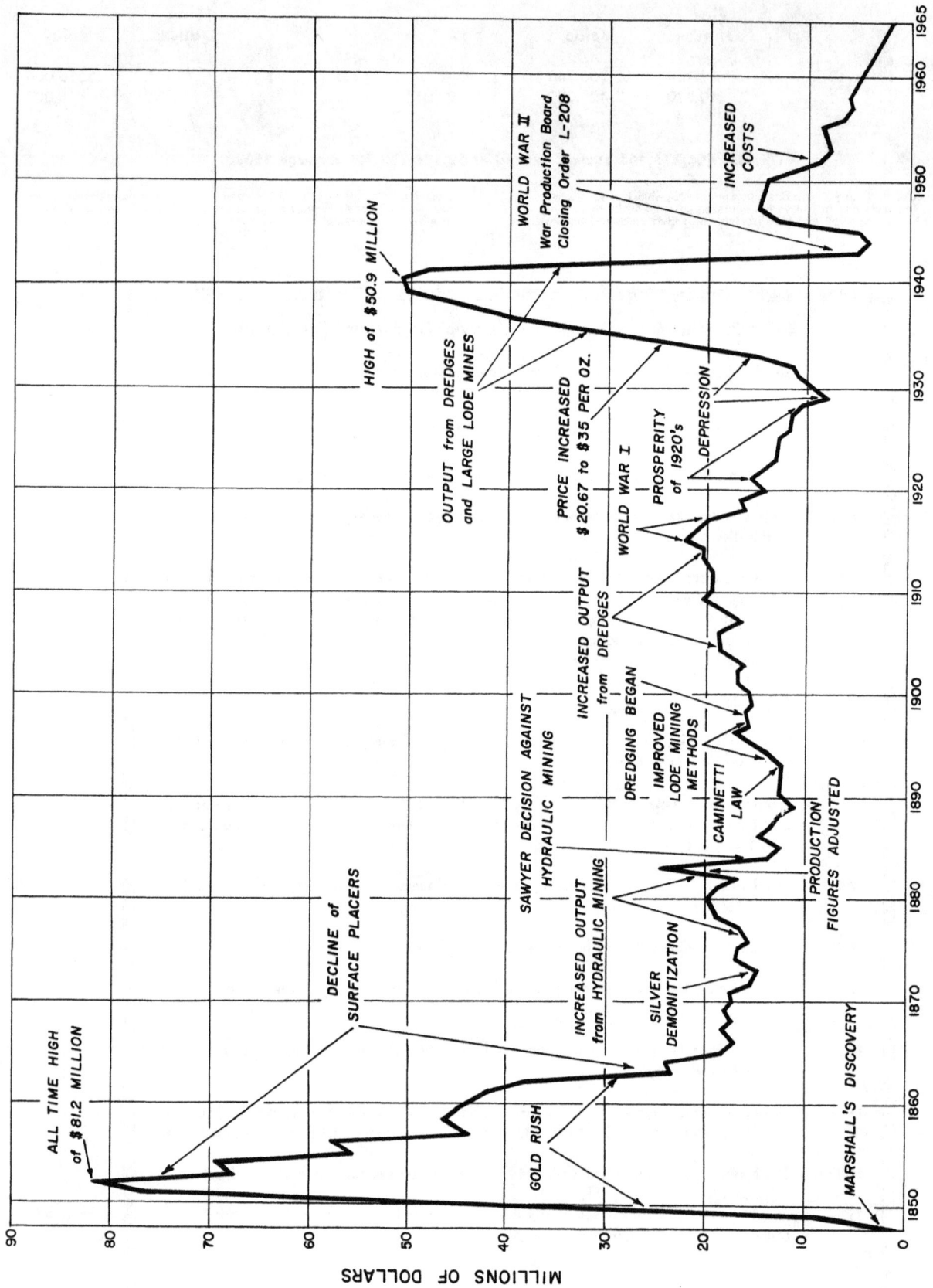

YEARS

Figure 1. Gold Production in California, 1848–1965.

MILLIONS OF DOLLARS

Table 2.   Significant Dates in the History of Gold Mining in California.—Continued

1855   The rich surface placers were largely exhausted by this date, and river mining accounted for much of the state's output until the early 1860s. All of the rivers in the gold regions were mined.

1859   The famous 54-pound Willard nugget was found at Magalia, Butte County.

1859   The Comstock silver rush began in Nevada. This development caused a large exodus of gold miners from California. However, it stimulated gold and silver prospecting in eastern and southeastern California.

1864   By this time California's gold rush had ended. The rich surface and river placers were largely exhausted; hydraulic mines were the chief sources of gold for the next 20 years.

1868   The first air drills were introduced. However, widespread use of air drills in mining did not come for another 30 years.

1876   The stampede to the Bodie district in Mono County began. This rush lasted until about 1888.

1880   Hydraulic mining reached its peak in the state. Vast systems of reservoirs, tunnels, ditches, flumes, and pipelines supplied water to these operations.

1883   Gold production figures began to be collected for the calendar year instead of the fiscal year.

1884   Sawyer Decision. In the case of Woodruff vs. North Bloomfield Gravel Mining Company, Judge Lorenzo Sawyer issued a decree prohibiting the dumping of debris into the Sacramento and San Joaquin Rivers and their tributaries. Action against other hydraulic mines soon followed. A few mines constructed tailings storage dams and continued to operate, but hydraulic mining has not been important in the Sierra Nevada since. For a few years drift mines partially made up for the loss in output of surface placer gold.

1890   Beginning about this time and continuing for several decades, great improvements were made in mining and milling methods. These changes enabled many more lode deposits, especially large but low-grade accumulations, to be profitably worked. The improvement of air drills, explosives, and pumps, and the introduction of electric power lowered mining costs greatly. The introduction of rock crushers, increase in size of stamp mills, and new concentrating devices, such as vanners, lowered milling costs. Cyanidation was introduced in 1896 and soon replaced the chlorination processes.

1893   The Caminetti Act was passed creating the California Debris Commission. This commission licenses hydraulic mining operations in the Sierra Nevada. It is empowered to assess such mines to build debris dams.

1893   Gold was discovered in Goler Gulch in the El Paso Mountains in eastern Kern County. This led to other discoveries in the area and the influx to the Rand district, which began in 1895.

1898   The first successful bucket-line dredge was started on the lower Feather River near Oroville. Gold dredging soon became a major industry that continued for more than 65 years.

1904   The lost, high-grade Tightner vein was rediscovered at Alleghany in Sierra County. Large amouts of rich ore were taken from this vein, and mining activity, reviving in this district, continued until 1965. This was the last district in the state where gold mining was the chief industry.

1916   The general prosperity that began during World War I and continued until 1929, with accompanying high costs, caused a decrease in gold output.

1922   Argonaut disaster. A fire on the 3350-foot level at the Argonaut mine in the Jackson district, Amador County, caused the loss of 47 lives.

Continued on p. 8

Table 2.   Significant Dates in the History of Gold Mining in California.—Continued

1929   Peak of post World War I boom. Lowest point in gold production since 1849.

1930   Gold production started to rise because of the depression and resulting low operating costs.

1933–35 The price of gold increased from $20.67 to $35 per fine ounce. This rise ultimately resulted in a large increase in gold output and in much greater exploration activities.

1940   Gold output totaled nearly $51 million. This was the most valuable annual output since 1856. Thousands of miners were employed in the quartz mines at Grass Valley, Alleghany, Nevada City, Jackson, Sutter Creek, Jamestown, Mojave, and French Gulch. There were many active bucket-line dredges, and dragline dredges became important producers of placer gold.

1942   World War II caused a precipitous drop in gold output. War Production Board Limitation Order L-208, issued on October 8, caused the gold mines to be shut down.

1944   Gold production touched the lowest point since 1848.

1945   Order L-208 was lifted, effective July 1. Some of the bucket-line dredges resumed operations, but only a few important lode mines at Grass Valley, Alleghany, and Sutter Creek were reopened. Production increased slightly for 4 years.

1950   Gold output resumed its decline because of rising costs and depletion of dredging gound. This trend was accelerated by the Korean War.

1953   The Central Eureka mine at Sutter Creek, the last major operating lode mine in the Mother Lode belt, was shut down.

1956   The mines of Empire-Star Mines Ltd., and Idaho-Maryland Mines, Inc., at Grass Valley were shut down. The industry of gold mining completed nearly 106 years of operation in this locality.

1960   Gold output fell below $5 million as the dredges continued to curtail operations.

1962   The last dredge of the Folsom field in Sacramento County was shut down, ending more than 60 years of operation. One of the last active lode-gold mines in California, the Sixteen-to-One in the Alleghany district, curtailed operations.

1963   The three large dredges of the Yuba Mining Division, Yuba Consolidated Industries—in the Hammonton district, Yuba County—were the only major sources of gold in the state. The small output from the substantial number of part-time prospectors, pocket miners, snipers, and skin divers did not offset the decrease in output from larger commercial operations. Several mines in the Alleghany district obtained U.S. Government exploration loans.

1964   The Brush Creek mine, a substantial source of gold in the Alleghany district, Sierra County, ceased operations.

1965   Governor Edmund G. Brown signed Senate Bill 265 designating gold as California's official state mineral. The Sixteen-to-One mine at Alleghany, Sierra County, was shut down at the end of the year. This was the last lode mine in the state that had been operated on a sustained basis.

1967   Two of the three remaining dredges at Hammonton were shut down.

1968   The last gold dredge at Hammonton was shut down on October 1. This was the last sustained commercial gold-mining operation in California.

1968   The U.S. Treasury suspended purchases of newly-mined gold. The free market price rose to $44 an ounce early in 1969, falling by November to $38.50, because of greater stability in international currencies.

Table 3.　Estimated Gold Production by Counties, 1848–1965.

County	Production (in millions)	County	Production (in millions)
Alpine	$4	Plumas	$105
Amador	200	Riverside	7
Butte	150	Sacramento	135
Calaveras	150	San Bernardino	20
Del Norte	2	San Diego	5
El Dorado	110	San Joaquin	5
Fresno	4	Shasta	60
Humboldt	5	Sierra	150
Imperial	12	Siskiyou	100
Inyo	14	Stanislaus	14
Kern	65	Trinity	75
Lassen	4	Tulare	1
Los Angeles	5	Tuolumne	190
Madera	6	Ventura	3
Mariposa	60	Yuba	145
Merced	17		
Mono	38		
Nevada	440		
Placer	120		

Other counties that have yielded some gold—all less than $1 million worth—are Colusa, Mendocino, Modoc, Monterey, Napa, Orange, San Francisco, San Luis Obispo, San Mateo, Santa Barbara, Santa Clara, Santa Cruz, Sonoma, Tehama, and Yolo Counties.

Famous Gold Nuggets

Many large and spectacular fragments of native gold have been found in California. Most of these were taken during the gold rush and contributed further to the excitement in the mining camps and "stampedes" to the various "diggings". Few of these nuggets exist today; most were melted down soon after they were discovered. Although the term "nugget" is technically restricted to water-worn gold fragments in alluvial deposits, it frequently is used to describe chunks of vein gold not far removed from the point of origin. It is not used to describe high-grade "pockets" or small but rich ore shoots—these are listed in the next section.

The largest piece of native gold that is believed to have been found in California was the 195-pound mass taken at Carson Hill in 1854. The largest true nugget was the Willard, Dogtown or Magalia nugget, which was found at Magalia in 1859. A celebration was held and the nugget was melted in Oroville soon afterward. It weighed 54 pounds troy. Replicas are owned by the Division of Mines and Geology and the Paradise Chamber of Commerce. An annual celebration is held in Magalia commemorating the discovery.

Other spectacular nuggets found in California were:
• The 50-pound slab from Knapp's Ranch, Tuolumne County.
• The 28-pound Holden Chispa nugget from Holden's Gardens in Sonora.
• A 28-pound nugget from Sullivan Creek, Tuolumne County.
• The 426- and 532-ounce nuggets from French Ravine, Sierra County.
• The gold-quartz boulder that held more than $8000 in gold from Pilot Hill, El Dorado County.
• A 360-ounce oblong smooth piece of native gold from Sullivan Creek, Tuolumne County.

• A 150-pound quartz-gold mass from Wood's Creek, Tuolumne County, that yielded 75 pounds of gold.
• A 52-pound mass of gold quartz from the Diltz mine, Whitlock district, Mariposa County.

To find a large nugget was not a blessing for all men. The Second Report of the State Mineralogist (California State Mining Bureau, 1882) tells of a French immigrant who took a piece worth more than $5,000 from Spring Gulch, in the Columbia district. The report relates (p. 149): "The discovery of this nugget proved to be a great misfortune, for the finder became insane the following day and was sent to Stockton. The French Consul recovered the nugget or the money obtained for it, and sent it to his family in France."

The first discovery of a spectacular gold specimen in California was in the summer of 1848, when a young soldier of Stevenson's Regiment found a 25-pound nugget on the banks of the Mokelumne River. Later that same year, General E. F. Beale took it to New York, where it caused much excitement. In 1865 a beautiful cluster of gold crystals weighing 201 ounces was found in the Grit mine at Spanish Dry Diggins, El Dorado County. It was sent to New York, where it was purchased by a Mr. Fricot, who had formerly lived in Grass Valley. Later this specimen was presented to the California Division of Mines and Geology by the Fricot family and is now displayed in the Division's mineral exhibit in the Ferry Building in San Francisco. A photograph of this specimen is the frontispiece of this bulletin.

The most famous nugget of all, that found by Marshall at Sutter's Mill in 1848 and which led to California's gold rush, weighed less than a quarter of an ounce. It is not known if this nugget still exists. A flake of gold which Captain Folsom sent to Washington in 1848 and described as Marshall's first piece is now in the Smithsonian Institution.

Table 4. Large Nuggets and Gold Masses From California.

Source	Date	Weight**
Carson Hill district	1854	195 pounds*
Wood's Creek, Sonora district	1848?	75 pounds*
Willard nugget, Magalia district	1859	54 pounds
Monumental mine, Sierra City district	1869	1893 ounces*
Monumental mine, Sierra City district	1860	1596 ounces*
Diltz mine, Whitlock district	1932	52 pounds*
Knapp's Ranch, Columbia district	1850s	50 pounds
French Ravine, Sierra County	1855	532 ounces
French Ravine, Sierra County	1851	426 ounces
Pilot Hill, El Dorado County	1867	426 ounces
Sullivan Creek, Columbia district	1849	408 ounces
Gold Hill, Columbia district	1850s	360 ounces
Holden Chispa nugget, Sonora district	1850s?	28 pounds
Mokelumne River, Amador County	1848	25 pounds
Downieville, Sierra County	1850	25 pounds
Polar Star claim, Dutch Flat district	1876	288 ounces*
Columbia district, Tuolumne County	1853	283 ounces*
Minnesota, Alleghany district	1850s?	266 ounces
Spring Gulch, Columbia district	1850s	250 ounces
Michigan Bluff, Placer County	1864	226 ounces
Fricot nugget, Spanish Dry Diggings (crystallized gold)	1865	201 ounces
Remington Hill, Nevada County	1855	186 ounces
Live Yankee claim, Alleghany district	1854–62	Twelve nuggets, 30–170 ounces
Smith's Flat, Sierra County	1864	140 ounces
Remington Hill, Nevada County	1869?	107 ounces
Little Grizzly Diggings, Sierra County	1869	107 ounces
Oregon claim, Alleghany district	1856–62	Several nuggets, 30–100 ounces
Hope claim, Alleghany district	unknown	94 ounces
Campo Seco, Calaveras County	1854	93 ounces
French Ravine, Sierra County	1860	93 ounces
Smith's Flat, Sierra County	1861	80 ounces
Lowell Hill, Nevada County	1865	58 ounces
Ruby mine, Alleghany district	1930s, 1940s	Several nuggets up to 52 ounces

* Mass of gold and quartz but mostly gold.

** In troy ounces or pounds.

Famous High-Grade Pockets

A considerable number of rich, small ore pockets or pocket shoots have been developed in mines in some lode-gold districts. Many of these pocket shoots were in districts commonly referred to as "high-grade" belts. The richest and most famous in California is the Alleghany district in Sierra County. Much of the output of this district has been from small but rich pockets. Other noted high-grade districts are the Sonora, West Point, Soulsbyville, Kinsley, Whitlock, Spanish Flat, and Kelsey-Garden Valley districts. A number of other lode-gold districts, such as the Grass Valley, Nevada City, Sierra City, French Gulch, Cargo Muchacho, Bodie and several Mother Lode districts, have yielded appreciable amounts of high-grade ore.

High-grade pockets usually occur in the veins and consist of a mixture of vein material and free gold. In some pockets, sulfide minerals are abundant, but in others they are absent or nearly so. In a few districts telluride or silver minerals are associated with the gold. Table 5 lists some of the famous high-grade pockets or pocket ore shoots discovered in California.

Table 5. High-Grade Pockets.

Location	Date	Value
Original 16-to-1 mine, Alleghany district	1920s	Nearly $2 million
Original 16-to-1 mine, Alleghany district	1920s	Nearly $1 million
Original 16-to-1 mine, Alleghany district	1930s	$750,000
Oriental mine, Alleghany district	pre-1890	$734,000
Alhambra mine, Spanish Flat district	1939	$550,000
Four Hills mine, Sierra City district	1860s?	$250,000 to $500,000
Tightner mine (now part of the 16-to-1 mine) Alleghany district	1912	+$375,000
Tightner mine (now part of the 16-to-1 mine) Alleghany district	1904	$375,000
Keltz mine, West Point district	1860s	+$300,000
Bonanza mine, Sonora district	1879	$300,000
Original 16-to-1 mine, Alleghany district	1920s to 1950s	Several that yielded about $200,000
Kate Hardy mine, Alleghany district	1948	About $200,000
Rainbow (now part of the 16-to-1 mine), Alleghany district	1881	$116,000
Carson Hill mine, Carson Hill district	1850s	$110,000
Oriental mine, Alleghany district	1930s	$100,000
North Fork mine, Alleghany district	1870s	$100,000
Kenton mine, Alleghany district	1930s	$100,000
Angels mine, Angels Camp district	1910	$100,000
Green Emigrant mine, Ophir district	1867	$100,000
Finnegan mine, Carson Hill district	1867	$80,000 to $100,000
Red Star mine (now part of the 16-to-1 mine) Alleghany district	1870s	$80,000
St. Patrick mine, Ophir district	1872	$75,000
Plumbago mine, Alleghany district	1920s	$60,000

DISTRIBUTION OF GOLD

California can be divided into 11 well-recognized natural divisions or geomorphic provinces. Each of these provinces has distinctive physiographic and geological features, and in each the distribution of economic mineral disposits, including gold, follows certain definite patterns associated with the major geological structures and rock types. The provinces are: 1) Klamath Mountains, 2) Cascade Range, 3) Modoc Plateau, 4) Coast Ranges, 5) Great Valley, 6) Sierra Nevada, 7) Basin Ranges, 8) Mojave Desert, 9) Transverse Ranges, 10) Peninsular Ranges, and 11) Colorado Desert. Figure 2 shows these provinces and the distribution of the gold-bearing areas. Tables 6 to 9 list the principal gold mining districts, lode mines, hydraulic mines, and drift mines. The figure and tables appear on p. 12–14.

Most of California's gold production has come from four of the 11 geomorphic provinces. These provinces are the Sierra Nevada, which has been by far the most productive, the Klamath Mountains, Basin Ranges, and Mojave Desert. Lesser amounts have been mined in the Transverse and Peninsular Ranges of Southern California, the Modoc Plateau and the Coast Ranges. Gold has been recovered along the eastern margin of the Great Valley, where it was derived from the Sierra Nevada. Starting on p. 15, this book is divided into seven chapters dealing with the gold-bearing provinces: 1) Sierra Nevada, 2) Klamath Mountains, 3) Basin Ranges, 4) Mojave Desert, 5) Transverse and Peninsular Ranges, 6) Modoc Plateau, and 7) Coast Ranges. Each chapter has a general introduction, and the district descriptions follow in alphabetical order.

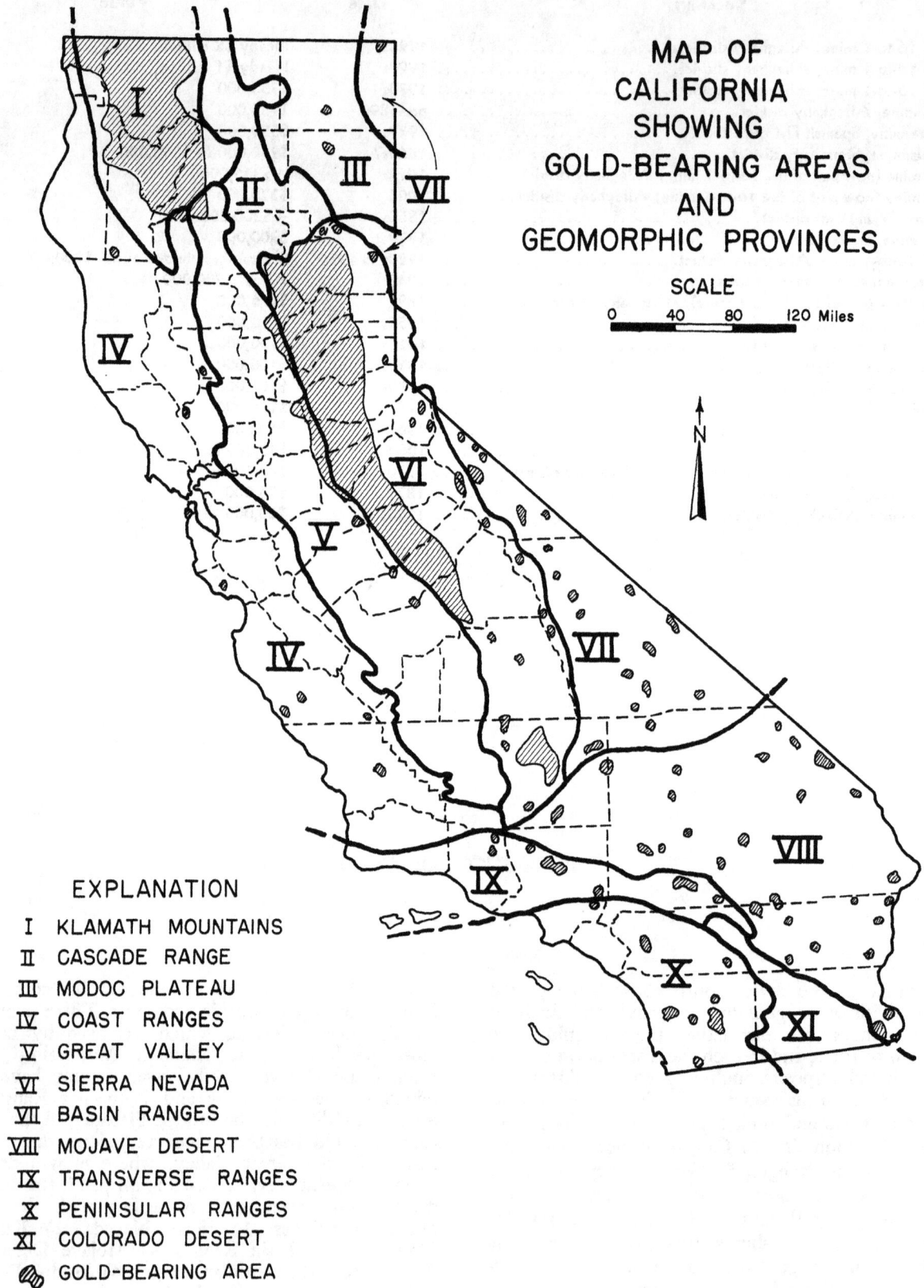

MAP OF
CALIFORNIA
SHOWING
GOLD-BEARING AREAS
AND
GEOMORPHIC PROVINCES

SCALE

0   40   80   120 Miles

N

EXPLANATION

I   KLAMATH MOUNTAINS
II   CASCADE RANGE
III MODOC PLATEAU
IV   COAST RANGES
V   GREAT VALLEY
VI   SIERRA NEVADA
VII BASIN RANGES
VIII MOJAVE DESERT
IX   TRANSVERSE RANGES
X   PENINSULAR RANGES
XI   COLORADO DESERT
     GOLD-BEARING AREA

Figure 2.

Table 6. Principal Gold Districts.

District	County	Type	Production* (in millions)
Grass Valley	Nevada	Lode	$300+
Jackson-Plymouth	Amador	Lode	180
Hammonton	Yuba	Dredge field	130+
Folsom	Sacramento	Dredge field	125
Columbia	Tuolumne	Placer	87
La Porte	Plumas	Placer	60+
Oroville	Butte	Dredge field	55
Nevada City	Nevada	Lode and placer	50+
Alleghany	Sierra	Lode and placer	50+
French Gulch	Shasta-Trinity	Lode; some placer	30+
Bodie	Mono	Lode	30+
Sierra City	Sierra	Lode	30
Angels Camp	Calaveras	Lode; some placer	30
Jamestown	Tuolumne	Lode	30
Placerville	El Dorado	Placer and lode	27+
Carson Hill	Calaveras	Lode	27
Magalia	Butte	Placer	25+
Big Oak Flat	Tuolumne	Placer; some lode	25+
Forest Hill	Placer	Placer	25+
Mojave	Kern	Lode	23
Iowa Hill	Placer	Placer	20+
Rand	Kern	Lode	20+
Soulsbyville	Tuolumne	Lode	20
Snelling	Merced	Dredge field	17
Poker Flat	Sierra	Placer	15+

* Most of these figures are rough approximations. Complete production records of most of the gold districts do not exist.

Table 7. Principal Lode-Gold Mines.

Mine	District	Production* (in millions)
Empire-Star group	Grass Valley	$130
Idaho-Maryland group	Grass Valley	70
Central Eureka	Jackson-Plymouth	36
Kennedy	Jackson-Plymouth	34.2
Carson Hill group	Carson Hill	26
Argonaut	Jackson-Plymouth	25.1
Sixteen-to-One	Alleghany	25+
Keystone	Jackson-Plymouth	24
Standard Cons.	Bodie	18.4
Utica	Angels Camp	17
Sierra Buttes	Sierra City	17+
Brown Bear	French Gulch	15+
Plymouth Cons.	Plymouth	13.5
Yellow Aster	Rand	12+
Lava Cap	Nevada City	12
Golden Queen	Mojave	10+
Plumas-Eureka	Johnsville	8+
Eagle-Shawmut	Jacksonville	7.4
Gwin	Paloma	7
Sheep Ranch	Sheep Ranch	7
Gladstone	French Gulch	6.9
Georgia Slide	Georgetown	6.5
App-Heslip	Jamestown	6.5
Bagdad-Chase	Stedman	6+
Rawhide	Jamestown	6

Table 8. Major Hydraulic Mines.

Mine	Location
Alpha	Washington district, Nevada County
Badger Hill	Badger Hill district, Nevada County
Blue Tent	Blue Tent district, Nevada County
Brandy City	Brandy City district, Sierra County
Buckeye Hill	Scotts Flat district, Nevada County
Cherokee	Cherokee district, Butte County
Cherokee	Badger Hill district, Nevada County
Chips Flat	Alleghany district, Sierra County
Craigs Flat	Eureka district, Sierra County
Deadwood	Last Chance district, Placer County
Depot Hill	Indian Hill district, Sierra County
Dutch Flat	Dutch Flat district, Placer County
Elephant	Volcano district, Amador County
French Corral	French Corral district, Nevada County
Gibsonville	Gibsonville district, Sierra County
Howland Flat	Poker Flat district, Sierra County
Indian Diggings	Indian Diggings district, El Dorado County
Indian Hill	Indian Hill district, Sierra County
Iowa Hill	Iowa Hill district, Placer County
La Grange	Weaverville district, Trinity County
La Porte	La Porte district, Plumas County
Last Chance	Last Chance district, Placer County
Liberty Hill	Lowell Hill district, Nevada County
Lost Camp	Emigrant Gap district, Placer County

Continued on p. 14   Continued on p. 14

Table 7. Lode-Gold Mines—Continued

Mine	District	Production* (in millions)
Soulsby	Soulsbyville	5.5
South Eureka	Jackson-Plymouth	5.3
Bunker Hill	Jackson-Plymouth	5.1
Cactus Queen	Mojave	5+
Fremont-Gover	Jackson-Plymouth	5+
Jumper	Jamestown	5
Princeton	Mt. Bullion	5
Providence	Nevada City	5
Royal	Hodson	5
Wildman-Mahoney	Jackson-Plymouth	5
Zeila	Jackson-Plymouth	5
Confidence	Confidence	4.2
Brush Creek	Alleghany	4+
Midas	Harrison Gulch	4+
Pine Tree-Josephine	Bagby	4+
Mt. Gaines	Hornitos	3.6
Black Oak	Soulsbyville	3.5
Plumbago	Alleghany	3.5
Original Amador	Jackson-Plymouth	3.5
Clearinghouse	Clearinghouse	3.3
Angels	Angels Camp	3.2
Black Bear	Liberty	3.1
Champion	Nevada City	3+
Siskon	Dillon Creek	3+
Hite	Hite Cove	3

*Many of these figures are rough estimates. Complete production records on many important gold mines do not exist.

Table 8. Hydraulic Mines—Continued

Mine	Location
Lowell Hill	Lowell Hill district, Nevada County
Malakoff	North Bloomfield district, Nevada County
Mayflower	Forest Hill district, Placer County
Michigan Bluff	Michigan Bluff district, Placer County
Minnesota	Alleghany district, Sierra County
Moore's Flat	Moore's Flat district, Nevada County
Morristown	Eureka district, Sierra County
North Columbia	North Columbia district, Nevada County
Omega	Washington district, Nevada County
Paragon	Forest Hill district, Placer County
Port Wine	Port Wine district, Sierra County
Poverty Hill	Poverty Hill district, Sierra County
Quaker Hill	Scott's Flat district, Nevada County
Red Dog	You Bet district, Nevada County
Relief	North Bloomfield district, Nevada County
Remington Hill	Lowell Hill district, Nevada County
Sawpit Flat	Sawpit Flat district, Plumas County
Scales	Poverty Hill district, Sierra County
Scott's Flat	Scott's Flat district, Nevada County
Smartsville	Smartsville district, Yuba County
Stewart	Gold Run district, Placer County
Texas Hill	Placerville district, El Dorado County
Todd Valley	Forest Hill district, Placer County
Whiskey Diggings	Gibsonville district, Sierra County
Yankee Jim's	Forest Hill district, Placer County
You Bet	You Bet district, Nevada County

Table 9. Major Drift Mines.

Mine	Location	Mine	Location
Bald Mountain	Alleghany district, Sierra County	Lyons	Placerville district, El Dorado County
Bald Mountain Extension	Alleghany district, Sierra County	Magalia	Magalia district, Butte County
Big Dipper	Iowa Hill district, Placer County	Morning Star	Iowa Hill district, Placer County
Blue Lead	Bangor district, Butte County	Morris Ravine	Morris Ravine district, Butte County
Calaveras Central	Angels Camp district, Calaveras County	Mountain Gate	Damascus district, Placer County
Emma	Magalia district, Butte County	Occidental	Iowa Hill district, Placer County
Feather Fork	Gibsonville district, Plumas County	Pacific Slab	Last Chance district, Placer County
Glenn	Duncan Peak district, Placer County	Pershbaker	Magalia district, Butte County
Hepsidam	La Porte district, Plumas County	Royal	Magalia district, Butte County
Hidden Treasure	Damascus district, Placer County	Ruby	Alleghany district, Sierra County
Hook-and-Ladder	Placerville district, El Dorado County	Startown	Last Chance district, Placer County
Indian Springs	Magalia district, Butte County	Tiedemann	Kentucky Flat district, El Dorado County
Live Yankee	Alleghany district, Sierra County	Vallecito Western	Vallecito district, Calaveras County

SIERRA NEVADA PROVINCE

Geology

The Sierra Nevada, the dominant mountain range in California, is approximately 400 miles long, with steep multiple scarps on its eastern flank and a gentle western slope. It has been the source of the bulk of the state's gold production and contains the richest and the greatest number of districts.

The main mass of the Sierra Nevada is a huge batholith of granodiorite and related rocks that is intrusive into metamorphosed rocks of Paleozoic and Mesozoic age. The metamorphic rocks occur largely along the western foothills and in the northern end of the range. They are complexly folded and faulted and consist of a number of major rock units. The principal units are the slates, phyllites, schists, quartzites, hornfels, and limestones of the Calaveras Formation (Carboniferous to Permian); the Amador Group (Middle and Upper Jurassic) of metasedimentary and metavolcanic rocks; the Mariposa Formation (Upper Jurassic), much of which is slate; schists, phyllites, and quartzites of the Kernville Series (Jurassic or older) in the southern Sierra Nevada; and a vast amount of undifferentiated pre-Cretaceous greenstones and amphibolites.

In addition, there are numerous intrusions of basic and ultra-basic rocks, many of which are serpentinized. The serpentine bodies apparently have been structurally important in the localization of some gold-bearing deposits and often are parallel to or occur within the belts of gold mineralization. Also, there are numerous dioritic and aplitic dikes that are closely associated with gold-bearing veins.

Lode Deposits

Much of the gold mineralization is in the belt of metamorphic rocks that extends along the western foothills and in the northern end of the range, although some important districts are in granitic rocks. Some are associated with small intrusions or stocks related to the Sierra Nevada batholith. The richest as well as the largest number of lode-gold deposits are in the northern and central portions of the range. In the Butte-Plumas County area at the northern end of the Sierra Nevada, the gold belt is nearly 70 miles wide. Continuing south it narrows and dies out almost completely in the Fresno-Tulare County area but appears again in Kern County in the southern end of the range. There are a few widely separated districts along the steep eastern flank of the range.

The most productive lode-gold districts in the northern end of the Sierra Nevada have been the Alleghany, Crescent Mills, Downieville, Forbestown, Graniteville, Grass Valley, Johnsville, Nevada City, and Sierra City districts. In the central portion the most productive and best-known districts are in the Mother Lode gold belt. Although the entire foothill region of the Sierra Nevada is sometimes loosely termed the "Mother Lode Country," technically the Mother Lode is a 120-mile-long system of linked or en echelon gold-quartz veins and mineralized schist

and greenstone that extends from the town of Mariposa, north and northwest to northern El Dorado County (see fig. 4).

The most production portion of the Mother Lode has been the 10-mile segment between Plymouth and Jackson in Amador County. Other major sources of gold in the Mother Lode have been the Angels Camp, Bagby, Carson Hill, Coulterville, Georgetown, Greenwood, Jacksonville, Jamestown, Kelsey, Mount Bullion, Nashville, and Placerville districts.

Although the terms "East Gold Belt" and "West Gold Belt" have been arbitrarily coined to describe the gold deposits east and west of the Mother Lode, each contains extensive systems of gold-bearing veins (see fig. 4). Unfortunately few systematic studies have been made of these belts. The principal sources of gold in the East Gold Belt have been the Grizzly Flat, West Point, Sheep Ranch, Soulsbyville, Confidence, Clearinghouse, Hite Cove and Kinsley districts. The most important in the West Gold Belt have been the Ophir, Shingle Springs, Hunter Valley, Hodson, and Hornitos districts. To the southeast in Madera and Fresno Counties there are some gold districts, but they have been much less productive than those to the north.

In the southern Sierra Nevada, in Kern County, considerable quantities of lode gold have been mined in the Cove district and from scattered areas to the west and south that include the Keyesville, Clear Creek and Loraine districts. Gold has been mined from a few districts along the east flank of the Sierra Nevada, the most productive having been the Bishop Creek district, Inyo County, and the Homer, Mammoth and Jordan districts in Mono County. Appreciable quantities of by-product gold have been recovered from the Sierra Nevada copper belts in the western foothills (see separate section below) and the Plumas County copper districts. Some has been recovered from tungsten mines on the east flank of the Sierra Nevada.

Placer Deposits

The alluvial or placer deposits of the western Sierra Nevada have contributed more than 40 percent of California's total gold output. They are divisible into the Tertiary (older) deposits, which consist predominantly of quartzitic gravels, and the Quaternary (younger), which are in and adjacent to the present stream channels. The Tertiary channel deposits have been mined by hydraulic and drift mining, while the greatest yield from the Quaternary deposits has been from dredging. The flush production of the gold rush was from Recent surface placers that were mined by small-scale methods. These surface placers have largely been exhausted.

The most productive Tertiary channel deposits have been in the Magalia, Cherokee, and Bangor-Wyandotte districts of Butte County; the La Porte and Sawpit Flat districts of Plumas County; the Smartsville district of Yuba County; the Gibsonville, Downieville, Pov-

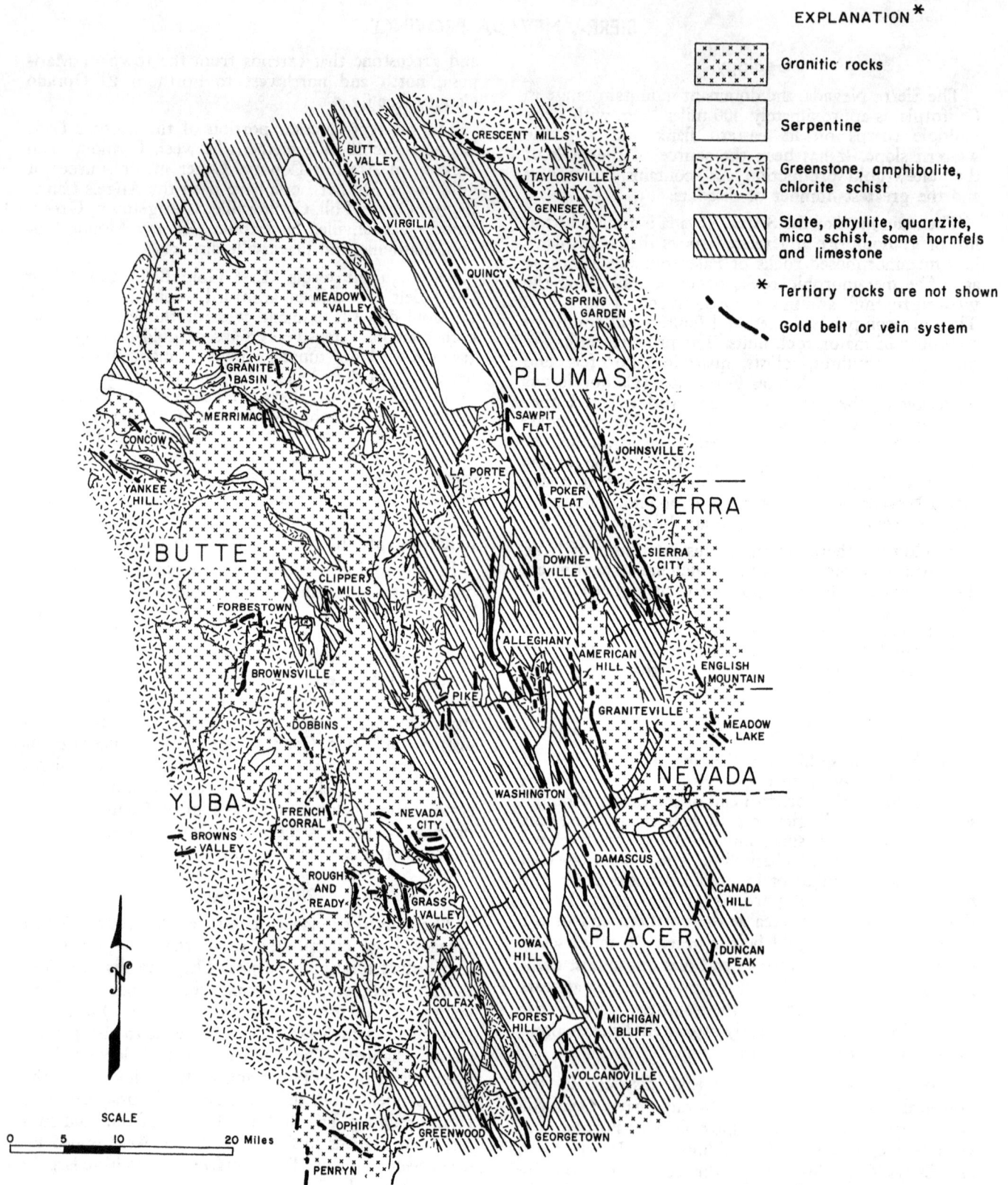

Figure 3. Map of Major Rock Units and Lode Gold Belts, Northern Sierra Nevada.

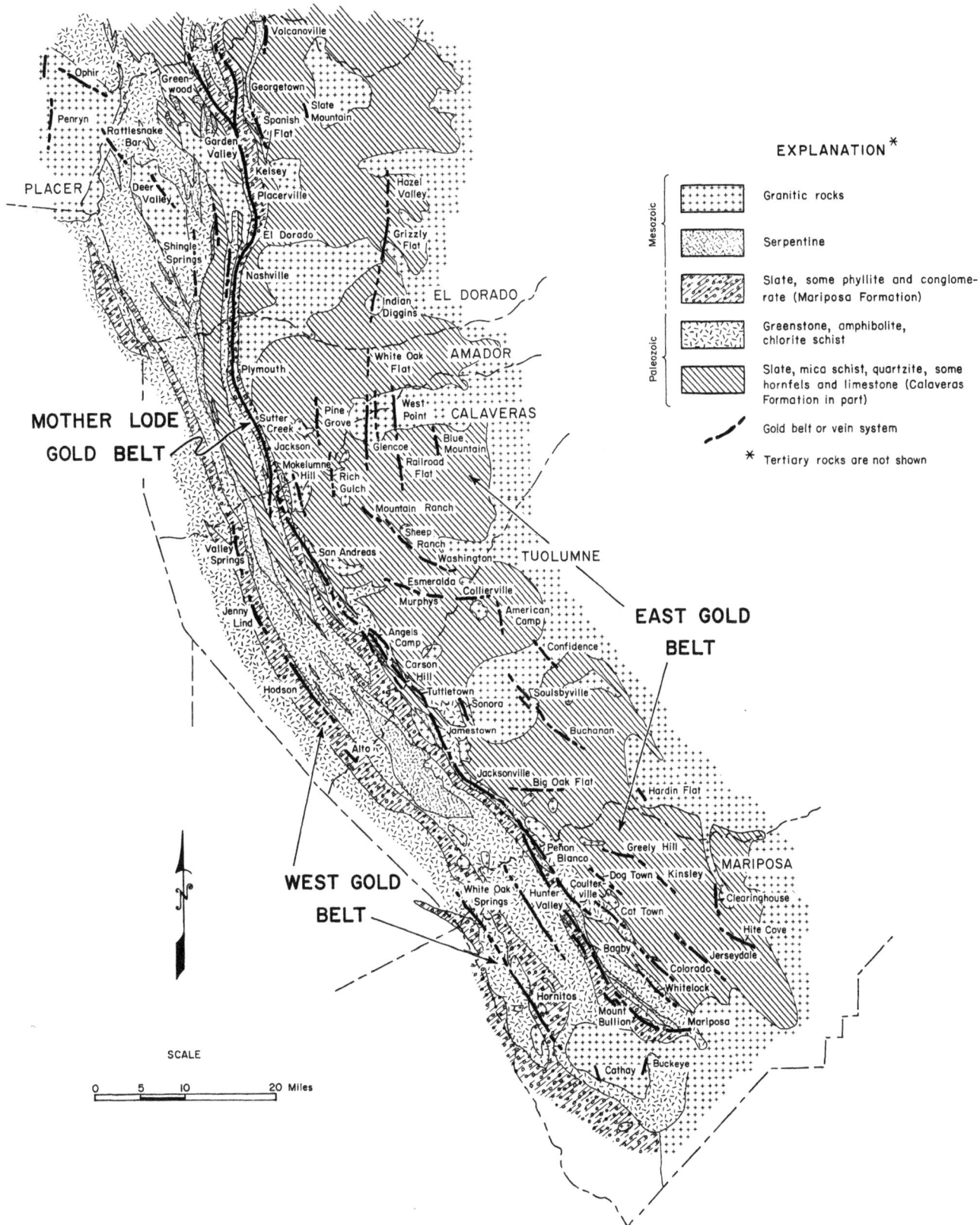

Figure 4. Map of Major Rock Units and Lode-Gold Belts, Central Sierra Nevada. The Mother Lode and the related East and West gold belts are shown.

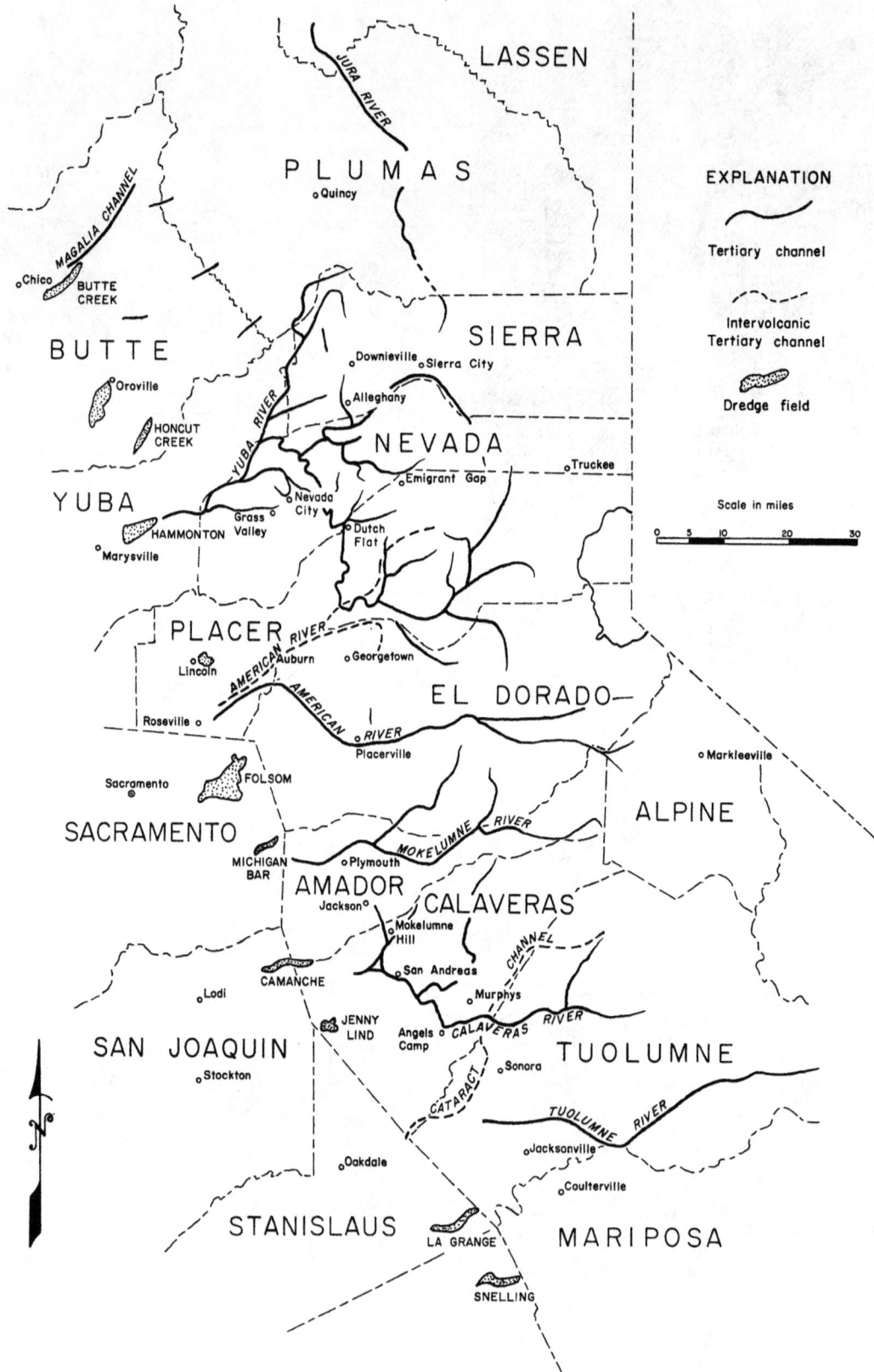

Figure 5. Map of Tertiary Channels and Dredge Fields, Sierra Nevada. *After Lindgren, 1911, and Jenkins, 1935.*

Photo 4. Locomotive, Bald Mountain Drift Mine, Alleghany District. This early steam locomotive, at the mine in Forest, Sierra County, was one of the few used in California gold mines. The photo dates back to possibly the 1870s. *Photo courtesy of Calif. State Library.*

erty Hill, Poker Flat, Brandy City, and Alleghany districts in Sierra County; the North Bloomfield, North Columbia, North San Juan, French Corral, Scotts Flat, You Bet and Washington districts, Nevada County; Dutch Flat, Gold Run, Forest Hill, Iowa Hill, Damascus, Last Chance and Michigan Bluff districts, Placer County; Placerville district, El Dorado County; Fiddletown and Volcano districts, Amador County; Mokelumne Hill and Vallecito districts, Calaveras County, and the Columbia district of Tuolumne County.

All of the major streams and their tributaries that flow across the gold-bearing areas have been placer-mined, many of them several times. The rivers that have yielded the most gold have been the Feather, Yuba, and American Rivers, but large quantities have been recovered from the Bear, Cosumnes, Mokelumne, Stanislaus, Tuolumne, Merced, and Kern Rivers and some from the Chowchilla, Fresno, Kings, White, and San Joaquin Rivers. The two greatest dredging fields are the Hammonton district on the lower Yuba River, Yuba County, and the Folsom district adjacent to and south of the Lower American River in Sacramento County. Other major dredging fields were on Butte and Honcut Creeks and the lower Feather River at Oroville in Butte County; Lincoln, Placer County; Michigan Bar, Sacramento County; Camanche in Calaveras and San Joaquin Counties, La Grange in Stanislaus County and Snelling in Merced County.

Alleghany

Location. Alleghany is in southwestern Sierra County. This district is in a belt of gold mineralization that extends from Goodyear's Bar, south and southeast through Forest, Alleghany, Chip's Flat, and Minnesota. This gold-bearing belt continues south to the Washington district in Nevada County. The Downieville and American Hill districts are to the east, and the Pike district is to the west.

History. The streams in the area were placer-mined soon after the beginning of the gold rush, and the Forest diggings were discovered in the summer of 1852 by some sailors. Some of these sailors were "Kanakas" or Hawaiians who also had deserted their ships in San Francisco. Forest, first known as Brownsville and then Elizaville, got its present name in 1853. The Bald Mountain and other drift mines were highly productive from then until around 1885. Hydraulic mining was done at Minnesota and Chip's Flat during these years. The town of Alleghany was named for Alleghany, Pennsylvania.

Quartz mining was reported to have begun in the district in 1853 at the German Bar and Irelan mines. Although the quartz mines were moderately productive until the 1870s, drift mining was the principal source of gold then. The rediscovery of the Tightner vein in 1904 by H. F. Johnson (erroneously given as 1907 in many reports) led to the revival of lode mining, which continued until 1965.

Photo 5.   Brush Creek Mine, Allegany District. This 1954 view to the south shows the mine yard and the portal of the main adit at the mine in Sierra County.

Photo 6.   Kate Hardy Mine, Allegany District. This 1954 view of the Sierra County mine looks west.

Alleghany was the only town in California after World War II where gold mining was the principal segment of the economy. After 1960, production from the district, which had been averaging more than $500,000 per year, decreased greatly as more and more mining operations were curtailed. By 1963, the output was less than $100,000 per year. The Sixteen-to-One mine, the largest gold source in the district, curtailed normal operations late in 1962, and the Brush Creek mine, the second largest operation, was shut down in 1964. At the end of 1965 the Sixteen-to-One mine was completely shut down, ending an operation that had lasted more than 60 years. Intermittent operations have continued at several mines, such as the Kate Hardy, Oriental, El Dorado-Plumbago, and Mugwump mines. Several of the mines received Federal exploration loans. Skin divers are active in the streams of the area.

Alleghany was the most famous high-grade gold mining district in California. The value of the total output is unknown, estimated at $50 million. Much of this production was from small but spectacularly rich ore bodies.

Geology. The district is underlain by north and northwest-trending beds of metamorphic rocks of the Calaveras Formation (Carboniferous to Permian), serpentine, and greenstone. In the vicinity of Alleghany and Forest this formation has been divided into six units: Blue Canyon Slate, Tightner Formation (chiefly amphibolite and chlorite schist), Kanka Formation (conglomerate, chert, and slate), Relief Quartzite, Cape Horn Slate, and the Delhi Formation (phyllite and slate). These rocks have been invaded by many basic and ultra-basic intrusions; the ultra-basic rocks have been largely serpentinized. Mariposite-bearing rock, locally known as "bluejay," is commonly adjacent to the serpentine. Also present are fine to medium-grained dioritic dikes. The higher ridges are capped by andesite and basalt, which in places overlies auriferous Tertiary channel gravels.

Photo 7. Oriental Mine, Alleghany District. This 1954 view of the Sierra County mine looks north. The mine was active in 1969.

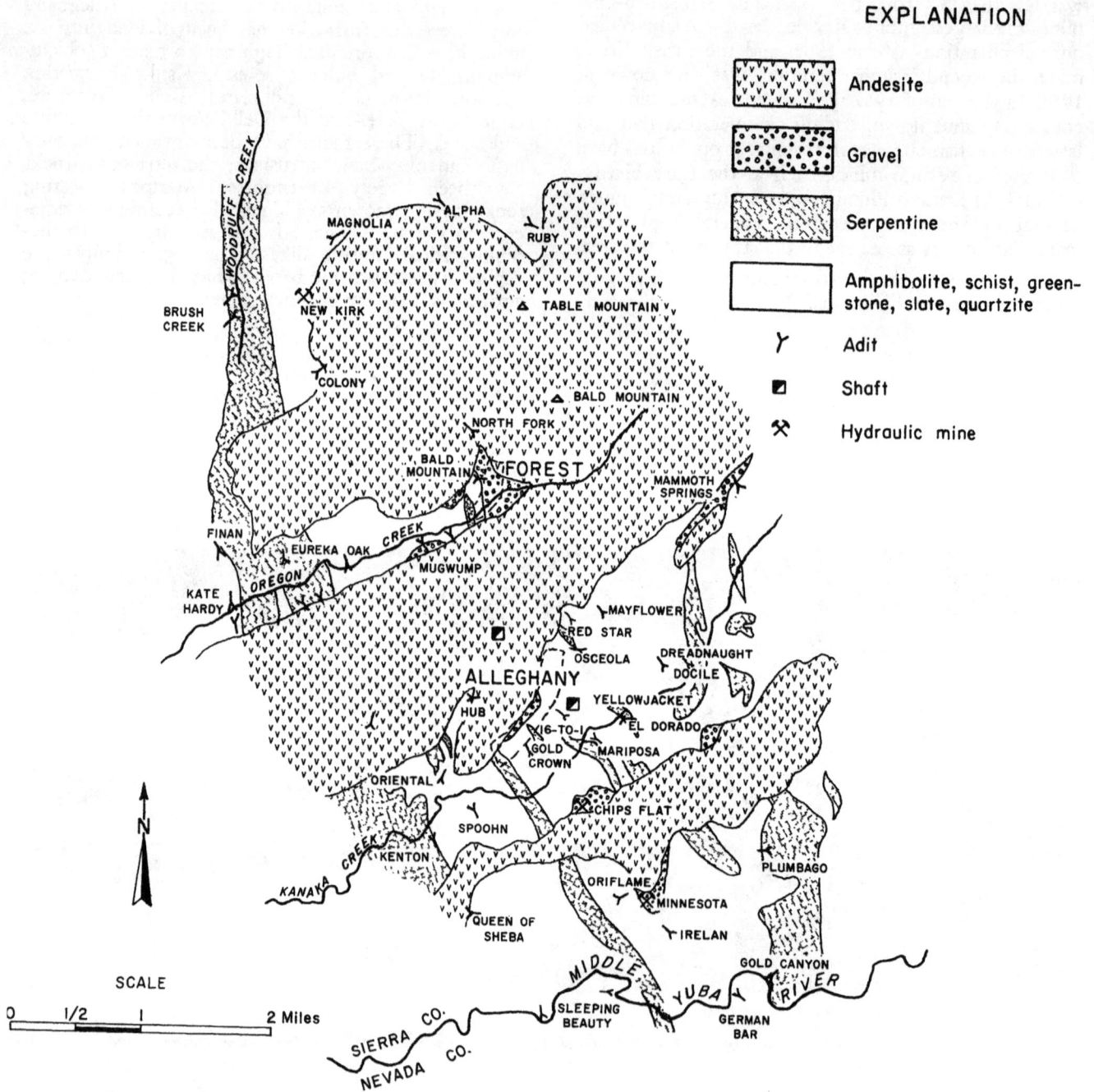

Figure 6. Geologic Map of Alleghany District, Sierra County. The locations of mines are shown. *After Ferguson and Gannett, 1932, and Carlson and Clark, 1956.*

Photo 8. Sixteen-to-One and Gold Crown Mines, Alleghany District. This 1954 view of the two mines looks east. The original Sixteen-to-One mine is in the background, Gold Crown in the foreground.

Ore deposits. The gold-quartz veins strike in a northerly direction, dip either east or west, and usually range from two to five feet in thickness. They occupy minor reverse faults, and occur in all of the rocks of the Calaveras Formation, and in the greenstone. The largest number of mines are in amphibolites of the Tightner Formation. The most characteristic features of the ore deposits are the extreme richness, erratic distribution and small size of the ore shoots. They range from small masses of gold and quartz yielding a few hundred dollars to ore bodies that have yielded hundreds of thousands of dollars. One ore body at the Sixteen-to-One mine, which had a pitch length of 40 feet, contained nearly $1 million, while another at the Oriental mine about 14 feet long yielded $734,000.

The gold occurs in the native state commonly with arsenopyrite but only small amounts of other sulfides. In a few places pyrite is abundant. The numerous serpentine bodies and associated mariposite rock are structurally important in the localization of the ore bodies. The quartz veins tend to fray or bend near serpentine, and it is in these frayed or bend portions of the veins that the high-grade ore bodies are often found. High-grade ore also is found in vein junctions or in sheared portions of the veins.

Channel gravels. A major tributary of the Tertiary Yuba River extended south from Rock Creek through Forest and Alleghany and then southeast through Chip's Flat and Minnesota to Moore's Flat in Nevada County. This is commonly known as the "Great Blue Lead" or Forest channel. It was uniformly rich except where cut by later channels. The largest gold producers were the Ruby, Live Yankee, and Bald Mountain drift mines, where many coarse nuggets were recovered. During the late 1930s a number of fist-sized gold nuggets were recovered from the Ruby mine. These were displayed for many years in the Sierra County exhibit at the California State Fair in Sacramento.

Mines. Lode: Brush Creek $4 million+, Dreadnaught $50,000 to $100,000, Docile $100,000 to $200,000, Eclipse $20,000 to $50,000, El Dorado $325,000, German Bar $200,000, Gold Canyon $750,000 to $1 million, Gold Crown, Golden King $250,000, Irelan $350,000 to $500,000, Kate Hardy $700,000, Kenton $1 million to $1.25 million, Mariposa $50,000, Morning Glory $80,000 to $100,000, Mugwump (both lode and placer) $50,000, North Fork (both lode and placer)

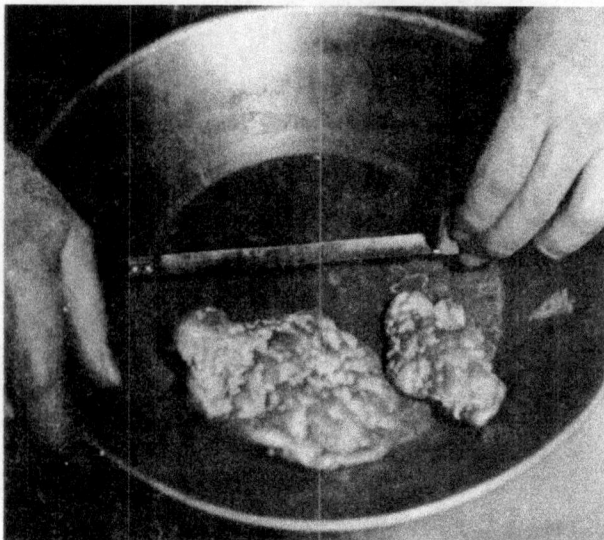

Photo 9. Large Nuggets, Ruby Drift Mine, Alleghany District. The pair of nuggets was found in the Sierra County mine in 1938. *Photo courtesy of L. L. Huelsdonk, Downieville.*

$125,000, Oriflamme, Ophir *, Oriental $2.85 million, Osceola *, Plumbago $3.5 million, Rainbow * $2.5 million, Rainbow Extension*, Red Ledge, Red Star-Osceola * $200,000, Rising Sun $58,000, Shannon, Sixteen-to-One $25 million+, South Fork (both lode and placer), Spoohn, Tightner *, Twenty One *, Wyoming, Yellowjacket. Drift: Bald Mountain $3.1 million, Bald Mountain Extension $500,000 to $1 million, Gold Star $250,000+, Highland & Masonic $300,000+, Live Yankee $750,000 to $1 million, Ruby $1 million+.

Bibliography

Averill, C. V., 1942, Mines and mineral resources of Sierra County; California Div. Mines Rept. 38, pp. 17–48.

Carlson, D. W., and Clark, W. B., 1956, Lode gold mines of the Alleghany-Downieville area, Sierra County: California Journal of Mines and Geology, vol. 52, no. 3, pp. 237–272.

Clark, W. B., and Fuller, W. P., Jr., 1968. The Original Sixteen-to-One Mine: California Div. Mines and Geology Mineral Information Service, vol. 21, no. 5, pp. 71–75 and 78.

Cooke, H. R., Jr., 1947, The Original Sixteen-to-One gold quartz vein, Alleghany, California: Econ. Geology, vol. 42, no. 3, pp. 211–250.

Ferguson, H. G., 1915, Lode deposits of the Alleghany district, California: U. S. Geol. Survey, Bull. 580, pp. 153–182.

Ferguson, H. G., and Gannett, R. W., 1932, Gold-quartz veins of the Alleghany district, California: U. S. Geol. Survey Prof. Paper 172, 139 pp.

Lindgren, W., 1900, U. S. Geol. Survey Geol. Atlas of the U. S., Colfax folio 66.

Lindgren, Waldemar, 1911, The Tertiary gravels of the Sierra Nevada of California: U. S. Geol. Survey Prof. Paper 73, p. 142.

Logan, C. A., 1923, Quartz mining in the Alleghany district: California Mining Bureau, Report 18, pp. 499–519.

Logan, C. A., 1929, Alleghany District: California Mining Bureau Report 25, pp. 156–159.

MacBoyle, E., 1920, Sierra County, Alleghany mining district: California Mining Bureau, Report 16, pp. 1–5.

Simkins, W. A., 1923, The Alleghany district of California: Pacific Mining News of Eng. and Min. Journal, vol. 2, pp. 288–291.

Turner, H. W., 1897, Downieville folio, U. S. Geol. Survey Geol. Atlas of the U. S., folio 37, 8 pp.

* Now part of the Sixteen-to-One mine.

Alto

This district is in the southwest corner of Calaveras County about six miles south of Copperopolis. It was named by the author for the Alto mine, a major source of gold in the area. The region was first placer-mined during the gold rush, when the gravels underlying nearby Tuolumne Table Mountain were worked by drifting. The Alto mine was discovered in 1886 and operated on a large scale until 1907. It has an estimated total output of $1 million. The district is underlain by slate of the Mariposa Formation (Upper Jurassic) with some interbeds of massive greenstone. The gold occurs in thin quartz veins or with disseminated pyrite in the greenstone. Placer gold was recovered from Eocene quartzitic gravels overlying the slate. Some of the gravels are capped by latite porphyry of Tuolumne Table Mountain.

Bibliography

Clark, W. B., and Lydon, P. A., 1962, Calaveras County, Alto mine: California Div. of Mines and Geology County Report 2, pp. 37 and 40.

Lowell, F. L., 1919, Alto gold mine: California Min. Bur. Rept. 16, p. 629 (erroneously shown as being in Stanislaus County).

Taliaferro, N. L., and Solari, A. J., 1948, Geologic map of the Copperopolis quadrangle: California Div. Mines colored quadrangle map (Pl. I of Bull. 145).

American Camp

Location. This district is in northwestern Tuolumne County in the general vicinity of American Camp Station, which is about eight miles northeast of Columbia. It includes the Italian Bar, French Camp, Star Ridge, Grant Ridge, and Cedar Ridge areas. The famous Columbia placer-mining district adjoins it on the southwest.

History. The streams in the district were first mined during the gold rush. The town of Italian Bar was the principal settlement at that time. It later was destroyed by fire. Lode mining began about 1860, and continued almost steadily until around 1900. There was some mining in the district again in the 1920s and 1930s, and there has been minor prospecting since. The Grant mine was prospected for uranium in 1953–54.

Geology. The district is underlain predominantly by argillite, quartzite, siliceous schist, and limestone of the Calaveras Formation (Carboniferous to Permian). There are also a few small granodiorite stocks. Fine-grained diorite and aplite dikes are common and are often associated with the gold-quartz veins.

Ore deposits. Numerous gold-quartz veins and stringers contain small to medium-sized ore shoots. The veins strike either west or north-north-east. In places the ore is rich. The ore contains free gold associated with varying amounts of sulfides, especially galena and chalcopyrite. There are several patches of Eocene channel gravels that have been mined by hydraulicking. At the Grant mine black uraninite associated with gold occurs in quartz.

Mines. Argentum Consolidated, Black Bear, Contention, Gold Ridge, Grant, Gray Eagle, Ham and Birney $100,000, Hazel Bell, Indian Girl, Keltz $300,000,

Photo 10. Angels Mine, Angels Camp District. This eastward view of the Calaveras County mine was taken in about 1914. *Photo courtesy of Hillcrest Studio, Angels Camp.*

Lucky Strike, Mountain Lily, Noonday, Rifle, Sonnet, Star, Tiffany, Volunteer.

Bibliography

Goldstone, L. P., 1890, Tuolumne County, Keltz mine: California Min. Bur. Rept. 10, pp. 755–757.

Logan, C. A., 1928, Tuolumne County, gold quartz mines: California Div. Mines, Rept. 24, pp. 8–21.

Turner, H. W., and Ransome, F. L., 1898, Big Trees folio: U. S. Geol. Survey Geol. Atlas of the U. S., folio 51, 8 pp.

American Hill

Location. The American Hill district is in southwestern Sierra County about five miles east of the town of Alleghany. It is both a lode and placer district, but the placer deposits have been more important. It includes the Cornish House area.

Geology. The district is chiefly underlain by slate. Serpentine and amphibolite are to the west, and granodiorite is just to the east. Extensive gravel deposits are part of a tributary to the Forest channel of the Tertiary Yuba River, which extends in a southwest direction through the district. The northern part of the district is covered by andesite. Dioritic dikes often are associated with the gold-quartz veins.

Ore deposits. The gravels are quartzitic and contain coarse gold. One of the gravel deposits is as much as 300 feet thick and covered with clay and sand. The quartz veins are lenticular and occur either in the slate or near the slate-granodiorite contact. The ore bodies contain free gold with pyrite, arsenopyrite, and galena. Sometimes carbon is found in cavities in the quartz. Some high-grade pockets have been found, but most of ore averages less than ½ ounce per ton.

Mines. Placer: American Hill, Bear Creek, Excelsior, Mable Mertz, Yellow Jacket. Lode: Comet, Ironsides, Jim Crow, Lonesome Pine, Pilgrim, Von Humboldt.

Bibliography

Lindgren, Waldemar, 1900, Colfax folio: U. S. Geol. Survey Geol. Atlas of the U. S., folio 66, 10 pp.

Lindgren, Waldemar, 1911, Tertiary gravels of the Sierra Nevada: U. S. Geol. Survey Prof. Paper 73, pp. 142–143.

MacBoyle, Errol, 1920, Sierra County, American Hill district: Calif. Min. Bur. Rept. 16, pp. 5–6.

Angels Camp

Location. This district is in southwest Calaveras County in the vicinity of the town of Angels Camp and Altaville. It is an important part of the Mother Lode belt and lies between the San Andreas district to the northwest and the Carson Hill district to the southeast. It is both a lode and placer district, but the lode mines have been more productive.

History. The streams in the area were mined shortly after the beginning of the gold rush. The town was founded in 1848 and named for Henry Angel, who had established a trading post here. Rich surface ores were mined in the oxidized zones in the 1850s, and most of the important veins were discovered at that time. By 1885 Angels Camp had become one of the major gold-mining districts in the state. The Utica Mining Company was organized in the middle 1850s, and for the next 40 years the Utica mine was a major source of gold. From 1893 to 1895 this mine yielded more than $4 million worth of gold. All of the major mines were shut down during World War I. There was some activity in the district again during the 1930s, and the Calaveras Central and Altaville drift mines have been intermittently prospected during the past 15 years. This district has an estimated total output of at least $30 million, and it may be considerably more.

The colorful jumping frog jubilee held each year at nearby Frogtown, the county fairgrounds, is based on Mark Twain's tale, *The Jumping Frog of Calaveras County.* He is supposed to have first related the story at the Angels Hotel.

Geology. This district is underlain by a series of northwest-striking beds of amphibolite and chlorite schist, phyllite, greenstone, and metagabbro (see fig. 7). To the east and west are slate, impure quartzite,

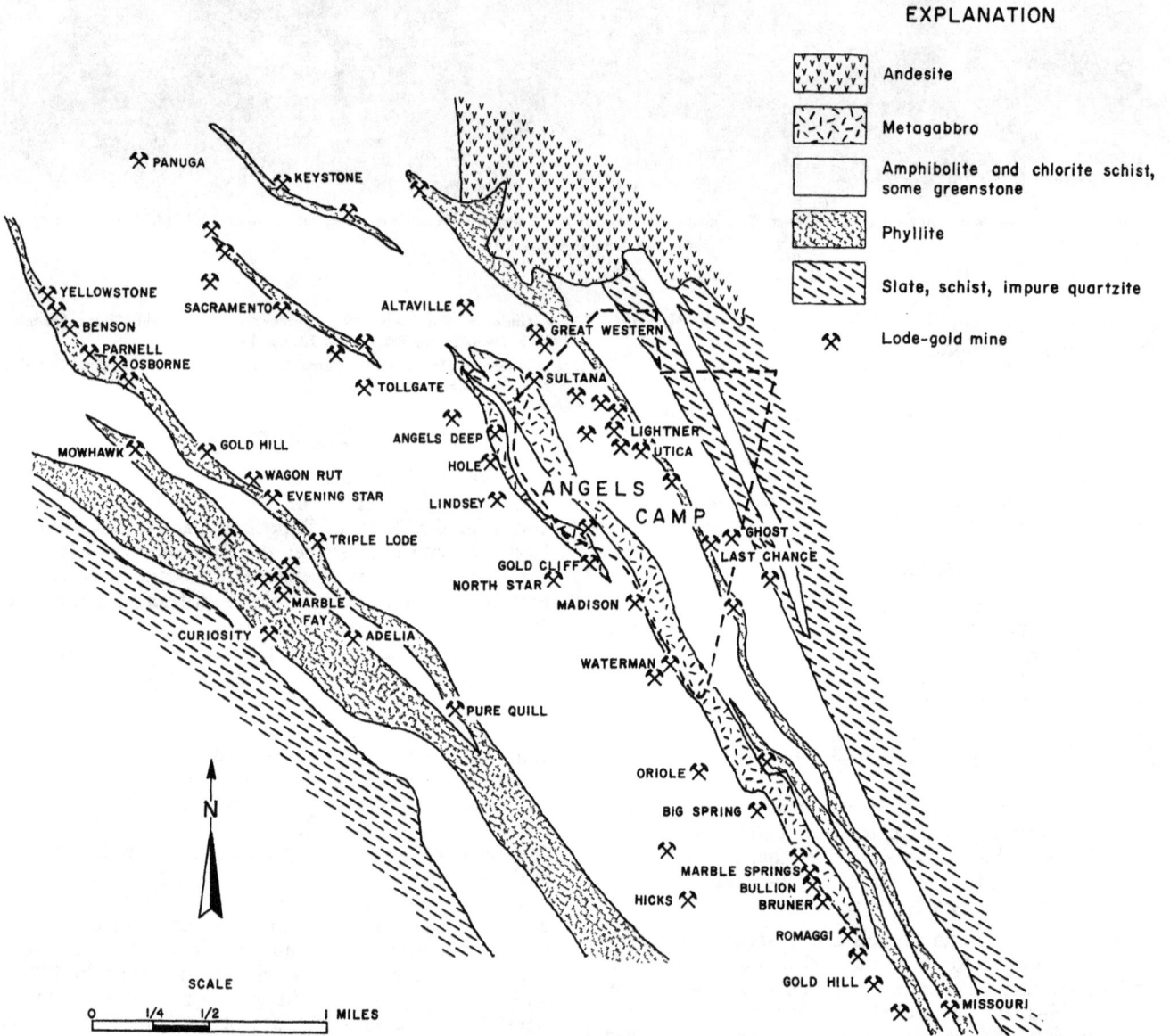

EXPLANATION

- Andesite
- Metagabbro
- Amphibolite and chlorite schist, some greenstone
- Phyllite
- Slate, schist, impure quartzite
- Lode-gold mine

Figure 7. Geologic Map of Angels Camp District, Calaveras County. The map shows the central portion of the district and the locations of lode-gold mines. *After Clark and Lydon, 1962, figure 2.*

Photo 11. Gold Cliff Mine, Angels Camp District. This view, in the 1870s or 1880s, shows the open cut at the Calaveras County mine. The wallrock is massive greenstone and amphibolite. *Photo courtesy of Tuolumne County Museum.*

and micaceous schist. In the north and northeast part of the district is the west and northwest-trending Tertiary Central channel, which is a tributary of the ancestral Calaveras River. The channel gravels are cemented, contain abundant quartz, and are overlain by rhyolite tuff and andesite.

Ore deposits. The lode deposits consist of massive quartz veins, zones of parallel quartz stringers and bodies of mineralized schist and greenstone. The ore contains disseminated free gold and auriferous pyrite. Usually the gold is in fine particles, although occasional high-grade pockets containing coarse gold have been found. Calcite, talc, ankerite, and sericite, are

commonly present in the ore. The milling-grade ore usually averaged 1/7 to 1/5 ounce of gold per ton, but the ore bodies were scores of feet in thickness, hundreds of feet in length, and were mined to depths of several thousand feet.

The deposits occur either in the amphibolite and chlorite schist or phyllite. There are three principal vein systems (see fig. 7). In the system on the west the veins are in phyllite. In the center one the veins are along the west margin of a northwest-trending belt of metagabbro. In the eastern system, which contains the famous Utica mine, the ore deposits are in amphibolite and greenstone.

Photo 12. Utica Mine, Angels Camp District. This northeast view, in about 1900, shows the surface plant at the Calaveras County mine. The north shaft is at left and the south shaft, at right. The Utica was the most productive mine at Angels Camp. *Photo courtesy of Hillcrest Studio, Angels Camp.*

Mines. Lode: Altaville, Angels $3.25 million+, Angels Deep $100,000+, Belmont-Osborn $100,000+, Benson, Big Spring, Bruner, Bullion, Chaparral Hill, Cherokee, Evening Star, Ghost, Gold Cliff $2,834,-000+, Gold Hill $100,000+, Great Western, Hardy, Keystone, Last Chance, Lightner $3 million+, Lindsey, Madison $1 million+, Marble Springs, Mohawk, Mother Lode Central $100,000+, Oriole, Parnell, Pure Quill, Santa Ana, Sultana $200,000+, Tollgate, Triple Lode, Tulloch, Utica $17 million, Vonich, Whittle, Wagon Rut, Waterman, Yellowstone. Drift: Altaville, Calaveras Central 22,000 ounces+.

Bibliography

Averill, C. V., 1946, Calaveras Central mine: California Div. Mines Bull. 135, pp. 235–247.

Clark, W. B., and Lydon, P. A., 1962, Calaveras County, gold: California Div. Mines and Geology, County Rept. 2, pp. 32–93.

Crawford, J. J., 1894, Utica-Stickles mine: California Min. Bur. Rept. 12, pp. 98–99.

Eric, J. H.; Stromquist, A. A., and Swinney, C. M., 1955, Geology of the Angels Camp and Sonora quadrangles: California Div. Mines Spec. Rept. 41, 55 pp.

Fairbanks, H. W., 1890, Geology of the Mother Lode region: California Min. Bur., Rept. 10, pp. 59–62.

Julihn, C. E., and Horton, F. W., 1938, Mines of the southern Mother Lode region: U. S. Bur. Mines Bull. 413, pp. 21–94.

Knopf, Adolph, 1929, Gold Cliff mine: U. S. Geol. Survey Prof. Paper 157, pp. 71–72.

Logan, C. A., 1935, Mother Lode gold belt—Calaveras County: California Div. Mines Bull. 108, pp. 125–152.

Ransome, F. L., 1900, Mother Lode district folio, California: U. S. Geol. Survey Geol. Atlas of the U. S., folio 63, 11 pp.

Storms, W. H., 1900, The Mother Lode region—Calaveras County: Calif. Min. Bur. Bull. 18, pp. 100–127.

Tucker, W. B., 1916, Calaveras County, Angels, Gold Cliff, Lightner, and Utica mines. California Min. Bur., Rept. 14, pp. 68–69, 81–82, 89–90, 110–112.

Turner, H. W., 1894, Jackson folio, California: U. S. Geol. Survey Geol. Atlas of the U. S., folio 11, 6 pp.

Badger Hill

Location. This placer-mining district is in north-central Nevada County. The North Columbia district adjoins it on the east and the North San Juan district on the west. It was hydraulicked extensively from the 1850s through the 1880s. Later, Chinese miners reworked the tailings. It includes the Cherokee diggings. This district was recently studied by the U.S. Bureau of Mines and U.S. Geological Survey as part of their heavy metals programs.

Geology. The deposits are part of the Tertiary North Bloomfield-North Columbia-North San Juan channel. The gravels are thick, the lower part being quartzitic and the upper part containing abundant sand and clay. The lower gravels were rich. The upper gravels yielded 10 to 15 cents per yard. Bedrock is slate and phyllite, with granodiorite lying to the west. The value of the total output for the district is unknown, but it has been estimated at several million dollars. In 1891 the U.S. Army Engineers estimated that 10 million yards had been excavated and 33 million remained.

Bibliography

Lindgren, Waldemar, 1911, Tertiary gravels of the Sierra Nevada: U. S. Geol. Survey Prof. Paper 73, pp. 121–130.

Lindgren, Waldemar, and Turner, H. W., 1895, Smartsville folio, California: U. S. Geol. Survey Geol. Atlas of the U. S., folio 18, 6 pp.

Bagby

Location. This district is in western Mariposa County in the vicinity of the towns of Bagby and Bear Valley. It is in the Mother Lode gold belt.

History. The streams were placer-mined early in the gold rush, and the Pine Tree and Josephine veins were discovered in 1849. Part of the area was on the Las Mariposa Spanish land grant of General John C. Fremont. Bagby was first known as Benton Mills, named by Fremont for Senator Thomas Hart Benton, his father-in-law. It was renamed in the 1890s for A. Bagby, a hotel owner. The town was a stop on the Yosemite Valley Railroad, which once extended up the Merced River canyon to Yosemite National Park.

Gold mining activity continued until around 1875. There was mining in the district again in the early 1900s. The Pine Tree-Josephine mine was worked on a major scale from 1933 to 1944, and the Red Bank mine has been active in recent years. Part of the area, including the old town of Bagby, was inundated by the Exchequer Reservoir in 1967.

Geology. In this district the Mother Lode gold belt is about 1½ miles wide. It is underlain by northwest-striking beds of slate, phyllite, and meta-sandstone of the Mariposa Formation (Upper Jurassic), with greenstone and green schist both to the west and southeast (see fig. 18, page 95). A belt of serpentine extends northwest through the central portion of the district and is structurally important in relation to some of the gold-bearing veins.

Ore deposits. There are several northwest-trending vein systems that consist of quartz veins and stringers with brecciated slate, schist, and associated bodies of pyritic ankerite and mariposite-quartz rock. These vein systems are often scores of feet in thickness. The ore contains free gold, pyrite, and arsenopyrite with small amounts of chalcopyrite, galena, millerite, sphalerite, and niccolite. Milling ore averaged 1/7 to 1/2 ounce of gold per ton. In places high-grade ore is abundant. The ore shoots had stoping lengths of up to 600 feet, and the veins were mined to an inclined depth of 1500 feet.

Mines. Dolman, French $116,000+, Jumper, Juniper, Live Oak, Mexican I $50,000, Oso $50,000+, Pine Tree-Josephine $4 million+, Queen Specimen, Red Bank $100,000+, Specimen.

Bibliography

Bowen, O. E., Jr., 1957, Mariposa County, Pine Tree and Josephine mine: California Journal of Mines and Geology, vol. 53, pp. 151–155.

Costello, W. O., 1921, Mariposa County, Bagby district: California Min. Bur. Rept. 17, pp. 91–92.

Knopf, Adolph, 1929, Mariposa Estate mines: U. S. Geol. Survey Prof. Paper 157, pp. 83–84.

Logan, C. A., 1935, Pine Tree and Josephine mines: California Div. Mines Bull. 108, pp. 186–189 and plate X.

Ransome, F. L., 1900, Mother Lode district folio: U. S. Geol. Survey Geol. Atlas of the U. S., folio 63, 11 pp.

Storms, W. H., 1900, The Mother Lode region—Mariposa County: Calif. Min. Bur. Bull. 18, pp. 143–146.

Bangor-Wyandotte

Location. This district is in southeastern Butte County about 10 miles southeast of Oroville. It is an extensive area of placer deposits that occur in the vicinity of the old towns of Bangor and Wyandotte. The Honcut dredging district is just to the southwest.

History. The district was originally mined during the gold rush. Extensive drift and hydraulic mining was done from the middle 1850's through the 1890's. Bangor, founded in 1855 by the Lumbert brothers and named for their home town in Maine, was an important mining and staging center. Numerous Chinese were in the district from the 1870s through the 1890s. This and the adjoining Honcut district were dredged in the early 1900s and again in the 1930s.

Geology. Tertiary gravels covering a broad 3- by 8-mile area are believed to represent an ancient delta of the Tertiary Yuba River. The channel flowed northwest through Bangor and then west. Farther west are shore gravels of Pleistocene age.

The channel gravels were mined by drifting, the shore gravels by hydraulicking, and the Recent gravels by dredging. The channel gravels are as much as 150 feet thick, well rounded, well cemented, and consist of intrusive and metamorphic rock fragments with 10 to 20 percent quartz. The gold was flaky and often rusty. A number of coarse nuggets were found, one of which was reported to have weighed 14 pounds. Bedrock consists mainly of greenstone and amphibolite.

Drift mines. Bangor, Blue Lead, Catskill, Gray Lead, Turner.

Bibliography

Crawford, J. J., 1894, Bangor mine: California Min. Bur., Rept. 12, pp. 80–81.

Lindgren, Waldemar, 1895, Smartsville folio, California: U. S. Geol. Survey Geol. Atlas of the U. S., folio 18, 6 pp.

Lindgren, Waldemar, 1911, U. S. Geol. Survey Prof. Paper 73, pp. 122–123.

Logan, C. A., 1930, Butte County, Blue Lead Mining Company: California Div. Min. Rept. 26, p. 387.

Bidwell Bar

Location. Bidwell Bar is in southeastern Butte County about 10 miles northeast of Oroville and west of the junction of the South and Middle Forks of the Feather River. The district includes the Hurleton, Stringtown and Enterprise areas to the east.

History. Gold was discovered here in 1848 by General John Bidwell, soon after Marshall's discovery at Coloma. News of this rich find spread, and there was a general rush to the Feather River region. Bidwell Bar, Long's Bar, Thompson Flat, Potter's Bar, Adamstown, and other settlements were soon established. All of these towns have long since disappeared, as the gravels were exhausted in a few years and the miners moved elsewhere. The old suspension bridge and a few remaining buildings later became a state park. Much of the area was inundated by Oroville Lake.

Geology. The district is underlain by amphibolite in the west and granite in the east. Most of the gold was obtained from Recent and Pleistocene gravels in and adjacent to the river. There are a few narrow gold-quartz veins.

Bibliography

Turner, H. W., 1898, Bidwell Bar folio: U. S. Geol. Survey Geol. Atlas of the U. S., folio 43, 6 pp.

Big Creek

Location. Big Creek is in eastern Fresno County about 10 miles northeast of the town of Trimmer and 50 miles northeast of Fresno. Superficial placer mining was done here during the early days and sporadic lode mining from the 1890s to around 1915.

Geology. A few narrow quartz veins contain free gold and often abundant sulfides. The country rock is granite and schist. A number of contact metamorphic zones in the region contain tungsten and some rare barium minerals, including sanbornite. The chief gold sources were the Contact and Hancock mines.

Bibliography

Bradley, W. W., 1919, Fresno County, Contact mine: California Min. Bur. Rept. 16, pp. 441–443.

Big Dry Creek

Location. This district is in northeastern Fresno County about 10 miles northeast of Clovis and two miles north of Academy. The area was placer-mined on a small scale during the gold rush and lode-mined from the 1870s to around 1900. Apparently very little mining has been done here since.

Geology. The area is underlain by granitic rocks with thin bands of slate and mica schist. A serpentine belt lies to the east. A number of narrow north-trending quartz veins contain free gold and varying amounts of sulfides. The veins occur near or at the contacts between granitic and metamorphic rocks. The ore bodies are shallow, none having been mined to a depth greater than 150 feet. Copper also has been mined here.

Mines. Blue Rock, Confidence, Crystal Springs, Defiance, Monte Cristo, Rip Van Winkle, Thorn.

Bibliography

Goldstone, L. P., 1890, Fresno County, Big Dry Creek Mining district: California Min. Bur. Rept. 10, pp. 193–194.

Irelan, Wm., Jr., 1888, Big Dry Creek mining district: California Min. Bur. Rept. 8, pp. 208–209.

Big Oak Flat

Location. The Big Oak Flat district is in the East Gold Belt of the Sierra Nevada in southwestern Tuolumne County. It includes the Groveland, Deer Flat, and Second Garrotte areas to the east.

History. This district was first mined shortly after the beginning of the gold rush by James D. Savage. It was named for an oak tree that had a trunk with a diameter of 11 feet. The placer deposits here were highly productive, those at Big Oak and Deer Flat having been credited with a production of $25 million. Lode mining began soon afterward, and continued steadily until World War I. The town of Garrotte was renamed Groveland by later residents who replaced the name given in 1850, when a thief was hanged there. But a place to the east, named Second Garrotte (after 1850) for a similar reason, has kept its 19th Century name. This area has prospered from tourist trade that originates along the Big Oak Flat road, which serves Yosemite National Park. The Hetch Hetchy Railroad, which was used in the construction of the San Francisco Water Department's Hetch Hetchy Reservoir to the east, served the area for some years. Some lode mining was done in the district again in the 1930s, and there has been intermittent prospecting and development work at several mines since.

Geology. The area is underlain by a northwest-trending belt of schist, argillite, and quartzite of the Calaveras Formation (Carboniferous to Permian), intruded by several small granodiorite stocks and flanked on the west by amphibolite, slate, and serpentine of the Mother Lode belt.

Ore deposits. A considerable number of north, northwest and west-trending quartz veins are found in both the metamorphic rocks and the granodiorite. The ore shoots usually are limited in extent, but often are rich. The ore contains free gold and often abundant sulfides, especially galena. Most of the veins are only a few feet thick. The surface placers mined during the gold rush were extremely rich.

Mines. Bicknell, Champion, Contact, Criss Cross, Del Monte, Goodnow, Kanaka, Long Fellow $100,-000+, Mack, Mississippi, Mohrman $40,000, Mt. Jefferson, National, Nonome, Red Jacket, Rhode Island, Venus, Wide West.

Bibliography

Logan, C. A., 1949, Tuolumne County, lode gold: California Journal of Mines and Geology, vol. 45, pp. 54–73.

Ransome, F. L., 1900, Moth Lode district folio: U. S. Geol. Survey Geol. Atlas of the U. S., folio 63, 11 pp.

Tucker, W. B., 1916, Tuolumne County, Big Oak Flat and Groveland district: California Min. Bur. Rept. 14, p. 136.

Turner, H. W., and Ransome, F. L., 1897, Sonora folio: U. S. Geol. Survey Geol. Atlas of the U. S., folio 41, 7 pp.

Bishop Creek

This district is on the east flank of the Sierra Nevada in northwestern Inyo County. It is about 15 miles southwest of Bishop. The principal source of gold in the district has been the Cardinal or Wilshire-Bishop Creek mine, which was worked on a large scale during the early 1900s and again from 1933 to 1938. It has an estimated total production of more than $1 million. The ore deposits occur in a zone of quartzite, which is enclosed in granitic rocks. The ore bodies are up to eight feet thick and contain fine free gold and fairly abundant pyrite, pyrrhotite, chalcopyrite, sphalerite and arsenopyrite. The deposits have been developed to a depth of 600 feet.

Bibliography

Tucker, W. B., and Sampson, R. J., 1938, Inyo County, Cardinal Gold Mining Company: California Div. Mines Rept. 34, pp. 389–390.

Waring, C. A., and Huguenin, Emile, 1919, Inyo County, Wilshire-Bishop Creek mine: California Min. Bur. Rept. 15, p. 85.

Blue Mountain

Blue Mountain is in eastern Calaveras County about 10 miles southeast of West Point. A number of narrow quartz veins exist in granodiorite, schist, and hornfels. The ore bodies are small and irregular but contain extremely abundant sulfides. The quartz often has a dark color. The principal properties have been the Black Wonder, Gold King, and Heckendorn mines.

Bibliography

Clark, W. B., and Lydon, P. A., 1962, Calaveras County, lode gold mines: California Div. Mines and Geology County Rept. 2, pp. 32–76.

Blue Tent

Blue Tent is northeast of the Nevada City district in western Nevada County. The area was mined years ago by hydraulicking, but little or no mining has been done since. The gravels are part of the Tertiary channel that extends north-northwest from You Bet through Scotts Flat and Quaker Hill to North Columbia. Although the gravels here are extensive, they were reported not to have been very remunerative. Lindgren (1911) estimated that 15 million cubic yards had been removed and 90 million yards remain, much of it barren clay and sand. The gravel next to bedrock was reported to have yielded about 50 cents in gold per yard. To the east and south the gravels are overlain by andesite. Bedrock is phyllite and slate.

Bibliography

Lindgren, Waldemar, 1911, Tertiary gravels of the Sierra Nevada: U. S. Geol. Survey Prof. Paper 73, p. 143.

Lindgren, Waldemar, 1900, Colfax folio: U. S. Geol. Survey Geol. Atlas of the U. S., folio 66, 10 pp.

Brandy City

Location. This district is along the extreme west margin of Sierra County about 12 miles west of Downieville and 10 miles by road north of Camptonville. It is on the ridge between Canyon Creek on the north and the North Fork of the Yuba River on the south. It is principally a placer-mining district and was extensively hydraulicked from the 1850s until the early 1890s. Some work was done again from the early 1900s to the 1930s. The value of the total production of the district is unknown, but it probably amounts to several million dollars' worth of gold.

Geology. The main channel of the Tertiary North Fork of the Yuba River, which is also known as the La Porte channel, extended in a south-southwest direction from Poverty Hill into this district and thence to Indian Hill. Pay gravel here is 700 to 800 feet wide, up to 200 feet thick, and overlain by 40 to 60 feet of andesite. The upper 130 feet of gravel are mostly pebbles; the lower portion is bouldery and well cemented.

The entire deposit is reported to have averaged 25¢/yard in gold at the old price of gold, and near bedrock it was reported to have contained as much as $2.50/yard in gold. Bedrock is slate and serpentine with amphibolite both to the east and west.

Bibliography

Lindgren, Waldemar, 1911, U. S. Geol. Survey Prof. Paper 73, p. 101.

MacBoyle, Errol, 1918, Sierra County, The Brandy City district: California Min. Bur. Rept. 16, pp. 6–8, 66.

Taylor, George F., The Brandy City hydraulic mines, Sierra County: Eng. and Min. Jour., vol. 89, June 4, 1910, pp. 1152–1153.

Turner, H. W., 1898, Bidwell Bar folio, California: U. S. Geol. Survey Geol. Atlas of the U. S., folio 43, 6 pp.

Brown's Valley

Location. Brown's Valley is in north-central Yuba County about 15 miles northeast of Marysville. The Hammonton dredge field is just to the south. The area was mined during the gold rush, when very coarse placer gold was recovered. It was named for a prospector who reportedly recovered $12,000 from a quartz vein in a few weeks in 1850. Much lode mining was done here in the 1860s and 1870s, and intermittent activity continued through World War I. The Dannebroge mine was reopened in 1945 and has been intermittently active since. The lode mines are estimated to have yielded $3 million to $5 million.

Geology. The central portion of the district is underlain by a northwest-striking belt of fine-grained greenstone that is classified as metadiabase and andesite porphyry. Amphibolite lies to the east, valley alluvium to the west, and gravels of Yuba River to the south.

Ore deposits. A number of nearly west-striking quartz veins in greenstone dip either north or south. The veins range from one to 10 feet thick. The ore contains free gold, pyrite, some galena, and chalcopyrite. Much of the recovered values have been free gold, but the sulfide concentrates often run high. Stoping lengths average around 150 feet, and the veins have been mined to inclined depths of 1700 feet.

Mines. Cleveland, Dannebroge, Hibbert and Burris, Jefferson, Pennsylvania, Too Handy.

Bibliography

Lindgren, Waldemar, 1895, Smartsville folio; California: U. S. Geol. Survey Geol. Atlas of the U. S., folio 18, 6 pp.

O'Brien, J. C., 1952, Yuba County, Dannebroge mine: California Jour. Mines and Geology, vol. 48, no. 2, pp. 149–151.

Preston, E. B., 1890, Brown's Valley: California Min. Bur. Rept. 10, pp. 798–799.

Waring, C. A., 1919, Yuba County, Brown's Valley: California Min. Bur. Rept. 15, p. 422.

Brownsville

Location. The Brownsville district is in northeastern Yuba County about 35 miles northeast of Marysville and 27 miles southeast of Oroville. It includes the lode-gold deposits here and in the Hansonville-Rackerby area, and the New York Flat placer "diggings". The town was named for I. E. Brown, who established a sawmill here in 1851.

Geology and ore deposits. The chief rocks in the area are greenstone with several gabbrodiorite intrusions. Amphibolite and some slate lie to the east. A number of north- and a few west-striking quartz veins commonly occur near intrusive-metamorphic contacts. The veins are one to seven feet thick and contain free gold with varying amounts of pyrite and other sulfides. The ore shoots are limited in size. Milling ore usually yielded $\frac{1}{7}$ to $\frac{1}{3}$ ounce of gold per ton, but some high-grade pockets were encountered.

Mines. Abbott, Arbucco, B. A. C., Beaver, Beehive (Mt. Hope) $100,000, Easy Money, Golden Key, Horseshoe, Manzanita, Napa and Oro, Ora Lewa, R. C., Rogers, Seaborg and Davis, Spanish, Twentieth Century Wonder, William Arthur.

Bibliography

Lindgren, Waldemar, 1895, Smartsville folio: U. S. Geol. Survey Geol. Atlas of the U. S., folio 18, 6 pp.

Waring, C. A., 1919, Yuba County, gold-quartz mines: California Min. Bur. Rept. 15, pp. 443–456.

Buckeye

This is a small lode-gold district in southwestern Mariposa County about eight miles south of the town of Mariposa. The principal gold sources in the district were the Granite King and Live Oak mines, which were worked from the 1870s through the early 1900s and again from 1938 to 1941. The gold-quartz veins are two to five feet thick, strike to the northeast, and occur in granodiorite. The ore contains native gold, pyrite, galena, and sphalerite. Milling ore averages about $\frac{1}{4}$ ounce of gold per ton with occasional pockets of high-grade ore.

Bibliography

Bowen, O. E., Jr., 1957, Mariposa County, Granite King and Live Oak mines: California Jour. Mines and Geology, vol. 53, pp. 104–105.

Butte Creek

Location. Butte Creek is in north-central Butte County. It is chiefly a dredging district that extends along Butte Creek from about three miles southeast of Chico northeast to Centerville and Helltown, a distance of almost 12 miles. The Magalia district is contiguous on the northeast. The streams were placer-mined during the gold rush, and hydraulic mining and some drift mining of Tertiary gravels followed. The creek was worked with primitive power shovels and washing plants in the early 1900s. It was dredged from around 1902 to the early 1920s, again in the 1930s, and from 1945 to 1949.

Geology. The deposits consist mainly of stream and bench gravels in and along Butte Creek. They range from a few hundred feet wide at the upper end to nearly a mile at the lower end. The gravels are coarse, well rounded, and consist of andesite with some chert and minor quartz. Also there are Tertiary shore and bench gravels in the Centerville and Helltown areas. Dredging depths ranged from 13 to 35 feet, with an uneven bedrock. The last operations were reported to have yielded as much as 35 cents of gold per yard.

Dredging concerns. Butte Creek Cons., 1909–16, one bucket-line; Lancha Plana, 1941–49, one bucket-line; Pacific Gold Dredging Co., 1902–17?, two bucket-line; Piedmont Dredging Co., 1941, one Becker-Hop-

kins; Thurman and Wright, one dragline; Yuba Cons., 1941, one Becker-Hopkins.

Bibliography

Lindgren, Waldemar, 1911, U. S. Geol. Survey Prof. Paper 73, pp. 84–86.

Waring, C. A., 1919, Butte County, Gold dredging: California Min. Bur. Rept. 15, pp. 187–193, 194, 197.

Winston, W. B., 1910, Gold dredging in California, Butte Creek district: California Min. Bur. Bull. 57, pp. 159–162.

Butt Valley

Location. This district is in northwestern Plumas County. It is an extensive area that lies between Lake Almanor on the north and the Virgilia-Twain area on the south and southwest. The district includes the Caribou and Seneca areas. The Crescent Mills district is just to the east, and the Meadow Valley district lies to the south.

History. During the gold rush vast amounts of placer gold were recovered from the Feather River and its tributaries. The valley was named for Horace Butts, a successful early-day miner. The town of Seneca was an important center at that time. The Butt Valley Reservoir was constructed by the Great Western Power Company in 1921. Lode-gold mining, which originally began in the 1850s, continued through the 1930s. Since then there has been intermittent prospecting at a few properties such as the Sunnyside mine. Skin divers and weekend prospectors have been active along the streams.

Geology. The district is underlain by a series of metamorphosed sediments ranging from Silurian to Triassic in age. Slate is most abundant, but also present are sandstone, limestone, quartzite, and conglomerates. In addition, there are greenstones and serpentine.

Ore deposits. The gold-quartz veins occur in brecciated zones in all of the formations, but they are most abundant in the slates and greenstones. The ore contains free gold and varying amounts of sulfides, including arsenopyrite. Milling ore averaged usually only a few dollars per ton, but the ore bodies were extensive. Some of the veins and vein zones are 20 or more feet thick. An appreciable number of high-grade pockets were found. In places the placer deposits, both Tertiary and Recent in age, were extremely rich.

Mines. Lode: Chico Star, Dean, Del Monte, Duncan, Elizabeth Cons., Halstead, Hazzard, Horseshoe, Justice, Lictum, Plumas Amalgamated, Reising, Rich Gulch Cons., Savercool, Seneca Cons., Shenandoah, Summit, Virgilia, White Lily $225,000+. Placer: Barker Hill, Cameron $100,000+, Dominion, Ellis, Lot, Providence Hill, Red Rock, Sunnyside $150,000+, Swiss, Dutch Hill $575,000+.

Bibliography

Averill, Charles V., 1937, Plumas County, gold: California Div. Mines Rept. 33, pp. 103–124.

Logan, C. A., 1928, Plumas County, gold: California Div. Mines and Mining, Rept. 24, pp. 285–310.

MacBoyle, Errol, 1920, Plumas County, Butt Valley district: California Min. Bur. Rept. 16, pp. 1–4.

Calaveritas

Location. This is a placer-mining district in central Calaveras County, in the vicinity of the old town of Calaveritas. It includes the Old Gulch area to the north. At one time there were many Mexican miners in the area.

Geology. The district is underlain by slate, graphitic schist, quartzite, limestone, green schist and granodiorite. Numerous gravel patches, remnants of the west-trending Fort Mountain channel of Tertiary age, were mined by hydraulicking and sluicing during the early days. The recent stream gravels were worked by drag-line dredges during the 1930s. There are a few narrow gold-quartz veins.

Mines. Barnhardt, Calaveritas Hill Consolidated, Oro Fino, Railroad Hill, Richie Hill.

Bibliography

Clark, L. D., 1954, Geology and mineral deposits of the Calaveritas quadrangle: California Div. Mines Spec. Rept. 40, 23 pp.

Clark, W. B., and Lydon, P. A., 1962, Calaveras County, placer gold: California Div. Mines and Geology County Report 2, pp. 76–93.

Camanche-Lancha Plana

Location. This district extends along the Mokelumne River from the vicinity of Lancha Plana, Amador County, west through the Camanche-Wallace area, Calaveras County, to Clements, San Joaquin County, a distance of about 12 miles.

History. The creeks were first worked during the early part of the gold rush, and hydraulic mining of the terrace gravels followed. The town of Camanche was named for a town in Iowa. Later, the Chinese mined the river and reworked the old tailings. From 1904 to 1923, the river was dredged on a large scale by the American Dredging Company. Dragline dredging was done during the 1930s and bucket-line dredging from then until 1951. Much of the region was recently inundated by the Camanche reservoir. The output from dredging is estimated to be about $10 million.

Geology. The deposits consist of unconsolidated gravels in and adjacent to the Mokelumne River and floodplain deposits. Tightly packed shore gravels of Eocene age are found near Wallace. In the west portion of the dredging field, the gravels ranged from six to 50 feet deep while, in the east, they were six to 25 feet deep. Bedrock is clay and volcanic rock. Values recovered by dredging ranged from 10 to 25 cents per yard. The gold was fine grained and 850 to 900 in fineness.

Operations. American Dredging Co., 1904–23, three bucket-lines; Camanche Placers Ltd., 1935–, one bucket-line; Gold Gravel Products, 1935; Gold Hill Dredging Co., 1936–51, two bucket-lines; Lancha Plana Gold Dredging Co., 1926–40, one? bucket-line; Wallace Dredging Co., 1935–40, two bucket-lines.

Bibliography

Clark, W. B., 1955, San Joaquin County, Gold Hill dredges: California Jour. Mines and Geology, vol. 51, no. 1, pp. 37–39.

Logan, C. A., 1927, Calaveras County, American Dredging Company: California Min. Bur. Rept. 23, p. 198.

Turner, H. W., 1894, Jackson folio, California: U. S. Geol. Survey Geol. Atlas of the U. S., folio 11, 6 pp.

Tucker, W. B., 1919, Oro Water, Light, and Power Company dredges: California Min. Bur. Rept. 15, pp. 127–128.

Winston, W. B., 1910, Gold dredging in California, Calaveras County: California Min. Bur. Bull. 57, pp. 205–208.

Campo Seco-Valley Springs

Campo Seco and Valley Springs are in northwestern Calaveras County. Substantial amounts of gold have been produced in this district from a variety of mineral deposits. These deposits include Recent stream gravels, quartz-rich gravels of Eocene age, narrow quartz veins, and massive copper and zinc sulfide deposits of the noted Penn mine, from which gold was recovered as a by-product.

The older quartzitic gravels were mined by hydraulicking and ground sluicing in the early days and later by draft mining. The Recent gravels were worked by dragline dredges. The Penn mine is credited with a production of 60,000 ounces of gold.

The principal rocks underlying the district are green schist, slate, serpentine and greenstone. The quartz-rich gravels which overlie these rocks occur in patches and are the remnants of an Eocene river channel system. Some of these gravels are capped by andesite. The massive sulfide deposits at the Penn mine and the gold-quartz veins are in greenstone and schist.

Bibliography

Clark, W. B., and Lydon, P. A., 1963, Calaveras County, placer gold: California Div. Mines and Geology County Report 2, pp. 76–84.

Heyl, G. R., et al., 1948, Copper in California, Penn zinc-copper mine: California Div. Mines Bull. 144, pp. 61–84.

Turner, H. W., 1894, Jackson folio: U. S. Geol. Survey Geol. Atlas of the U. S. folio 11, 6 pp.

Camptonville

Camptonville, in northeast Yuba County and western Sierra County, adjoins the Pike district on the east. Rich gold discoveries were made here in 1850–51. The town was named in 1854 for Robert Campton, the local blacksmith. The Pelton wheel, long used in mining machinery and electric generators, was invented here by Lester Pelton. The old town is now a tourist attraction. A number of moderate-sized deposits of Tertiary channel gravel were mined by hydraulicking. These include the deposits at Young's Hill, Weed's Point, and Galena Hill. They are part of the Tertiary Yuba River that extended southwest from Indian Hill to this district and then to North San Juan. Bedrock is amphibolite, slate, greenstone, and granodiorite.

Bibliography

Lindgren, Waldemar, and Turner, H. W., 1895, Smartsville folio: U. S. Geol. Survey Geol. Atlas of the U. S., folio 18, 6 pp.

Canada Hill

Location. This district is in eastern Placer County 25 miles northeast of Forest Hill and about 10 miles

Photo 13. Carson Hill Mine and Mill, Carson Hill District. This northward view of the Calaveras County mine and large mill was taken in the 1920s. *Photo courtesy of Hillcrest Studio, Angels Camp.*

south of Cisco. It includes the Sailor Flat, Robertson Flat, Sailor Canyon and New York Canyon areas.

Geology. The Canada Hill channel, a branch of the Tertiary American River, is believed to have flowed northeast and east across Sailor Canyon and then southeast to join the southwest-flowing main Tertiary channel near French Meadows. The channel is steep, narrow, and has not been too productive. The gravels are usually angular and poorly washed. Bedrock is slate and quartzitic schist of the Blue Canyon Formation (Carboniferous) and schist of the Sailor Canyon Formation (Lower Jurassic). The gravels are capped by rhyolite and andesite. There are some gold-quartz veins that contain abundant sulfides.

Mines. Beauty, Carmac, Lost Emigrant, Merz, Monumental, Pacific Blue Lead, Placer Queen, Reed, Sailor Canyon, Trinidad, Walker, X-Ray.

Bibliography

Lindgren, Waldemar, 1900, Colfax folio: U. S. Geol. Survey Geol. Atlas of the U. S., folio 66, 10 pp.

Lindgren, Waldemar, 1897, Truckee folio: U. S. Geol. Survey Geol. Atlas of the U. S., folio 39, 8 pp.

Lindgren, Waldemar, 1911, Tertiary gravels of the Sierra Nevada: U. S. Geol. Survey Prof. Paper 73, pp. 157–158.

Logan, C. A., 1936, Gold mines of Placer County, placer mines: California Div. Mines Rept. 32, pp. 49–96.

Waring, C. A., 1919, Placer County, placer mines: California Min. Bur. Rept. 15, pp. 325–386.

Carson Hill

Location. Carson Hill is on the Mother Lode belt in southwestern Calaveras County. The district consists of that portion of the Mother Lode that extends from Carson Flat southeast through Carson Hill to the town of Melones on the Stanislaus River. It has also been known as the Melones district.

History. Carson Hill was named for James H. Carson, a soldier who came to California in 1847 and who discovered gold at nearby Carson Creek in 1848. Lode gold was first discovered in 1850 at the Morgan mine and many miners soon came to the area. By 1851, the town of Melones had a population of 3,000 to 5,000. It was named for the melon seed-shaped gold nuggets found here. The district was extremely productive then, much of the mineral coming from fantastically rich surface pockets. Gold to the amount of $110,000 was exposed by one blast, and in 1854, a 195-pound mass of gold, the largest ever taken in California, was found here. Telluride minerals were recovered in quantity, but most were lost in unsuccessful attempts to extract the gold.

The gold production from the district declined in the late 1850s. Large-scale mining of low-grade ore bodies began in 1889 at the Calaveras mine. The Melones mine was worked on a major scale from 1895 to 1918. The Morgan, Calaveras, and Melones mines were consolidated in 1918 and worked as a unit until 1926. They were operated again from 1933 until 1942. This was one of the richer portions of the Mother Lode, the Carson Hill group alone having yielded an estimated total of $26 million. Part of Carson Hill will be inundated by the New Melones Reservoir.

Geology. The district is underlain by a series of northwest-striking beds of phyllite, amphibolite, green schist, and serpentine. Widespread hydrothermal alteration has changed much of the serpentine to extensive bodies of mariposite-ankerite-quartz rock. Slate of the Mariposa Formation (Upper Jurassic) lies to the west and metasediments of the Calaveras Formation (Carboniferous to Permian) to the east.

Ore deposits. The deposits consist of thick, massive, and often barren quartz veins, with adjacent large bodies of auriferous schist and pyritic ankerite-mariposite-quartz rock containing numerous thin quartz seams and stringers. Milling ore usually was low in grade, but the ore bodies were extensive. The famous Hanging Wall ore body, which consisted of auriferous schist, averaged ½ ounce of gold per ton and had dimensions of 175 x 4500 x 15 feet. Much rich high-grade ore from surface pockets was recovered in the 1850s. Tellurides, which included calaverite, sylvanite, hessite, and petzite, were recovered in quantity during the early days, near the surface. Both calaverite and melonite, a rare nickel telluride, were first found and described from this district.

Mines. Carson Hill Mines $26 million (Calaveras, Finnegan, Melones, Morgan, Reserve, Stanislaus Mines), Carson Creek $1 million, Hardy, Santa Ana, Tulloch.

Bibliography

Burgess, John A., 1937, Mining methods at the Carson Hill mine, Calaveras County, California: U. S. Bur. Mines Inf. Circ. 6940, 15 pp.
Clark, W. B., and Lydon, P. A., 1962, Calaveras County, Carson Hill mines: California Div. Mines and Geology, County Rept. 2, pp. 44–50.
Eric, J. H., Stromquist, A. A., and Swinney, C. M., 1955, Geology and mineral deposits of the Angels Camp and Sonora quadrangles, Calaveras and Tuolumne Counties, California: California Div. Mines Spec. Rept. 41, 55 pp.
Knopf, Adolph, 1929, Mines on Carson Hill: U. S. Geol. Survey Prof. Paper 157, pp. 72–77.
Logan, C. A., 1935, Carson Hill mines: California Div. Mines Bull. 108, pp. 129–137.
Moss, F. A., 1927, The geology of Carson Hill: Eng. and Min. Jour., vol. 124, no. 26, pp. 1010–1012.
Ransome, F. L., 1900, Mother Lode district folio: U. S. Geol. Survey Atlas of the U. S., folio 63, 11 pp.
Storms, W. H., 1900, Melones Consolidated mines: California Min. Bur. Bull. 18, pp. 121–122.
Turner, H. W., 1894, Jackson folio, California: U. S. Geol. Survey Geol. Atlas of the U. S., folio 11, 6 pp.
Young, G. J., 1921, Gold mining at Carson Hill: Eng. and Min. Jour., vol. 112, pp. 725–729.

Cathey

The Cathey or Cathay district is in southwestern Mariposa County about 10 miles southwest of Mariposa and near the town of Catheys Valley. The area was first placer-mined during the gold rush, and lode mining began in the 1850s. The principal mines were the Francis, Moore Hill, and Rich mines, which were last worked in the 1930s. A number of north-trending quartz veins up to 9 feet thick, in granodiorite and schist, contain native gold and sulfides. The sulfides, which include pyrite, arsenopyrite, and galena, sometimes are extremely abundant.

Bibliography

Bowen, O. E., Jr., 1957, Mariposa County, lode gold mines: California Jour. Mines and Geology, vol. 53, pp. 72–186.

Photo 14. Suction Dredge, Calaveras County. Floating suction dredges, like this one on the Stanislaus River near Melones in 1966, have been active in streams of the gold-mining districts in recent years.

Photo 15.   Cherokee Hydraulic Mine, Cherokee District. This view of the Butte County mine dates back probably to the 1870s. *Photo courtesy of Calif. State Library.*

Cat Town

This district is in northwestern Mariposa County at the site of the old mining camp of Cat Town. Although the Kinsley-Greeley Hill district adjoins it on the north and the Coulterville portion of the Mother Lode is on the west, this district is on a separate northwest-trending vein system. It may be on the same belt of mineralization as the Whitlock district to the southeast (see fig. 4). The height of mining activity here was in the 1880s and 1890s. The Gold Bug mine and several others were active in the 1930s, and there has been some prospecting since. Also the area was recently prospected for molybdenum.

The deposits consist of gold-bearing quartz stringers in schist, slate, metachert, and greenstone. The values usually occur in small but rich pockets and are closely associated with albitite and diorite dikes. The deposits have not been mined to depths of greater than 100 feet. The principal gold sources have been the Black Bart, Gold Bug, and White Porphyry mines.

Bibliography

Bowen, O. E., Jr., 1957, Mariposa County, Gold Bug and White Porphyry mines: California Jour. Mines and Geology, vol. 53, no. 102 and 181–183.

Turner, H. W., and Ransome, F. L., 1897, Sonora folio: U. S. Geol. Survey Geol. Atlas of the U. S., folio 41, 7 pp.

Cherokee

Location and history. This district is in central Butte County, 12 miles north of Oroville on the north side of Table Mountain and in the vicinity of the town of Cherokee or Cherokee Flat. It was so named for a party of Cherokee Indians who migrated here in the 1850s to mine gold. It has also been known as the Spring Valley district. Most of the ouput has come from the single large hydraulic mine, which is estimated to have yielded about $15 million. The town reached its heyday in the middle 1870s when it had a population of about 700.

Geology. The Tertiary placer deposits are associated with a west-trending channel. In this area the channel is in a trough about 700 feet wide. The sequence from bottom to top of the hydraulic pit is as follows: irregular greenstone gravel 5–10 feet thick that is lean in gold and contains local black clay streaks and minor basalt blocks; a rich 20- to 30-foot layer of coarse fresh blue gravel with large greenstone blocks, coarse and fine gold, small diamonds, and minor platinum (this layer yielded as much as several dollars per yard); several feet of decomposed gravel; 50 feet of sand and quartzitic gravel, the lower part of which yielded 25 cents per yard; 200 feet of clayey sand; and 50 to 75 feet of massive basalt.

Between 400 and 500 small diamonds were recovered from the gold-bearing gravels at the Cherokee mine. Several of the stones were more than two carats in weight and of good quality, but most were small and had a pale-yellow tinge. This is the best-known diamond-bearing locality in California.

Bibliography

Creely, R. S., 1965, Geology of the Oroville Quadrangle: California Div. Mines and Geology Bull. 184, 86 pp.

Irelan, William, 1886, Spring Valley hydraulic mine: California Min. Bur. Rept. 6, pp. 24–25.

Lindgren, Waldemar, 1911, Tertiary gravels of the Sierra Nevada: U. S. Geol. Survey Prof. Paper 73, pp. 86–87.

Miner, J. A., 1890, Spring Valley hydraulic gold mine: California Min. Bur. Rept. 10, pp. 124–125.

Preston, E. B., 1893, Channel of Spring Valley Hydraulic Mining Company: California Min. Bur. Rept. 11, pp. 155–157.

Weatherbe, D'Arcey, 1906, A hydraulic mine in California: Min. and Sci. Press, vol. 93, Sept. 8, 1906, pp. 296–298.

Chinese Camp

Chinese Camp is in western Tuolumne County about 10 miles southwest of Sonora. It was a placer-mining center settled by Chinese miners in 1849. Much work was done in the 1850s, and the piles of soil and gravel turned over in the search for gold can still be seen in nearly every gulch. The old mining town of Chinese Camp is quite well-preserved.

Much of the placer gold was recovered from extremely rich quartzitic gravels of Eocene age. Some gold also was mined from gravels of late Tertiary age. Most of the early-day work was done by hydraulicking and ground sluicing; during the 1930s there was some dragline dredging. Bedrock consists of serpentine, greenstone, and slate. There are a few gold-quartz veins in the area. The value of the total output from placer mining is estimated at $2.5 million.

Bibliography

Turner, H. W., and Ransome, F. L., 1897, Sonora folio, California: U. S. Geol. Survey Geol. Atlas of the U. S., folio 41, 7 pp.

Chowchilla

In recent years small amounts of placer gold have been mined from the lower Chowchilla River west of Raymond in western Madera County. The gold has been recovered by several small- to medium-sized floating suction dredges, which consist of pontoons that carry suction pumps and gold-recovery equipment. The river gravels range from 10 to as much as 35 feet in depth. This area was first placer-mined during the gold rush, when the Grub Gulch district to the northeast was active.

Bibliography

Logan, C. A., 1950, Madera County, gold: California Jour. Mines and Geology, vol. 46, pp. 453–456.

Clear Creek

Location. The Clear Creek or Havilah mining district is in east-central Kern County, about 26 miles east-northeast of Bakersfield and five miles south of Bodfish. It is an extensive region that includes the Red Mountain and Walker Basin areas. It is also a tungsten district.

History. Gold was discovered in Clear Creek in 1863 or 1864 by Claude de la Borde, George McKay, Benjamin Mitchell, and Hugh McKeadney. The town of Havilah was established in 1865 and soon became an important center with a population of at least 3000. It was the seat of Kern County from 1867 until 1874. The area declined in the 1880s, but was intermittently active for many years afterward. Some work has been done in recent years at the Joe Walker and Rand mines.

Geology. The area is underlain by quartz diorite with roof pendants of metasedimentary rocks in the north and south portions of the district. A body of gabbro lies to the northeast. The gold deposits are mostly confined to the quartz diorite west of Havilah and in the Walker Basin. They consist of quartz veins up to six feet thick, which contain free gold and varying amounts of sulfides.

Mines. Friday, Jackpot, Joe Walker $600,000+, Porter, Rand group $125,000, Rochfort, Southern Cross, Washington.

Bibliography

Dibblee, T. W., Jr., and Chesterman, C. W., 1953, Breckenridge Mountain quadrangle: California Div. Mines Bull. 168, 56 pp.

Troxel, B. W., and Morton, P. K., 1962, Kern County, Clear Creek district: California Div. Mines and Geology County Rept. 1, pp. 25–27.

Tucker, W. B., and Sampson, R. J., 1933, Kern County, Joe Walker mine: California Div. Mines Rept. 29, pp. 310–311.

Clearinghouse

Location and history. Clearinghouse is in central Mariposa County a few miles west of El Portal and Yosemite National Park. It was named for the Clearinghouse mine, the largest source of gold in the district. The Merced River, which flows through the area, was extensively placer-mined during the gold rush. The Clearinghouse mine was discovered in 1860 and worked on a fairly large scale until about 1880. There was mining activity again during the early 1900s and 1930s, and there has been intermittent prospecting and development work since. Substantial quantities of limestone and barite and some tungsten are found here. At one time the Yosemite Valley Railroad extended through the area to El Portal, the line's eastern terminus. From El Portal, passengers were taken by stage to Yosemite Valley.

Geology and ore deposits. The principal rocks underlying the district are graphitic schist, slate, quartzite, and hornfels of the Calaveras Formation (Carboniferous to Permian). There are some granitic dikes, and the main mass of the Sierra Nevada granitic batholith is a few miles to the east.

The gold deposits occur in north-striking quartz veins that usually range from one to five feet in thickness. The ore contains free gold and often abundant sulfides. Appreciable amounts of high-grade ore have been found here, and milling-grade ore commonly contained one ounce or more of gold per ton. Several of the veins have been developed to inclined depths of about 1200 feet.

Mines. Clearinghouse $3.35 million, Gold Star, Old Timer, Rutheford and Cranberry, South Cranberry, Uncle Jim, West Rutherford.

Bibliography

Bowen, O. E., Jr., 1957, Mariposa County, lode mines: California Jour. Mines and Geology, vol. 53, pp. 72–187.

Laizure, C. McK, 1928, Mariposa County, gold lode mines: California Div. Mines and Mining Rept. 24, pp. 79–122.

Clipper Mills

Clipper Mills is in southeastern Butte and northeastern Yuba County, about eight miles east-northeast of Forbestown. Several moderate-sized deposits of Tertiary channel gravels have been mined by drifting and hydraulicking. The Pratt and Gentle Anna drift mines and the Pittsburg Hill hydraulic mines have been the principal placer-gold sources. A few narrow gold-quartz veins have yielded small but rich pockets. The district is underlain by slate, serpentine, and gabbro-diorite.

Bibliography

Lindgren, Waldemar, 1911, Tertiary gravels of the Sierra Nevada: U. S. Geol. Survey Prof. Paper 73, pp. 99–100.

Turner, H. W., 1898, Bidwell Bar folio: U. S. Geol. Survey Geol. Atlas of the U. S., folio 43, 6 pp.

Coarsegold

Location. This district is in east-central Madera County in the vicinity of the town of Coarsegold about 30 miles northeast of Madera. It is in the central portion of a 20-mile long belt that extends from Grub Gulch on the northwest to Fine Gold on the southeast. It includes part of the area once known as the Potter Ridge district. Coarse and heavy placer gold was recovered from shallow deposits during the gold rush. Lode mining began in 1853, when the Texas Flat mine was discovered. The district has been intermittently mined and prospected ever since.

Geology. The area is underlain by medium to coarse-grained granodiorite with some narrow belts of slate and schist. A number of north-trending quartz veins that contain free gold and varying amounts of sulfides are enclosed in both the metamorphic and grantic rocks. The veins were mined to depths of 1000 feet.

Mines. Baker, Balfron, Daisy Bell, Five Oaks, Golden Road, Melvin, Morning Star, New Citizen, Texas Flat $200,000+.

Bibliography

Irelan, William, 1888, Texas Flat mine: California Min. Bur. Rept. 8, pp. 212–213.

Laizure, C. McK., 1928, Madera County, gold: California Div. Mines Rept. 54, pp. 324–328.

McLaughlin, R. P., and Bradley, W. W., 1916, Madera County, Texas Flat mine: California Min. Bur. Rept. 14, p. 551.

Colfax

Location. The Colfax district is in southwestern Placer County in the vicinity of the town of that name. It includes the deposits in the Weimar-New England Mills area and placer-mining areas along the American and the Bear Rivers. The Colfax district also has been know as the Illinois district.

History. Placer mining began here soon after the beginning of the gold rush. The locality was first known as Alder Gulch and later Illinoistown. The Rising Sun mine was discovered in 1866. Colfax, which was named for U. S. Senator Schuyler Colfax who visited the place in 1865, was an important center during the construction of the Central Pacific Railroad. Considerable mining continued in the district until about 1900. Small-scale placer mining, some done by skindivers, has continued until the present time.

Geology. Slates of the Mariposa Formation (Upper Jurassic) are found in the central portion of the district, and slate of the Cape Horn Formation (Carboniferous) crops out to the east. A gabbro intrusion lies to the north and northwest, a diabase body to west, and serpentine lenses and amphibolite to the east and southeast. A fairly extensive patch of Tertiary channel gravel is exposed on the north side of Colfax Hill. The extensive quartz gravels in the Bear River are hydraulic mine tailings from the You Bet and Lowell Hill districts.

Ore deposits. The ore bodies occur in a number of northeast-striking quartz veins that usually range from two to five feet in thickness. The ore contains free gold and often abundant sulfides. Considerable high-grade ore was found near the surface during the early days.

Mines. Lode: Annie Laurie, Big Oak Tree $100,000+, Bauer, Black Oak, Brushy Creek, Chubb, Hinchy, Last Chance, Live Oak Ravine, Rising Sun $2 million+, Victory, Whiskey Tunnel. Placer: Bear River Ext., Bear River Tunnel, Burnt Flat, Collins, Rocky Bar.

Bibliography

Chandra, D. K., 1961, Geology and mineral deposits of the Colfax and Foresthill quadrangles: California Div. Mines Spec. Rept. 67, 50 pp.

Irelan, William, Jr., 1888, Rising Sun and Big Oak Tree mines: California Min. Bur. Rept. 8, pp. 462–464.

Lindgren, Waldemar, 1900, Colfax folio, California: U. S. Geol. Survey Goel. Atlas of the U. S., folio 66, 10 pp.

Logan, C. A., 1936, Gold mines of Placer County, Rising Sun mine: California Div. Mines Rept. 32, pp. 34–35.

Collierville

Location. Collierville is in the East Gold Belt of the Sierra Nevada in south-central Calaveras County and north-central Tuolumne County. It is about five miles due east of Murphys.

Geology. The area is underlain by west-striking beds of siliceous slate, banded quartzite, and mica schist, with several limestone lenses. Granodiorite stocks up to two miles in diameter lie in the northeast and southeast. A number of west-striking and north-dipping or northeast-striking southeast-dipping quartz veins in the metamorphic rocks contain free gold and often abundant sulfides. The quartz is white to dark gray in color. Some high-grade pockets were recovered here, and tellurides have been reported.

Mines. Collier, Dorsey, Golden Sulphuret, Louise, North Chimney Rock, Sailor Boy, True Business.

Bibliography

Clark, W. B., and Lydon, P. A., 1962, Calaveras County, gold: California Div. Mines and Geology, County Rept. 2, pp. 32–93.

Turner, H. W., and Ransome, F. L., 1898, Big Trees folio, California: U. S. Geol. Survey Geol. Atlas of the U. S., folio 51, 8 pp.

Coloma

Location. This district is in the vicinity of the old mining town of Coloma in western El Dorado County. It is on the South Fork of the American River about eight miles northwest of Placerville.

History. Although this is a relatively small district, it is significant, for it was here that James W. Marshall made his historic gold discovery. In August 1847, Captain John Sutter, grantee of a large Mexican land grant in the vicinity of present-day Sacramento, signed a contract with Marshall to erect a sawmill on the American River. Work commenced in September 1847. The mill was almost finished on January 24, 1848, when Marshall, inspecting the mill tailrace, noted several small flakes of what appeared to be gold. Work on the mill stopped, and more flakes were recovered. These were taken to Sutter at Sacramento for more tests, which proved beyond a doubt that it was gold. Attempts were made to keep the discovery a secret, but the news quickly leaked out.

Soon Sonorans from the Los Angeles placer-mining districts arrived, the vanguard of the thousands of gold seekers who came from all directions to Coloma. The surface placers here were soon exhausted, and the miners went elsewhere. Coloma was named for a nearby Southern Maidu Indian Village. Early spellings were "Colluma" and "Culoma". Some gold dredging was done on the American River here during the 1930s and 1940s.

Marshall never was associated with a really successful mining venture and died in 1885 in the nearby town of Kelsey, a poor man. The Marshall Monument, where he is buried, was dedicated in 1890. Part of the old town, the mill site and the monument joined the California state park system in 1927. A replica of Sutter's mill was recently constructed at the park. Also at the park is a museum containing many items of early-day mining equipment.

Geology and ore deposits. The central portion of the district is underlain by a granodiorite intrusion. It is surrounded by slate, mica schist, amphibolite, and several north-trending lenticular bodies of serpentine.

Most of the gold values were obtained from gravels in the American River or from terrace gravels along the bank. A few narrow gold-quartz veins crop out

and several contact-metamorphic copper-gold deposits are found along the margin of the granodiorite.

Bibliography

Clark, W. B., and Carlson, D. W., 1956, El Dorado County, gold: California Jour. Mines and Geology, vol. 52, pp. 400–437.

Lindgren, Waldemar, and Turner, H. W., 1894, Placerville folio, California: U. S. Geol. Survey Geol. Atlas of the U. S., folio 3, 3 pp.

Cutter, Donald W., 1948, The discovery of gold in California: California Div. Mines Bull. 141, pp. 13–17.

Columbia

Location. This famous placer-mining district is in north-central Tuolumne County, in the vicinity of the old mining town of Columbia, five miles north of Sonora. It includes the Yankee Hill, Sawmill Flat, Squabbletown, Brown's Flat, and Springfield areas. The Sonora district is just to the south and the American Camp district lies to the northeast.

History. Columbia was one of the richest and most famous placer-mining districts in California. Early in 1850 a group of Mexican miners who had been forced off their claims at Sonora struck it rich here. Americans moved in and in turn forced them to leave. For a short period, the district was known as Hidreth's Diggings and American Camp, but it soon became "Columbia, Gem of the Southern Mines". During the 1850s and early 1860s, the diggings were enormously productive, the output averaging $100,000 or more per week. Columbia was one of the largest cities in California at this time, with an estimated population of 25,000 to 30,000. The district declined in the late 1860s, but small-scale mining continued until recently. The central portion of the old town became a state park in 1945 and is now a popular tourist attraction. Many of the famous old buildings have been restored. The value of the total production of the district has been estimated to be at least $87 million, and some have put the figure as high as $150 million.

Geology. Columbia lies in a preserved Tertiary valley with pre-volcanic features. It is a flat valley that is underlain chiefly by crystalline limestone and dolomite of the Calaveras Formation (Carboniferous to Permian). The limestone has numerous deep potholes and cavities, which contained enormously rich gravel. Several very large nuggets and gold masses were taken here, including one that weighed over 50 pounds and several weighing more than 300 ounces. Slow degradation of the area in pre-volcanic times tended to concentrate coarse gold in this flat basin. It is south of the main Tertiary Stanislaus River. Vertebrate fossils were found in the gravels. In the early-day mining operations, the gravels were hoisted from the potholes and washed through sluices and long toms on raised platforms.

Bibliography

Haley, C. S., 1923, Gold placers of California: California Min. Bur. Bull. 92, p. 148.

Lindgren, Waldemar, 1911, Tertiary gravels of the Sierra Nevada: U. S. Geol. Survey Prof. Paper 73, pp. 212–213.

Logan, C. A., 1928, Tuolumne County, placer mines: California Min. Bur. Rept. 24, pp. 32–43.

Photo 16.   Placer Mining, Columbia District. This early photo was taken in Columbia, Tuolumne County, near the Wells-Fargo Express Company office. *Photo courtesy of Tuolumne County Museum.*

Photo 17.   Hummocky Limestone, Columbia District. Rich, gold-bearing gravels occurred in deep cavities in limestone bedrock, like these at Columbia, Tuolumne County.

Photo 18. Virginia Mine, Coulterville District. This photo of the Mariposa County mine was taken in 1919. *Photo by W. O. Castello.*

Louderback, G. D., 1933, Notes on the geologic section near Columbia, California, with special reference to the occurrence of fossils in the auriferous gravels: Carnegie Inst. Washington, Publ. no. 440, pp. 7–13.

Ransome, F. L., 1897, Mother Lode district folio: U. S. Geol. Survey Geol. Atlas of the U. S., folio 63, 11 pp.

Turner, H. W., and Ransome, F. L., 1898, Big Trees folio: U. S. Geol. Survey Geol. Atlas of the U. S., folio 51, 10 pp.

Storms, W. H., 1900, Confidence mine: California Min. Bur. Bull. 18, pp. 136–137.

Tucker, W. B., 1916, Tuolumne County, Confidence mine: California Min. Bur. Rept. 14, p. 142.

Turner, H. W., and Ransome, F. L., 1898, Big Trees folio: U. S. Geol. Survey Geol. Atlas of the U. S., folio 51, 8 pp.

Confidence

Location and History. The Confidence district is in west-central Tuolumne County between Twain Harte and Long Barn and about 15 miles northeast of Sonora. It is in the Sierra Nevada East Gold Belt and adjoins the Soulsbyville district on the northeast (see fig 24, p. 122). This region was first placer-mined during the gold rush, and lode mining began shortly afterward. The Confidence mine was worked on a large scale until around 1912. Some of the mines were active in the 1930s, and the district was later prospected for tungsten.

Geology. The mines in this district are located chiefly in a southwest-extending arm or "peninsula" of granitic rocks of the main Sierra Nevada batholith (see fig. 24). The principal rock is granodiorite; slate, mica schist, phyllite, and quartzite lie to the south. Fine-grained dioritic dikes are present and commonly are associated with the gold-quartz veins. Also present are small bodies of tungsten-bearing tactite, which occur as roof pendants in the granitic rocks.

Ore Deposits. The ore bodies occur in north to northwest-striking quartz veins that usually range from one to five feet in thickness. The ore contains free gold and often abundant sulfides, especially galena, which often is associated with gold. The Confidence vein was mined to an inclined depth of 1200 feet.

Mines. Casa Madera, Corona, Confidence $4.25 million, Excelsior $420,000, Fair Oaks, Gem, Geraldine, Green $200,000, Humbug, Lucky Strike, Morning Glory, Red Cloud, Ripperton, Ryan, Thunderbolt, Too Far North, Wall Street.

Bibliography

Logan, C. A., 1949, Tuolumne County, Confidence mine: California Jour. Mines and Geology, vol. 45, pp. 61–62.

Preston, E. B., 1893, The Confidence mine: California Min. Bur. Rept. 11, p. 503.

Coulterville

Location. This district is in southwestern Tuolumne County and northwestern Mariposa County. It is that portion of the Mother Lode gold belt that extends from the vicinity of the McAlpine mine southeast through Peñon Blanco and the town of Coulterville to Virginia Point, a distance of about 10 miles.

History. The surface portions of the veins and the streams were worked during the gold rush. The town was named for George Coulter, who opened a store there in 1849. Quartz-mining began about 1852 with the discovery of the Malvina and Mary Harrison veins. Considerable lode mining was done from the 1860s through the 1890s, when many of the mines belonged to the Cook estate. There was some activity from the early 1900s until 1942. In recent years some work has been done at the Virginia, McAlpine and a few other mines.

Geology. In this district, the Mother Lode system is associated with an extensive northwest-trending lenticular body of serpentine. In a number of places the serpentine has been hydrothermally altered to mariposa-quartz-ankerite rock. Slate and greenstone lie to the west, and amphibolite schist and greenstone are to the east. Also present are smaller amounts of chlorite schist, phyllite, and metadiorite and aplitic dikes.

Ore Deposits. The ore deposits consist of pyrite-bearing bodies of mariposite-quartz-ankerite rock, with numerous parallel quartz stringers and adjacent massive quartz veins. The gold occurs in the free state or with the pyrite and usually is found in the quartz stringers, although in places there are values in the massive veins. A number of high-grade pockets have been found in this district, some of which came from the gossan that overlies some of the deposits. Tellurides also have been found here. Milling ore usually ranges from $\frac{1}{5}$ to $\frac{1}{2}$ ounce of gold per ton. The ore shoots had stoping lengths of up to 400 feet and were mined to inclined depths of more than 1200 feet. The massive veins form prominent ridges; Peñon Blanco Ridge is visible for many miles.

Mines. Mariposa County: Adelaide, Big Lode, Champion I $150,000 to $200,000, Flyaway group, Louisa $100,000, Malvina $1 million, Mary Harrison $1.5 million, Midas, Oro Rico, Potosi, Tyro $110,-000+, Virginia $824,000. Tuolumne County: McAlpine $100,000+.

Bibliography

Bowen, O. E., Jr., 1957, Mariposa County, Virginia Mine: California Jour. Mines and Geology, vol. 53, nos. 1–2, pp. 176–179.

Castello, W. O., 1921, Mariposa County, Coulterville district: California Min. Bur. Rept. 17, pp. 92–93.

Logan, C. A., 1935, Mother Lode gold belt—Mariposa County: California Div. Mines Bull. 108, pp. 180–190.

Lowell, F. L., 1916, Mariposa County, Mary Harrison quartz mine: California Min. Bur. Rept. 14, p. 588.

Ransome, F. L., 1900, Mother Lode district folio: U. S. Geol. Survey Atlas of the U. S., folio 63, 11 pp.

Storms, W. H., 1900, The Mother Lode region—Mariposa County: California Min. Bur. Bull. 18, pp. 142–147.

Cove

Location and History. The Cove district is in northeastern Kern County between the towns of Kernville and Isabella on the west side of the Isabella Reservoir.

The upper Kern River here was placer-mined during the 1850s. The Big Blue vein was discovered by Lovely Rogers in 1860, and much activity followed during the 1870s and early 1880s. The mines were worked intermittently from then through the 1930s, the Big Blue group having been operated on large scale from 1934 until 1943. There has been only minor activity since. The main mines of the district have been consolidated into the Big Blue-Sumner group, which includes the Lady Belle group. The district has an estimated output valued at $8 million.

Geology and Ore Deposits. The principal mines are in granodiorite. To the east and south are schist, phyllite, quartzite, and marble of the pre-Cretaceous Kernville Series. Aplite dikes are often associated with the gold-bearing veins.

The ore deposits consist of extensive vein systems as much as 150 feet wide that occur in shear zones in granodiorite. The ore consists of quartz with finely disseminated free gold, arsenopyrite, pyrite, chalcopyrite, and gelana. The milling ore usually averages 1/10 to 1/3 ounce of gold per ton with some high-grade streaks. Some scheelite also is present in the gold ore. The veins have been mined to depths of about 500 feet. There are two main vein systems: those of the Big Blue-Sumner and the Lady Belle groups.

Bibliography

Brown, G. C., 1916, Kern County, Cove district; Beauregard, Big Blue, Blue Gouge and Bull Run mines: California Min. Bur. Rept. 14, pp. 482 and 487–490.

Crawford, J. J., 1893, Big Blue mine: California Min. Bur. Rept. 11, p. 142.

Miller, William J., and Webb, Robert W., 1940, Descriptive geology of the Kernville quadrangle: California Div. Mines Rept. 36, pp. 343–378.

Prout, John W., Jr., 1940, Geology of the Big Blue group of mines, Kernville: California Div. Mines Rept. 36, pp. 379–421.

Troxel, B. W., and Morton, P. K., 1962, Kern County, gold: California Div. Mines and Geology County Rept. 1, pp. 92–196.

Tucker, W. B., and Sampson, R. J., 1940b, Mineral resources of the Kernville quadrangle: California Div. Mines Rept. 36, pp. 322–333.

Crescent Mills

Location. This district is in north-central Plumas County. It contains a northwest-trending belt of lode and placer deposits that extends from the vicinity of the town of Crescent Mills northwest through Greenville to Almanor, a distance of about 10 miles. At one time it was also known as the Cherokee district. Crescent Mills has been the most productive of the districts in the northern end of the Sierra Nevada.

Geology. This area is underlain by a series of northwest-trending beds of Paleozoic and Mesozoic metamorphic rocks. They are: pre-Devonian metarhyolite; Silurian metasedimentary rocks; Taylorsville Formation (Devonian) slate and sandstone; Taylor Meta-andesite (Carboniferous); Calaveras Formation (Carboniferous to Permian) metasedimentary rocks (four formations); greenstones (Carboniferous); Triassic metasedimentary rocks; Jurassic granodiorite and serpentinized basic and ultra-basic intrusives. Overlying the bedrock are patches of Tertiary and Quaternary gravels.

Ore Deposits. A number of northwest-striking quartz veins occur principally in the greenstones, slates, and granodiorite. Some of the veins are as much as 20 feet thick. The ore contains free gold, auriferous pyrite, and smaller amounts of other sulfides. The surface ores were especially rich, and numerous high-grade pockets were encountered during the early days. Milling-grade ore usually contained slightly less than ¼ ounce per ton, but the ore shoots had stoping lengths of up to 300 feet. Iron and copper also occur in the district.

Mines. Altona, Arcadia, Cherokee $250,000, Crescent $500,000+, Dagian, Droege $300,000+, Gold Stripes, Green Mountain $1 million to $2 million, Hobson, Indian Falls, Indian Valley $1.8 million+, Leete, Long Valley, Monitor, Mountain Lily, New York $400,000+, Pearless, South Eureka, Wardlow, Whitney.

Bibliography

Averill, C. V., 1928, Plumas County: California Min. Bur. Rept. 24, pp. 261–316.

Averill C. V., 1937, Mineral resources of Plumas County: California Jour. Mines and Geology, vol. 33, pp. 79–143.

Diller, J. S., 1905, Mineral resources of the Indian Valley region, California: U. S. Geol. Survey Bull. 260, pp. 45–49.

Diller, J. S., 1908, Geology of the Taylorsville region: U. S. Geol. Survey Bull. 353, 128 pp.

Irelan, William, 1888, Green Mountain mine, etc.: California Min. Bur. Rept. 8, pp. 479–481.

Lindgren, Waldemar, 1911, Tertiary gravels of the Sierra Nevada, U. S. Geol. Survey Prof. Paper 73, pp. 114–116.

MacBoyle, Errol, 1919, Plumas County, Crescent Mills district: California Min. Bur. Rept. 16, pp. 4–8.

Damascus

Location. This district is in east-central Placer County at the site of the old town of Damascus, about seven miles southeast of Dutch Flat. It includes the lode mines of the Pioneer-Humbug Bar area on the north and the extensive placer deposits that extend from Damascus south through Forks House to the Sunny South-Gas Hill area.

History. The streams in this area were originally mined during the gold rush, and drift mining began in the late 1850s. The Hidden Treasure drift mine was discovered in 1875 and was worked on a major scale through the early 1900s. There was some mining activity in the district again during the 1930s, and there has been intermittent prospecting since. This area has been quite productive; the drift mines alone have had a total output of more than $12 million and the lode mines several million dollars more.

Geology. An early Tertiary gravel channel extends from Damascus south to Gas Hill and eventually southwest to Michigan Bluff. An intervolcanic channel that apparently eroded away portions of the earlier channel enters the area from Westville to the east. The lower earlier "white" channel is more than 300 feet wide. It contains abundant quartz. The gravels in this early channel yielded 50 cents to $1.75 per ton at the old price, and the gold was coarse. The upper channels are commonly known as "blue" channels. Bedrock is slate and schist, and the gravels are capped by rhyolite and andesite. The channel has been mined by drifting almost continuously from Damascus south to Hidden Treasure, a distance of more than four miles. The quartz veins, which occur in slate, range from two to eight feet in thickness and contain free gold and often abundant sulfides. The ore usually is low to moderate in grade, but the ore shoots had stoping lengths of up to several hundred feet. The Pioneer mine was developed to a depth of 1400 feet.

Mines. Lode: American Eagle, Black Hawk, Central, Dover, Floyd, Lynn, Mars, North Star, Pioneer $1 million, Rawhide $300,000+, Southern Cross. Placer: Bullion, Cameron, Comet, Gas Hill, Golden River, Hermit, Hidden Treasure $4 million, Mountain Chief $700,000+, Mountain Gate, Rainbow, Tickell.

Bibliography

Browne, Ross E., 1890, The ancient rivers of the Forest Hill Divide: California Min. Bur. Rept. 10, pp. 435–465.

Crawford, J. J., 1894, Golden River and Hidden Treasure mines: California Min. Bur. Rept. 12, pp. 208–210.

Logan, C. A., 1936, Gold mines of Placer County, Hidden Treasure mine: California Div. Mines Rept. 32, pp. 65–66.

Lindgren, Waldemar, 1900, Colfax folio, California: U. S. Geol. Survey Geol. Atlas of the U. S., folio 66, 10 pp.

Lindgren, Waldemar, 1911, Tertiary gravels of the Sierra Nevada: U. S. Geol. Survey Prof. Paper 73, pp. 150–159.

Waring, C. A., 1919, Placer County, Damascus district and Placer County drift mines: California Min. Bur. Rept. 15, pp. 317 and 352–375.

Deer Creek

Deer Creek is in western El Dorado County and eastern Sacramento County. The creek, which flows southwest, was first placer-mined during the gold rush. In the 1930s and early 1940s, substantial amounts of gold were recovered here by dragline dredges. One operation was reported to have recovered about 4500 ounces of gold from 790,000 cubic yards of gravel.

Bibliography

Averill, C. V., 1946, Placer mining for gold in California: California Div. Mines Bull. 135, pp. 255–257.

Deer Valley

Location. This is a small lode-gold district in western El Dorado County about 10 miles northwest of Shingle Springs. The area was first worked during the gold rush, and it has been intermittently prospected since.

Geology. This district is on the north end of a large gabbro-diorite intrusion. Greenstones and serpentine lie to the east. There are a number of narrow gold-quartz veins containing small and shallow ore bodies. A few small high-grade pockets have been found. Sulfides, especially galena, which is associated with gold, are abundant.

Mines. Boneset, Delores, Morman Hill, Rose Kimberley.

Bibliography

Clark, W. B., and Carlson, D. W., 1956, El Dorado County, lode gold mines: California Jour. Mines and Geology, vol. 52, pp. 401–429.

Lindgren, Waldemar, and Turner, H. W., 1894, Placerville folio: U. S. Geol. Survey Geol. Atlas of the U. S., folio 3, 3 pp.

Diamond Mountain

Location and History. This district is in southern Lassen County about five miles south of Susanville. It is in the Diamond Mountain block, which lies at the extreme north end of the Sierra Nevada. Placer-mining began in the late 1850s, and several hundred thousand dollars in gold were soon recovered. Later some work was done by Chinese miners. Lode-mining began in the 1860s and continued sporadically through the early 1900s with some activity again in the 1930s. The output of the district is estimated at around $1 million.

Geology. The district is underlain by quartz diorite and granite, which is overlain in places by Tertiary gravel and andesite. The quartz veins are as much as 15 feet thick and occur in shear zones chiefly in the quartz diorite. The ore contains free gold and varying amounts of pyrite. Milling ore usually was low to medium in grade, but some high-grade pockets were found. The ore shoots were not large and none of the mines has been developed to any great depth. Some aplitic and basic dikes are associated with veins. Opal also is in the veins.

Mines. Arkansas $200,000+, Gold Belt (McDow) $20,000, Honey Bee $50,000+, Honey Lake, Harris-Mosgrove, Red Jacket (Gayman).

Bibliography

Averill, C. V., and Erwin, H. D., 1936, Lassen County, Diamond Mountain district: California Div. Mines Rept. 32, pp. 409–424.

Diller, J. S., 1908, Geology of the Taylorsville region, California: U. S. Geol. Survey Bull. 353, pp. 68–69.

Lindgren, Waldemar, 1911, Tertiary gravels in the Sierra Nevada, California: U. S. Geol. Survey Prof. Paper 73.

Preston, E. B., 1890, Lassen County: California Min. Bur. Rept. 10, pp. 273–276.

Tucker, W. B., 1919, Lassen County, Diamond Mountain mining district: California Min. Bur. Rept. 16, pp. 235–236.

Dobbins

Location. This district is in northeastern Yuba County in the vicinity of the town of Dobbins. The town was named for the Dobbins' brothers who settled herein 1849. It also has been known as the Indiana Ranch district.

Geology. The principal rock type here is granodiorite, with smaller amounts of quartz diorite and diorite intrusive into greenstone and slate. A number of north- and west-trending narrow quartz veins are found in granodiorite or along granodiorite-greenstone contacts. The ore contains free gold and varying amounts of sulfides and occasionally tellurides and scheelite.

Mines. California Mother Lode, Good Title, Higgins, Liberty, Lillian Francis, Red Cross, Red Ravine $100,000, Summit Hill, Templar.

Bibliography

Lindgren, Waldemar, 1895, Smartsville folio, California: U. S. Geol. Survey Geol. Atlas of the U. S., folio 18, 6 pp.

Downieville

Location. This is a lode and placer gold-mining district in west-central Sierra County in the general vicinity of the town of Downieville. It includes the Fir Cap Mountain, Craycroft, China Flat, and Slug Canyon areas and part of the Pliocene Ridge area. The Goodyear's Bar-Alleghany belt lies immediately to the west, the Sierra City district to the east, and the American Hill district to the south.

History. This area was prospected soon after the beginning of the gold rush. Major William Downie and his party arrived here in November, 1849. Soon a town was laid out, which was named for him early in 1850. The Downieville mining district was organized with "claims fixed at 30 feet per man". Many rich strikes were made; one claim 60 feet square yielded $80,000 in six months. At nearby Tin Cup Diggings, three men filled a tin cup with gold each day before quitting. A 25-pound nugget was found in the river upstream from the town in 1850. More than 5000 persons lived here in 1851. After the surface placers were exhausted, the river was mined, and hydraulic and drift mining became important. Mining continued almost steadily until World War II, and intermittent prospecting and skin diving for gold continues.

Geology. The district is underlain predominently by north-trending beds of phyllite, slate and quartzite of the Calaveras Formation (Carboniferous to Permian). To the west are greenstone, amphibolite, and serpentine. The higher ridges are capped by Tertiary andesite, which in places overlies rich Tertiary auriferous gravels. There are fairly extensive recent river and terrace gravels along the Yuba River and its branches.

Ore Deposits. A considerable number of gold-quartz veins occur chiefly in greenstone and slate. The veins range from one to 10 feet in thickness. The ore contains free gold and varying amounts of sulfides.

The milling-grade ore usually averages ¼ to ⅓ ounce per ton. Some ore shoots had stoping lengths of as much as several hundred feet. Some high-grade ore pockets have been taken from some of the mines. The Tertiary channel gravels are quartzitic, often well-cemented, and in places contain extremely coarse gold.

Mines. Lode: Alhambra, Bessler, Elcy, Finney (York) $75,000+, Gold Bluff $1.5 million, Gold Point (Grey Eagle) $100,000+, High Commission, Jumper, Mexican, Oro, Oxford $100,000+, Secret Canyon, Sierra Standard $75,000. Placer: Brown Bear, City of Six, Craycroft, Golden Hub, Kirkpatrick, Klondike, Mott and Mt. Vernon, Monte Carlo, New York, White Bear $200,000+, Wide Awake $100,-000+.

Bibliography

Averill, C. V., 1942, Sierra County, gold: California Div. Mines Rept. 38, pp. 17–48.

Carlson, D. W., and Clark, W. B., 1956, Lode gold mines of the Alleghany-Downieville area, Sierra County: California Jour. Mines and Geology, vol. 52, pp. 237–272.

Crawford, J. J., 1894, Gold Bluff mine: California Min. Bur. Rept. 12, p. 266.

Lindgren, Waldemar, 1911, Tertiary gravels of the Sierra Nevada: U. S. Geol. Survey Prof. Paper 73, p. 111.

Logan, C. A., 1929, Sierra County, Brush Creek, City of Six, and Grey Eagle mines: California Div. Mines Rept. 25, pp. 160, 161, and 165–166.

MacBoyle, Errol, 1920, Sierra County, Downieville mining district: California Min. Bur. Rept. 17, pp. 8–11.

Turner, H. W., 1897, Downieville folio; California: U. S. Geol. Survey Geol. Atlas of the U. S., folio 37, 8 pp.

Duncan Peak

Location and History. This is an extensive area of placer deposits in the general vicinity of Duncan Peak and the Greek Store guard station in eastern Placer County. It is 20 miles east of Forest Hill and six miles southeast of Last Chance. The area extends from just south of Duncan Peak south through Duncan Canyon to Ralston Ridge. The area was first mined in the early 1850s. The peak was named for Thomas Duncan, an early-day miner. During these early days, this region supported many Greek placer miners. Intermittent prospecting and development work has continued until the present time.

Geology and Ore Deposits. The placer deposits occur in a complex system of Tertiary channels that extends south and southwest to join the main west-trending Long Canyon channel in the Ralston Divide district. There are a number of tributaries and channel remnants, one of the main ones known as the Chalk Bluff channel. The deposits are up to several hundred feet wide and were extremely rich in places. Usually the gold is coarse. Quartz is sparse. The bedrock is quartzitic schist and slate, and the gravels are capped by andesite. There are some narrow gold-quartz veins in the district.

Mines. Bald Mountain, Barney, Blue Eyes, Dixie Queen, Glenn Cons., Gold Dollar, Golden Gate No. 1, Hard Climb, Hunted Hole, Miller's Defeat, Pine Nut, Pork and Brown, Red Star, Jack Robinson, Sauer Kraut, Savage, Trap Line, Yellow Jacket.

Bibliography

Lindgren, Waldemar, 1900, Colfax folio: U. S. Geol. Survey Geol. Atlas of the U. S., folio 66, 10 pp.

Lindgren, Waldemar, 1911, Tertiary channels of the Sierra Nevada: U. S. Geol. Survey Prof. Paper 73, pp. 152–153.

Logan, C. A., 1925, Ancient channels of the Duncan Canyon region: California Min. Bur. Rept. 21, pp. 275–280.

Logan, C. A., 1936, Gold mines of Placer County, placer mines: California Div. Mines Rept. 32, pp. 49–96.

Waring, C. A., 1919, Placer County, placer mines: California Min. Bur. Rept. 15, pp. 352–386.

Dutch Flat

Location. Dutch Flat is in north-central Placer County. This district includes the Alta and Towle areas. The Gold Run district lies just to the south, the You Bet district to the west, and the Lowell Hill district to the northeast.

History. Placer mining began here in 1849. The settlement was established by some Germans or "Dutch" in 1851. Hydraulicking began in 1857 and, during the following few years, the hydraulic mines were highly productive. Operations continued until 1883, when the mines were shut down by anti-debris injunctions. Some work was done in the district again in the 1890s and early 1900s. Logan (1936) estimated the district to have a total output of $4.5 million to $5 million although it may be more. The old town of Dutch Flat is well-preserved and is now a popular tourist attraction.

Geology. This district is located at the junction of several major channels of the Tertiary American River. One channel enters the area from the Lowell Hill district on the northwest, another from Lost Camp and Shady Run on the east, and a third from the Gold Run district on the south. The main channel then continues west and northwest through Little York, You Bet, Red Dog, and Hunt's Hill. It has been estimated that 90 to 105 million yards have been washed here. The gravels have a maximum depth of 300 feet, the lower 150 consisting of coarse blue gravel. The bottom gravels are well-cemented. Bedrock consists of slate, gabbro, quartzite, and amphibolite.

Bibliography

Lindgren, Waldemar, 1900, Colfax folio: U. S. Geol. Survey Geol. Atlas of the U. S., folio 66, 10 pp.

Lindgren, Waldemar, 1911, Tertiary gravels of the Sierra Nevada: U. S. Geol. Survey Prof. Paper 73, pp. 144–146.

Logan, C. A., 1936, Gold mines of Placer County, Dutch Flat district: California Div. Mines Rept. 32, pp. 56–58.

El Dorado

Location and History. This district is in west-central El Dorado County a few miles southwest of Placerville. It is in the Mother Lode gold belt and includes the Logtown area a few miles to the south of El Dorado. El Dorado, originally known as Mud Springs, was a camp on the Kit Carson emigrant trail before the beginning of the gold rush. There was much activity here during the gold rush and for some years afterward. The Church, Union and other mines were worked on a large scale during the 1890s and early 1900s. There was some work in the district again in the 1930s, but only minor prospecting since.

Geology. Slate of the Mariposa Formation (Upper Jurassic) occurs in the central portion of the district. Massive greenstone of the Logtown Ridge Formation (Upper Jurassic) is to the west, and granite rocks, schist and amphibolite lie to the east. The Mother Lode Belt here bends from the north to the northeast towards Placerville.

Ore Deposits. Most of the quartz veins in this district are confined to the slate, although a few are in greenstone and schist. They have north to northeast strikes and usually range from five to 10 feet in thickness. The ore contains free gold and pyrite and has an average content of about $\frac{1}{5}$ ounce of gold per ton. Considerable high-grade ore was recovered from shallow workings. Much fault gouge is present. The greatest depth of development is about 2,000 feet.

Mines. Bidstrup, Buena Vista, Church $1 million, Crown Point, Crusader, German, Griffith, Larkin, Martinez, McNulty, Ophir, Pochahantas, Red Wing, Starlight, Tullis, Union $5 million?

Bibliography

Clark, W. B., and Carlson, D. W., 1956, El Dorado County, lode gold mines: California Jour. Mines and Geology, vol. 52, pp. 401–429.

Lindgren, Waldemar, 1894, Placerville folio: U. S. Geol. Survey Geol. Atlas of the U. S., folio 3, 4 pp.

Logan, C. A., 1936, Mother Lode gold belt, Church and Martinez mines: California Div. Mines Bull. 108, pp. 21–22 and 30–31.

Storms, W. H., 1900, Church and Union mines: California Min. Bur. Bull. 18, pp. 91–92.

Tucker, W. B., 1919, El Dorado County, Church and Union mines: California Min. Bur. Rept. 15, pp. 283 and 299.

Emigrant Gap

Location and History. This district is in the east-central Placer and Nevada Counties in the vicinity of the towns of Emigrant Gap and Blue Canyon. It is both a lode and placer gold-mining district that was first worked during the gold rush. There has been intermittent prospecting and development work here since. The Zeibright mine was worked on a large scale during the 1930s. The Washington district is just to the northwest and the Westville district is to the south. The Zeibright mine camp is now a boy scout camp.

Geology. The district is chiefly underlain by thick beds of slate, schist and phyllite of the Blue Canyon Formation (Carboniferous). To the north and west these rocks are overlain by andesite and rhyolite tuff. There are several patches of auriferous Tertiary channel gravels.

Ore Deposits. There are a number of north-trending quartz veins in schist and slate that contain free gold, pyrite, and small amounts of other sulfides. The veins range from one to 10 feet in thickness and consist of a series of parallel quartz stringers. The milling-grade ore usually contained $\frac{1}{2}$ ounce per ton or less, but some of the ore shoots were extensive. Channel deposits at the Lost Camp hydraulic mine contain both cemented and free-washing quartzitic gravels. These gravels are overlain by sands and volcanic rocks. Some of the gold recovered from the gravels was coarse.

Mines. Lode: Red Rock, Texas, Van Avery, Zeibright $1 million+. Placer: Golden Channel, Golden Nugget, Lost Camp, Shell, Wild Yankee.

Bibliography

Lindgren, Waldemar, 1900, Colfax folio, California: U. S. Geol. Survey Geol. Atlas of the U. S., folio 66, 10 pp.

Lindgren, Waldemar, 1911, The Tertiary gravels of the Sierra Nevada: U. S. Geol. Prof. Paper 73, p. 146.

Logan, C. A., 1936, Gold mines of Placer County, Lost Camp mine: California Division of Mines Rept. 32, p. 68.

Logan, C. A., 1941, Nevada County, Zeibright mine: California Division of Mines Rept. 37, p. 431.

Waring, C. A., 1919, Placer County, Lost Camp mine: California Mining Bureau Rept. 15, pp. 376–377.

English Mountain

Location. This district is in northeastern Nevada County in the vicinity of English Mountain and Bowman Lake. The area was first mined during the 1860s, followed by activity at the English Mountain gold and copper mines during the 1890s and early 1900s. There has been minor prospecting since.

Geology. The area is underlain by slate, schist, hornfels and granite, all of which are cut by diorite and quartz-diorite dikes. The ore deposits consist of iron-rich zones containing quartz, free gold, galena, pyrite, and chalcopyrite. Molybednite is abundant in places. The principal gold sources have been the English Mountain Gold and Yellow Metal mines.

Bibliography

Lindgren, Waldemar, 1900, Colfax folio: U. S. Geol. Survey Geol. Atlas of the U. S., folio 66, 11 pp.

MacBoyle, Errol, 1919, Nevada County, English Mountain and Yellow Metal mines: California Min. Bur. Rept. 16, pp. 164 and 258.

Erskine Creek

Erskine Creek is a northwest-flowing tributary of the Kern River in northeast Kern County, a few miles to the southeast of Isabella. The district includes the area known as the Pioneer district. Gold and varying amounts of silver, antimony, tungsten, copper, and uranium have been recovered here. The principal sources of gold have been the Glen Olive mine, which has yielded $500,000, and the Iconoclast mine. Other properties include the Golden Bell, Laurel, Valley View, Faust, and King Solomon mines. Two northwest-trending roof pendants of pre-Cretaceous metamorphic rocks are surrounded by Mesozoic granitic rock. The ore deposits consist of quartz veins containing free gold and varying amounts of sulfides.

Bibliography

Troxel, B. W., and Morton, D. K., 1962, Kern County, Erskine Creek district: California Div. Mines and Geology, County Rept. 1, pp. 31–32.

Tucker, W. B., and Sampson, R. J., 1933, King Solomon mine: California Div. Mines Rept. 29, pp. 312–313.

Eureka

Location. This is an extensive area of scattered placers and a few lode deposits in northwestern Sierra County about eight miles northwest of Downieville. It includes the "diggings" not only at Eureka but also at Craig's Flat, Morristown, and Saddleback Mountain.

It is surrounded by a number of famous placer-mining districts: Downieville, Poker Flat, Port Wine, Poverty Hill, and Brandy City. The hydraulic mines here were worked on a major scale from the 1850s to the middle 1880s, and then intermittently on a small scale through the 1930s.

Geology. The principal Tertiary channel deposits are at Eureka, Craig's Flat, Morristown, and Monte Cristo, the most extensive being at Eureka. They are part of the Eureka channel, an indistinct branch of the Tertiary North Fork of the Yuba River. As in the other nearby placer-mining districts, the chief values were obtained from the lower quartzitic gravels. Some very coarse nuggets have been found here. Bedrock consists of slate and phyllite and several narrow belts of greenstone and serpentine. Several of the high ridges are capped by andesite. There are a few gold-quartz veins, the most productive having been at the Telegraph mine, which is on a slate-serpentine contact.

Bibliography

Lindgren, Waldemar, 1911, Tertiary gravels of the Sierra Nevada: U. S. Geol. Survey Prof. Paper 73.

Turner, H. W., 1897, Downieville folio: U. S. Geol. Survey Geol. Atlas of the U. S., folio 37, 8 pp.

Fairplay

Location and History. This district is in south-central El Dorado County about 20 miles southeast of Placerville. It includes the Slug Gulch and Cedarville areas. It is primarily a placer-gold district, but some copper has been mined here. The area was first settled in 1853 by N. Sisson and Charles Staples. The name, according to tradition, arose from an incident in which an appeal for fair play forestalled a fight between two miners.

Geology. The deposits here are part of an isolated Tertiary gravel channel that extends southwest from Slug Gulch toward Fairplay. The gravels are of various ages. Older benches of quartz gravel are rich, younger intervolcanic gravels are leaner. The gold is extremely coarse. At Slug Gulch the bedrock is limestone; elsewhere it is slate and schist; to the west is granodiorite. A few narrow gold-quartz veins are found in the district.

Bibliography

Clark, W. B., and Carlson, D. W., 1956, El Dorado County, placer deposits: California Jour. Mines and Geology, vol. 52, pp. 429–435.

Lindgren, Waldemar, and Turner, H. W., 1894, Placerville folio: U. S. Geol. Survey Geol. Atlas of the U. S., folio 3, 3 pp.

Lindgren, Waldemar, 1911, Tertiary gravels of the Sierra Nevada: U. S. Geol. Survey Prof. Paper 73, pp. 180–181.

Mining and Scientific Press, June 17, 1876.

Fiddletown

Location. This district is in northwestern Amador County in the general vicinity of the old mining town of Fiddletown, six miles east of Plymouth. It also has been known as the Oleta district.

History. Fiddletown was settled in 1849, reportedly by Missouri miners addicted to "fiddling." The district flourished in the 1850s when the drift and hydraulic mines were active. Attention was drawn to

Photo 19. Natomas Company Dredge No. 2, Folsom District. This 1921 view of the dredge, in Sacramento County, shows the double stackers and sand wheel. *Photo by C. A. Waring.*

the area by Bret Harte's short story, *An Episode at Fiddletown.* The name proved offensive to one of the distinguished residents and, in 1878, he succeeded in having the name changed by legislative enactment to Oleta. In 1937, the California Historical Society, with the approval of local residents, obtained the restoration of the original name. There was some drift mining and dragline dredging in the district in the 1930s and early 1940s.

Geology. Numerous patches of quartzitic gravels remain that were deposited by several channels of the Tertiary Cosumnes River. One channel comes in from the Coyoteville area from the northeast and another from Volcano from the southeast. Some of the gravels are capped by andesite. Bedrock is chiefly graphitic slate, metachert and schist of the Calaveras Formation (Carboniferous to Permian). Limestone lies to the east and granodiorite to the north. There are a few narrow gold-quartz veins in the district.

Bibliography

Carlson, D. W., and Clark, W. B., 1954, Amador County, gold: California Jour. Mines and Geology, vol. 50, pp. 164–200.

Haley, C. S., 1923, Gold placers of California: California Min. Bur. Bull. 92, pp. 146–147.

Lindgren, Waldemar, and Turner, H. W., 1894, Placerville folio: U. S. Geol. Survey Geol. Atlas of the U. S., folio 3, 3 pp.

Lindgren, Waldemar, 1911, Tertiary gravels of the Sierra Nevada: U. S. Geol. Survey Prof. Paper 73, p. 199.

Turner, H. W., 1894, Jackson folio: U. S. Geol. Survey Geol. Atlas of the U. S., folio 11, 6 pp.

Fine Gold

Location. This district is in east-central Madera County at the site of the old town of Fine Gold, about 35 miles northeast of Madera. It is in the south end of a 20-mile mineralized belt that extends southeast from the Grub Gulch and Coarsegold districts. It also includes the Quartz Mountain area. The district was first mined in the late 1850s. There has been intermittent prospecting in recent years, chiefly in the vicinity of Quartz Mountain.

Geology. The region is underlain by medium- to coarse-grained granodiorite cut by pegmatite and aplite dikes. Also present are thin beds of gneiss, schist and slate. There are a number of quartz veins containing small but sometimes rich ore shoots. The ore contains native gold, pyrite, and galena. Nearly all of the deposits are shallow. The veins range from less than one to more than 10 feet in thickness.

Mines. Ackers, Fresno Banner, Johnny, Little Johnny, Quartz Mountain, Standard, Waterloo, Zebra.

Bibliography

Irelan. William, Jr., 1888, Fine Gold Gulch district: California Min. Bur. Rept. 8, pp. 210–216.

McLaughlin, R. P., and Bradley, W. W., 1916, Madera County, Waterloo mine: California Min. Bur. Rept. 14, pp. 552–553.

Turner, H. W., 1896, Further contributions to the geology of the Sierra Nevada: U. S. Geol. Survey, 17th Ann. Rept., pt. 1, p. 695.

Watts, W. L., 1893, The Fine Gold mines: California Min. Bur. Rept. 11, pp. 215 and 216–217.

Folsom

Location. This district is in northeastern Sacramento County. It was mainly a dredging field that extended from the town of Folsom southwest along the south side of the American River to Fair Oaks, south through the town of Natoma to Nimbus and then west to the east border of what is now Mather Air Force Base. The dredged area is approximately 10 miles long in a southwest direction and up to seven miles wide. The Folsom district has also been known as the American River district.

History. The region around Folsom and Mormon Bar was extensively placer-mined during the gold rush, with minor lode mining. The area was originally settled in 1849 and first known as Negro Bar. The present town was laid out in 1855 by T. D. Judah for the Sacramento Valley Railroad and named for Captain J. L. Folsom, quartermaster of Stevenson's Regiment. Numerous Chinese worked the region from the 1860s through the 1890s. A primitive grab-dredger was active at Natoma in 1894. Bucket-line dredging began at Folsom in 1898 and soon became a major industry. Most of the dredging companies were merged into Natomas Consolidated of California in 1908. This firm, later known as the Natomas Company, was the principal operator in the district. The company designed and built its own dredges at extensive shops in the town of Natoma. In 1916, 11 active dredges yielded more than $2 million worth of gold. From 1927 to 1952, several other operators joined Natomas in dredging the district.

Dredging operations were curtailed during World War II but were resumed on a major scale shortly afterward. However, increasing costs, the depletion of dredging ground, and changing land values caused the dredging operations to be gradually curtailed. By 1960 there was only a single active dredge, and this was shut down in February 1962. Large portions of the dredged-over areas are now occupied by defense industries, such as the Aerojet-General Corporation and Douglas Aircraft Company plants, and by housing tracts. Folsom, one of the largest dredging fields in California, has a total output estimated at $125 million. Approximately one billion cubic yards of gravel were dredged by the Natomas Company.

Geology. Recent stream gravels lie in and adjacent to the present American River. To the south are sand and gravel deposits of the Victor Formation (Pleistocene) and silt, sand and some gravel of the Laguna Formation (Plio-Pleistocene). These are underlain by andesite of the Mehrten Formation (Pliocene). The paying gravels are either in or along the American River and near the lower contacts of the Laguna and Mehrten Formations. Digging depths ranged from 30 to 110 feet and recoveries from 10¢ to 20¢ per yard with gold valued at $35 per ounce. Minor amounts of platinum were recovered. There are a few narrow gold-quartz veins in greenstone east of the town of Folsom.

Dredging Operations. Capitol Dredging Co., 1927–52, four bucket-lines; General Dredging Co., 1938–51, three draglines; Gold Hill Dredging Co., 1933–37, one bucket-line; Lancha Plana Gold Dredging Co., 1940–49, one bucket-line. Natomas Co., 1909–62, fifteen bucket-lines; consolidated by Natomas Co. in 1908 were: Colorado Pacific Gold Dredging Co., El Dorado Gold Dredging Co., Folsom Development Co., Natoma Development Co., Syndicate Gold Dredging Co., Wilkes-Barre Dredging Co. (1916); later Natomas Co. production records, 1959, 7,894,592 cu. yds., 10.19¢/yd. yield, 9.56¢ cost per yard and 1958, 9.15¢/yd. yield, 8.73¢ cost per yard.

Bibliography

Carlson, D. W., 1955, Sacramento County placer mines, California Jour. Mines and Geology, vol. 51, pp. 134–142.

Crawford, J. J., 1896, Folsom district: California Min. Bur. Rept. 13, pp. 316–317.

Doolittle, J. E., 1905, Folsom district: California Min. Bur. Bull. 36, pp. 92–98.

Lindgren, Waldemar, 1894, Sacramento folio, California: U. S. Geol. Survey Geol. Atlas of the U. S., folio 5, 3 pp.

Lindgren, Waldemar, 1911, Tertiary gravels of the Sierra Nevada, Folsom dredge fields: U. S. Geol. Survey Prof. Paper 73, p. 222.

Logan, C. A., 1925, Sacramento County, Dredging: California Min. Bur. Rept. 21, pp. 12–14.

Waring, C. A., 1919, Gold dredging—Sacramento County: California Min. Bur. Rept. 15, pp. 405–415.

Winston, W. B., 1910, Gold dredging in California, Sacramento County, dredging: California Min. Bur. Bull. 57, pp. 176–204.

Forbestown

Location and History. The Forbestown district is in southeastern Butte County about 15 miles due east of Oroville. It includes the Feather Falls area. The Brownsville district is just to the south, and the Hurle-

Photo 20. Natomas Company Dredge No. 9, Folsom District. This is a 1914 photo of the dredge, in Sacramento County. *Photo by C. A. Waring.*

Photo 21. Natomas Company Dredge No. 8, Folsom District. This 1953 photo, taken in Sacramento County, shows the company's last active dredge in the Folsom field. No. 8 ceased operation in 1962.

ton and Bidwell Bar districts are to the west. The area was placer-mined during the gold rush. At that time the South Fork of the Feather River, which drains the area, yielded huge amounts of gold. The town was named for Ben F. Forbes who established a store here in 1850. The Gold Bank mine was operated on a major scale from 1888 until 1904. Moderate mining activity continued in the district through the 1930s, and there has been intermittent prospecting, some by skin diving, since.

Geology and Ore Deposits. The area is underlain by massive greenstone, amphibolite schist and fine-grained diorite. Granite lies to the north and gabbro diorite to the south. There are a number of north-trending quartz veins and stringers. The ore contains free gold and considerable amounts of sulfides; tellurides have been reported. The milling ore is usually low in value (1/5 oz. or less per ton), but the ore shoots often had stoping lengths of several hundred feet. The sulfide concentrates were valued up to $100 per ton.

Mines. Carlisle, Denver, Gold Bank $2 million, Golden Eagle, Golden Queen $372,000+, Miller, Shakespeare $137,000+, Southern Cross, Williams.

Bibliography

Crawford, J. J., 1894, Gold Bank mine: California Min. Bur. Rept. 12, p. 83.
Logan, C. A., 1930, Butte County, Forbestown Consolidated Gold Mines: California Div. Mines Rept. 26, pp. 373–376.
Miner, J. A., 1890, Gold Bank mine: California Min. Bur. Rept. 10, pp. 125–127.
Turner, H. W., 1898, Bidwell Bar folio, California: U. S. Geol. Survey Geol. Atlas of the U. S., folio 43, 6 pp.
Waring, C. A., 1919, Butte County, Gold Bank and Golden Queen mines: California Min. Bur. Rept. 15, pp. 216–217.

Forest Hill

Location. The Forest Hill district is in south-central Placer County in the general vicinity of the town of that name. This district is fairly large in area and includes not only the "diggings" at Forest Hill but those at Bath to the east, Todd Valley and Dardanelles to the southwest, and Yankee Jims to the northwest. The district is principally a placer-mining one, although there have been some productive quartz mines.

History. Gold was discovered here in 1850. By 1852 the area was highly productive. In that year the Jenny Lind mine was discovered, and hydraulic mining was introduced at Yankee Jims by Colonel McClure. The town was an important trading center in those days. By 1868 the mines in the vicinity of the town had yielded more than $10 million. Large-scale hydraulic mining continued until the early 1880s and drift mining until the early 1900s. There was appreciable activity in the district again in the 1930s and early 1940s, and a few mines, such as the Paragon and Three Queens, have been worked since. Forest Hill is now an important lumbering center. The total output of the district is estimated to be at least $25 million, and it may be considerably more.

Geology. The main early Tertiary channel of the Middle Fork of the American River enters the district from Michigan Bluff on the east. At Bath it turns north and then west and southwest and continues southwest through Forest Hill. At the Dardanelles mine west of Forest Hill, the channel swings northwest to Yankee Jims and then north to the Iowa Hill district. An intervolcanic channel extends west-southwest from Baker Ranch to north of Forest Hill. Another intervolcanic channel extends south-southwest between the above and Yankee Jims. The older quartzitic gravels near bedrock are coarse and well-cemented and have yielded the most gold. Much of the gravel is overlain by rhyolite and andesite. Bedrock is slate with some phyllite, schist, and serpentine. Some of the gold-quartz veins were rich, especially those that occur near serpentine. The veins are usually three to four feet thick and strike in a northwesterly direction. A number of small but rich pockets were found in the Three Queens mine, the principal lode mine in the district.

Mines. Placer: Baker Divide; Baltimore; Big Spring $150,000; Dardanelles $2 million+; Excelsior; Florida; Georgia Hill, Yankee Jim and Smiths Point, together $5 million; Grey Eagle; Homestake; Independent, New Jersey and Jenny Lind, together $2,-653,000; Mayflower $1 million; Maus; Paragon $2.65 million+; Peckham Hill and Todd Valley, together $5 million; Pond; San Francisco; Small Hope; Yankee Jims. Lode: Dry Hill, Eureka, Cons. International, Mitchell, Three Queens $100,000+.

Bibliography

Browne, Ross E., 1890, The ancient river beds of the Forest Hill Divide: California Min. Bur. Rept. 10, pp. 435–465.

Chandra, Deb K., 1961, Geology and mineral deposits of the Colfax and Forest Hill quadrangles, California: California Div. Mines Spec. Rept. 67, 50 pp.

Ellsworth, E. W., 1933, Tracing buried-river channel deposits by geomagnetic methods: California Div. Mines Rept. 29, pp. 244–250.

Irelan, William, Jr., 1888, Dardanelles, Baker Divide, and Breece and Wheeler mines: California Min. Bur. Rept. 8, pp. 464–468.

Jarman, Arthur, 1927, Forest Hill and south side of Forest Hill: California Min. Bur. Rept. 23, pp. 88 and 91–92.

Lindgren, Waldemar, 1900, Colfax folio: U. S. Geol. Survey Geol. Atlas of the U. S., folio 66, 10 pp.

Lindgren, Waldemar, 1911, Tertiary gravels of the Sierra Nevada: U. S. Geol. Survey Prof. Paper 73, pp. 149–151.

Logan, C. A., 1936, Gold mines of Placer County, Dardanelles, Mayflower, and Paragon mines: California Div. Mines Rept. 32, pp. 55, 69, and 73–75.

Waring, C. A., 1919, Placer County, Forest Hill district: California Min. Bur. Rept. 16, p. 317.

French Corral

Location. French Corral is in northwestern Nevada County about nine miles northwest of Nevada City. Much of the gold production came from hydraulic mines between here and Birchville to the northeast.

History. The town was named for a mule corral erected by a Frenchman, who was the first settler in the area. The principal period of gold mining was from the middle 1850s to the 1890s; there has been minor work since. Sometime before 1867, a 7½-carat diamond, the largest known to have come from California, was found here in a sluice box.

French Corral was the terminus of one of the first long-distance telephone lines in the United States. Installed by the Edison Company about 1878, it connected Birchville, North San Juan and North Bloomfield to Bowman or French Lake, in the high Sierra Nevada some 58 miles east. It was used primarily to send messages about the delivery of water to the hydraulic mines, but it also was used by Western Union to send other messages.

The total production of the district is unknown, but it has been estimated to be valued at between $3 million and $4 million. Lindgren, in 1911, estimated that 32.5 million yards of gravel had been removed and that 20 million remained; the U.S. Army Corps of Engineers, in 1891, estimated that the same amount had been removed, but that only 10 million yards remained.

Geology. A major channel of the Tertiary Yuba River entered the area from the northeast. It extends southwest for a distance of about four miles in this district. The gravels deposited by this channel are 150 to 250 feet thick and 600 or more feet wide. The gravels have yielded gold throughout, but the quartz-rich lower gravels were the richest. Bedrock is granodiorite with greenstone to the north. Also there are some gold-quartz veins and bodies of mineralized granodiorite and greenstone.

Bibliography

Lindgren, Waldemar, 1895, Smartsville folio, California: U. S. Geol. Survey Geol. Atlas of the U. S., folio 18, 6 pp.

Lindgren, Waldemar, 1911, Tertiary gravels of the Sierra Nevada: U. S. Geol. Survey Prof. Paper 73, pp. 123–125.

MacBoyle, Errol, 1919, Nevada County, French Corral mining district: California Min. Bur. Rept. 16, pp. 7–11.

Fresno River

Small amounts of gold are recovered intermittently from the lower Fresno River in western Madera County. There are several small floating suction dredges similar to those on the Chowchilla River to the north (see also the section on the Chowchilla district). This stream was first placer-mined during the gold rush, when the Coarsegold and Grub Gulch districts to the northeast were originally worked.

Bibliography

Logan, C. A., 1950, Madera County, gold: California Jour. Mines and Geology, vol. 46, pp. 453–456.

Friant

The Friant district is in northeastern Fresno County on the San Joaquin River, in the vicinity of the Friant Dam. Placer mining was carried on in the district during the early days in the vicinity of Fort Miller, a military post now inundated by the reservoir. Later this place was the terminus of a branch of the Southern Pacific Railroad and known as Pollasky. It was renamed in the 1920s for Thomas Friant of the White-Friant Lumber Company. About $200,000 worth of gold was recovered as a by-product from sand and gravel excavated for use in the construction of the dam in 1940–42. Since 1946, from $5000 to $25,000 worth of gold has been produced annually from the sand and gravel plants here. The gold occurs in the river gravels and small terrace deposits adjacent to the river. The gold is fine and flaky, recovered in riffles set below the fine screens in the sand-washing plants.

Bibliography

Bradley, Walter W., 1916, Fresno County, gold: California Min. Bur. Rept. 14, p. 440.

Logan, C. A., Braun, L. T., and Vernon, J. W., 1951, Fresno County, gold: California Jour. Mines and Geology, vol. 47, no. 3, pp. 503–504.

Genesee

Location. The Genesee district is in the southeast end of the Crescent Mills-Taylorsville-Genesee gold belt of east-central Plumas County. This is not a single belt but rather several parallel belts or zones of gold and copper mineralization. The well-known Walker copper mine is in this district.

History. Gold was first placer-mined in the streams during the gold rush, and lode mining began soon afterward. Genesee is believed to have been named by the Ingalls family for a valley in New York State. Mining activity continued almost steadily through the early 1900s. The Walker copper mine was worked on a major scale from about 1915 to 1942, and the concentrates were delivered to the Western Pacific Railroad at Spring Garden via a nine-mile aerial tramway. There has been intermittent gold and copper prospecting in the district since.

Geology. This area is underlain by the same series of Paleozoic and Mesozoic metamorphic rocks found in the Taylorsville district to the west and northwest (see Taylorsville district). Contact metamorphism, especially in the vicinity of the Walker mine, has altered the rocks into hornfels and schist. The gold-ore deposits consist of either quartz veins or zones of quartz stringers that contain free gold, limonite, and sulfides. A number of high-grade pockets have been found. There are several patches of auriferous Tertiary gravels.

Mines. Austrian Syndicate, Big Cliff, Blue Bell, Bullion, Calman, Cosmopolitan, Five Bears, Green Ledge, Gruss $460,000, Hinchman, Magpie, Mountain Lion, Native Son, Peter, Taylor (placer), Wards (placer).

Bibliography

Averill, C. V., 1937, Plumas County, copper and gold: California Div. Mines Rept. 33, pp. 93–124.

Diller, J. S., 1908, Geology of the Taylorsville region, California: U. S. Geol. Survey Bull. 353, 128 pp.

Diller, J. S., 1909, Mineral resources of the Indian Valley region: U. S. Geol. Survey Bull. 260, pp. 45–49.

MacBoyle, Errol, 1920, Plumas County, Genesee mining district: California Min. Bur. Rept. 16, pp. 12–18.

Georgetown

Location. The Georgetown district is in northwestern El Dorado County at the north end of the northeast segment of the Mother Lode belt. It extends from just north of Garden Valley north through Georgetown and the Georgia Slide area to the Middle Fork of the American River. It is both a lode- and placer-mining district.

History. Mining began here in 1849 by a party of placer miners from Oregon. The site was first known as Growlersburg, but was soon changed to Georgetown. It is reported to have been named for either George Ehrenhaft, who laid out the town, or George Phipps, a sailor-prospector. The placers were highly productive during the 1850s. The seam deposits at Georgia Slide were mined on a large scale by hydraulicking from 1853 to about 1895. There was

some activity during the early 1900s, and in the 1930s the Beebe and Alpine mines were worked on a fairly large scale. There has been minor prospecting and skin diving in the district since.

Geology. This district is in the northern end of the Mother Lode gold belt (see fig. 4). There is a two-mile wide north- and northwest-trending belt of Mariposa slate (Upper Jurassic) in the central portion of the district, with greenstone and green schist to the west and mica schist, slate, quartzite, amphibolite, and serpentine to the east. In places, especially at Georgia Slide, the bedrock is deeply weathered. Several patches of Tertiary gravel overlain by andesite are exposed on some hills in the northern part.

Ore Deposits. The ore deposits consist of thick zones of mineralized schist and slate that contain numerous quartz veins and veinlets. Where deeply weathered the gold became concentrated and such deposits were worked by placer-mining methods. These are known as "seam" deposits. Below the weathered zone they were mined as lode deposits. The seam deposit at Georgia Slide was 1000 feet long and 500 feet wide. Usually the milling ore yielded from $1/7$ to $1/5$ ounce to the ton, but many high-grade pockets were encountered. In addition, there are several wide quartz veins containing finely disseminated free gold and pyrite. These veins contained ore shoots with stoping lengths of up to 500 feet. The Tertiary gravel patches have yielded gold.

Mines. Lode-seam: Alpine $500,000+, Alma, Barney, Beebe $2 million, California Jack, Cove Hill, Georgia Slide $6 million, Mamaluke, Mount Hope. Placer: Anderson, Bottle Hill, Cary, Cement Hill, Holmes, Jones Hill, Little Chief, Mulvey Point, Patterson, Rowe, Shoemaker, Tanksley, Trimble.

Bibliography

Clark, W. B., and Carlson, D. W., 1956, El Dorado County, Seam deposits and Georgetown area: California Jour. Mines and Geology, vol. 52, pp. 431 and 435–437.

Irelan, William, Jr., 1888, Alpine mine: California Min. Bur. Rept. 8, pp. 167–168.

Lindgren, Waldemar, and Turner, H. W., 1894, Placerville folio, California: U. S. Geol. Survey, Geol. Atlas of the U. S., folio 3, pp.

Lindgren, Waldemar, 1911, The Tertiary gravels of the Sierra Nevada: U. S. Geol. Survey Prof. Paper 73, pp. 108–169.

Logan, C. A., 1935, Mother Lode gold belt of California, Alpine, Beebe, and Georgia Slide mines: California Div. Mines Bull. 108, pp. 15–19 and 46.

Preston, E. B., 1893, Georgetown: California Min. Bur. Rept. 11, pp. 202–204.

Gibsonville

Location. This placer-mining district is in northwestern Sierra County and southern Plumas County in the vicinity of the old town of Gibsonville. It is about six miles northeast of La Porte and 20 miles due north of Downieville. The district includes the Whiskey Diggings, Hepsidam, and Bunker Hill areas. The area was first worked during the gold rush, and considerable drift mining and some hydraulic mining followed in 1875–95. There has been intermittent prospecting since.

Geology. The deposits are located on the Tertiary North Fork of the Yuba River, which is also known as the La Porte channel. The channel extends in a southwest direction through the district. The channel gravels are as much as 1500 feet wide and capped by thick beds of andesite and clay. The gold was found mostly in quartz-rich gravels near bedrock and was mostly coarse-grained. Some ground yielded up to $3 per yard of gold at the old price. There are a number of drops in the channel, caused possibly by bedrock faulting. The bedrock consists of amphibolite with serpentine, slate, and schist lying to the east. The channel has been drift-mined almost continuously from Hepsidam southwest through Gibsonville to the Thistle shaft, a distance of about five miles. Although the district is reported to have had only a moderate output, extensive drifting and a number of rich pay streaks indicate that it must have been much more productive than originally believed.

Mines. Bellevue group, Bunker Hill, Empire, Feather Fork (Thistle), Garnet, Gibsonville Water and Mining (hydraulic), Homestake, Taber (Hal Taber), Union-Keystone $600,000, Washington (North American Cons.) (hydraulic), Whiskey (hydraulic).

Bibliography

Crawford, J. J., 1896, Feather Fork mine: California Min. Bur. Rept. 13, p. 376.

Lindgren, Waldemar, 1911, Tertiary gravels of the Sierra Nevada: U. S. Geol. Survey Prof. Paper 73, pp. 106–107.

MacBoyle, Errol, 1921, Sierra County, Gibsonville mining district: California Min. Bur. Rept. 17, pp. 11–13.

Turner, H. W., 1897, Downieville folio, California: U. S. Geol. Survey Geol. Atlas of the U. S., folio 37, 8 pp.

Weil, S. C., The ancient channel of Gibsonville: Min. and Sci. Press, vol. 91, July 29, 1905, p. 73.

Wiltsee, E. A., 1893, Gibsonville mining district: California Min. Bur. Rept. 11, pp. 418–419.

Globe

There are several small gold mines and prospects in an area known as the Globe mining district in central Tulare County about 15 miles east of Porterville. Most of the prospects are in the vicinity of Cow Mountain on the north side of the Tule River Indian Reservation or a few miles to the west. There was some prospecting here in the early 1900s, but apparently little or nothing has been done since. The deposits consist of shallow quartz veins in granite that contain fairly abundant pyrite and small amounts of free gold.

Bibliography

Tucker, W. B., 1919, Tulare County, gold: California Min. Bur. Rept. 15, pp. 912–915.

Gold Run

Location and History. This district is in north-central Placer County in the vicinity of and south of the town of Gold Run. Extensive Tertiary channel gravels extend from here south to Indiana Hill and the North Fork of the American River. Much of the output in the district has come from the vast Stewart hydraulic mine, which is traversed by U.S. Highway Interstate 80 across its north end. The area was first placer-mined in 1849, and the town was founded in 1854 by O. W. Hollenbeck. The town was originally called Mountain Springs. From 1865 to 1878 approxi-

mately $6,125,000 in gold was shipped from the express office here. Mining on a moderate scale continued until about 1915, with considerable production reported in 1908. There was minor work here in the 1920s and 1930s.

Geology. The deposits are located on a major Tertiary channel of the American River that enters the area from the south and continues north to Dutch Flat. The gravel deposits are more than a mile wide in an east-west direction, three miles long in a north-south direction, and up to 400 feet deep. The lower cemented blue gravel yielded as much as several dollars per yard. The upper gravels contain quartz with clay and sand and averaged 11 to 17 cents per yard, while the top gravels ran to four to five cents per yard. Bedrock is slate in the west portion and gabbroic rock to the east.

Bibliography

Anon. July 24, 1875 to Feb. 19, 1876, Hydraulic mining at Gold Run—The Blue Lead ancient river channel: Min. and Sci. Press.

Hobson, J. B., 1890, Gold Run district: California Min. Bur. Rept. 10, p. 427.

Jarman, Arthur, 1927, Gold Run: California Min. Bur. Rept. 23, pp. 81–86.

Lindgren, Waldemar, 1900, Colfax folio: U. S. Geol. Survey Geol. Atlas of the U. S., folio 66, 10 pp.

Lindgren, Waldemar, 1911, Tertiary gravels of the Sierra Nevada, Indiana Hill and Gold Run: U. S. Geol. Survey Prof. Paper 73, p. 145.

Logan, C. A., 1936, Gold mines of Placer County, Gold Run district: California Div. Mines Rept. 32, pp. 62–63.

Lydon, P. A., 1959, Geology along U. S. Highway 40: Mineral Information Service, vol. 12, no. 8, pp. 1–9.

Granite Basin

Location. The Granite Basin district straddles the Butte-Plumas County line about 30 miles northeast of Oroville and 30 miles west southwest of Quincy. It includes the Buckeye, Gold Lake, Milsap Bar, Soapstone Hill, and Merrimac areas. The area was placer-mined during and after the gold rush. There was some lode mining here in the 1930s, and there has been minor prospecting since.

Geology. Several granitic stocks are intrusive into slate, quartzite and limestone of the Calaveras Formation (Carboniferous to Permian) and amphibolite. Exfoliation has formed several round granitic domes, of which Bald Rock is the most prominent. Serpentine, some Tertiary basalt and a few patches of Tertiary auriferous gravel are present. The quartz veins usually occur in the granite. The veins are narrow, but the ore bodies often are rich. Pyrite and galena are abundant.

Mines. Lode: Hardquartz, Hose, Reynolds. Placer: Buckeye drift, Coquette, Horseshoe, Milsap Bar, Robinson.

Bibliography

Averill, C. V., 1937, Plumas County, Granite Basin Mining Company: California Div. Mines Rept. 33, pp. 108–109.

Heitanen, A. M., 1951, Metamorphic and igneous rocks of the Merrimac area. Geol. Soc. America Bull. Vol. 62, pp. 565–607.

Lydon, P. A., 1959, Geological section and petrography along the Poe tunnel, Butte County: California Div. Mines Spec. Rept. 61, 18 pp.

MacBoyle, Errol, 1920, Plumas County, Granite Basin mining district: California Min. Bur. Rept. 16, pp. 18–21.

Turner, H. W., 1898, Bidwell Bar folio: U. S. Geol. Survey Geol. Atlas of the U. S., folio 43, 6 pp.

Granite Springs

This district is in the northwest corner of Mariposa County and the southwest corner of Tuolumne County in the vicinity of Lake McClure and Don Pedro Reservoir. The mines apparently were last worked in the 1930s. The region is underlain chiefly by greenstone with some interbeds of graphitic slate. There are a considerable number of northwest-striking gold-quartz veins often associated with diorite dikes. Most of the deposits are relatively shallow.

Mines. Mariposa County: Anita, Burr $350,000, Florinita, Jackson, White Rock. Tuolumne County: Buzzard Roost, Diamond, Hedley, Oak Mesa, Solambo.

Bibliography

Bowen, O. E., Jr., 1957, Mariposa County, Burr mine: California Jour. Mines and Geology, vol. 53, p. 247.

Graniteville

Location and History. This district is in east-central Nevada County about 30 miles east of Nevada City. It is also known as the Eureka district. An extensive belt of gold mineralization in this region extends from the vicinity of the town of Graniteville south-southeast to the Emigrant Gap district, a distance of about 10 miles. The district includes the Gaston mine area, which sometimes has been classified as a separate district. The Alleghany-Washington gold belt lies a few miles to the west, and the American Hill district is to the north. The Graniteville district was first placer-mined during the gold rush, and quartz mining began soon afterward. Considerable mining activity continued from the 1860s until about 1900, and there was much activity again during the 1930s. The Ancho-Erie and a few other mines were worked for a short time after World War II.

Geology. The district is underlain by slate, schist, and phyllite of the Blue Canyon Formation (Carboniferous) in the west and granodiorite in the east. In addition, there are several patches of Tertiary gravel and several glacial moraines.

Ore Deposits. Three main north-striking vein systems run through the district. One in the western portion of the district is in slate and schist and contains the Culbertson, National, and Ancho-Erie mines. One to the east is in granodiorite and contains the Wisconsin, Baltic, and Iowa mines. The veins in the central system are along the slate-granodiorite contact. Properties in the central system include the Rocky Glen and Gaston mines. The quartz veins are as much as 15 feet thick. The ore contains free gold and varying amounts of auriferous sulfides. The milling ore usually averages ½ ounce per ton or less with very little high grade. Some of the ore shoots had stoping lengths of several hundred feet. The veins were mined to depths of as much as 500 feet. The Tertiary gravels at Graniteville and at Shands two miles to the west have yielded some gold.

Mines. Alpha, Ancho-Erie $1 million+, Anderson, Artic, Azalie, Baltic, Barren, Birchville, Blue Bell, Celina Flat, Cooley, Culbertson, Eagle Bird, Gaston $2 million?, German, Gold Bug, Hotwater, Iowa, IXL, Jim, Keller, Last Chance, Lindsay, Mountain View, National, Rainbow, Rattlesnake, Republic, Rocky Glen $300,000+, Star, Washington, Wisconsin, Yellow Metal, Yuba $2 million+.

Bibliography

Hobson, J. B., and Wiltsee, E. A., 1893, Eureka mining district: California Min. Bur. Rept. 11, pp. 308–310.

Irelan, William, Jr., 1888, Eureka district: California Min. Bur. Rept. 8, pp. 448–451.

Lindgren, Waldemar, 1900, Colfax folio, California: U. S. Geol. Survey Geol. Atlas of the U. S., folio 66, 10 pp.

Lindgren, Waldemar, 1911, Tertiary gravels of the Sierra Nevada: U. S. Geol. Survey Prof. Paper 73, p. 141.

Logan, C. A., 1930, Nevada County, Gaston mine: California Div. Mines Rept. 26, pp. 110 and 113.

Logan, C. A., 1941, Nevada County, Ancho and Erie group, Birchville mines: California Div. Mines Rept. 37, pp. 383 and 386.

MacBoyle, Errol, 1919, Nevada County, Graniteville mining district: California Min. Bur. Rept. 16, pp. 11–13.

Grass Valley

Location. This famous mining district is in western Nevada County in the immediate area of the town of Grass Valley. The Nevada City district adjoins it on the northeast and the Rough-and-Ready district is to the west.

History. Placer gold was first found in Wolf Creek in 1848 shortly after Marshall's discovery at Coloma. The earliest mining was done by David Stump and two companions who came from Oregon. The shallow placers were rich but were exhausted quickly. Gold-bearing quartz was discovered at Gold Hill in 1850 and soon afterward at Ophir, Rich, and Massachusetts Hills. Quartz mining soon developed into a major industry that was to last more than 100 years. The Gold Hill and Allison Ranch were the leading lode mines during the 1850s. Mining was curtailed somewhat during the Comstock rush of 1859–65, but the mines were productive again in the late 1860s. The camp declined in the 1870s, and by 1880 only the Empire and Idaho mines were active. In 1884 the North Star mine was reopened and activities increased; the North Star, Empire, Idaho-Maryland, Pennsylvania, and W.Y.O.D. all were highly productive. By 1900, the Idaho-Maryland mine had yielded a total of $12.5 million. From 1900 to 1925, the North Star and Empire mines were the largest producers, the Idaho-Maryland having been idle in 1901–19. By 1928, the North Star had had a total output valued at $33 million.

In 1929 the Empire and North Star groups were purchased by the Newmont Mining Corporation. This merger, which resulted in the Empire-Star Mines Co., included other important mines, such as the Pennsylvania, W.Y.O.D., and Sultana. From 1930 to 1941, the district was enormously productive. The 1930–40 output of Idaho-Maryland Mines Corp., which included the Idaho-Maryland and Brunswick mines, was $26.7 million. The Empire-Star group yielded 1,074,284 ounces of gold from 1929 to 1940. Nearly 4000 miners

Photo 22. Empire Mine, Grass Valley District. Mules frequently hauled ore cars, as in this underground scene at the mine in Nevada County. The photo was taken in about 1910.

were at work in the mines during the 1930s and early 1940s. The mines were shut down during World War II, but the Empire, Pennsylvania, North Star, and Idaho-Maryland reopened soon afterward. However, operations gradually decreased; the Idaho-Maryland stopped gold mining in 1956 and the Empire-Star group in 1957, closures that ended nearly 106 years of gold-mining operations in the Grass Valley district. Some tungsten ore was mined in 1954–57 at the New Brunswick unit of the Idaho-Maryland mine.

Grass Valley was the richest and most famous gold-mining district in California. The value of the total output of the lode mines is estimated to have been at least $300 million, and the placer mines yielded a few million dollars more worth of gold. The two largest operations, the Empire-Star and Idaho-Maryland groups, had total outputs of $130 million and $70 million, respectively. Many famous mining engineers and geologists worked in the Grass Valley district. A number of important inventions and improvements were made in mining and milling equipment in the Grass Valley gold mines. Many of the miners were of Cornish descent and were often known as "Cousin Jacks". For many years the town and the mines were served by the Nevada County Narrow Gauge Railroad,

Photo 23. End of a Shift, Empire Mine, Grass Valley District. A shift, circa 1900, ascends after a work tour at the Nevada County mine. Many of these miners were Cornishmen—"Cousin Jacks."

Photo 24.   Idaho-Maryland Mine, Grass Valley District. This photo, taken in about 1930, shows the Idaho shaft at the Nevada County mine. *Photo by Walter W. Bradley.*

Photo 25. New Brunswick Mine, Grass Valley District. This is a 1955 view of the Nevada County mine, which was a member of the Idaho-Maryland group. *Photo by D. W. Carlson.*

which extended north from Colfax. A few historic mine structures are still standing, but most of the extensive surface plants of the major mines have been dismantled. The old power house at the North Star mine and its 32-foot Pelton wheel are part of a Nevada County historical display.

Geology. An elongated body of granodiorite is in the central portion of the district (fig. 8). This body is five miles long in a north-south direction and ½ to two miles wide. It is intrusive into older metamorphic rocks and itself is cut by various dike rocks. Immediately east and west of the intrusion are dark green-

Photo 26. Scotia Mine, Grass Valley District. This scene shows the Nevada County mine in the 1940s. *Photo by Olaf P. Jenkins*

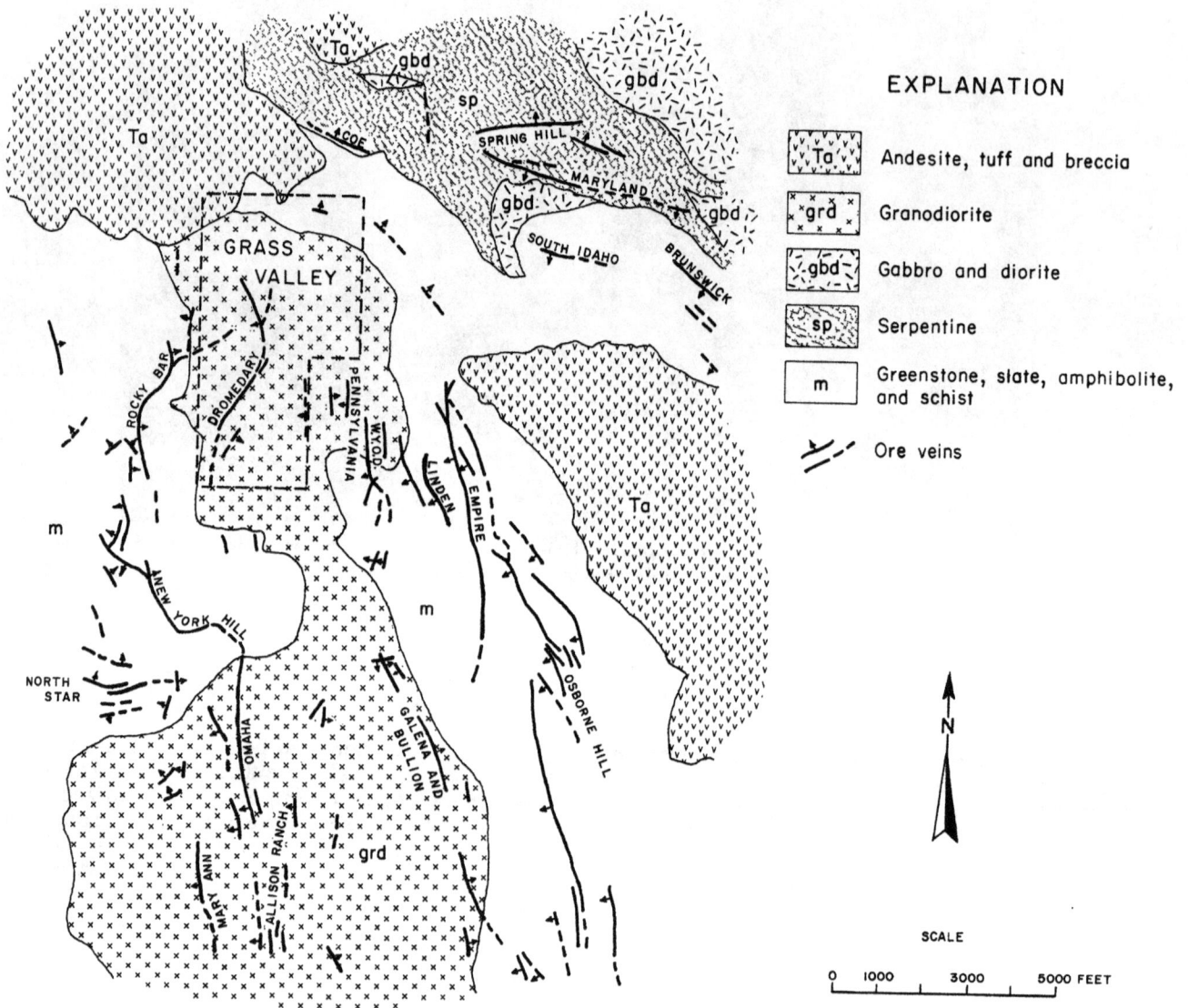

Figure 8. Geologic Map of Grass Valley District, Nevada County. The major veins and vein systems are shown. The names apply to veins, not mines. *After Johnston, 1940, plate 1, and Lindgren, Nevada City Special Folio, 1896.*

stones classified as metadiabase and metadiabase porphyry (so-called "porphyrites"), and continuing to the northeast are amphibolite schist, serpentine, gabbro and diorite, and slate. Just north of the granodiorite and to the southwest are slates, phyllite, quartzite, and schist of the Calaveras Formation (Carboniferous to Permian). A number of intermediate to basic dikes are present also, as well as a few aplite and granite porphyry dikes. Overlying part of the district to the east and to the northwest are Tertiary gravels, in turn largely overlain by andesite.

Ore Deposits. This is the most heavily mineralized and richest gold district in the state with a very large number of productive veins in a relatively small area. The veins fall into two major groups: 1) those of the granodiorite-greenstone area, which have gentle dips, and 2) those of the serpentine-amphibolite area, with

steep dips (see fig. 8). The veins of the granodiorite area are either in the granodiorite or in the adjacent greenstone, entering the granodiorite at depth. One group of veins strikes north and dips gently (about 35° on the average) either east or west. This group includes the Empire, Pennsylvania, Osborne Hill, Omaha, W.Y.O.D., and Allison Ranch veins. The other group of veins in the granodiorite strikes west or northwest and dips gently north. The North Star and New York Hill veins are included in this group. In the serpentine-amphibolite area the veins strike northwest and dip steeply southwest; a few dip northeast. These occur mostly in the amphibolite near or at the serpentine contact. The Idaho-Maryland, Brunswick, and Union Hill mines are here.

The veins usually range from one to 10 feet in thickness and consist of quartz with some calcite and ankerite. They fill minor thrust faults. Many veins contain several generations of quartz. There are numerous northeast-striking, vertical or steeply-dipping fractures or "crossings" that commonly are boundaries of ore shoots. The ore contains free gold and varying amounts of sulfides, chiefly pyrite. Present in smaller amounts are galena, chalcopyrite, arsenopyrite, sphalerite, and pyrrhotite. Galena is commonly associated with gold.

The ore shoots vary considerably in size and shape, and the distribution of gold within the shoots is erratic. Some have pitch lengths of up to several thousand feet, and the veins have been developed to inclined depths of as much as 11,000 feet. Much specimen ore has been found, but milling ore usually averaged from 0.25 to 0.5 ounce of gold per ton. Coarse-grained scheelite is present in several veins, notably in the Union Hill and New Brunswick mines.

Mines. Ajax, Alaska, Alcade, Allison Ranch $2.7 million, Alpha, Bella Union, Ben Franklin $750,000, Big Diamond, Black Hawk, Bow, Boundary, Buena Vista, Bullion, Cassidy (Linden), Centennial $500,-000+, Cheranne, Coe $500,000+, Conlan, Crown Point * $130,000+, Daisy Hill *, Dakota *, Diamond, East Star *, Empire *, Empire-Star group $130 million, Empire West *, Empress, Eureka † $5.7 million, Gaston, General Grant, Gladstone, Golden Center (Dromedary) $2.5 million+, Golden Gate, Golden Treasure, Gold Hill *, Gold Point †, Goodall, Granite Hill, Grant, Hartery $350,000, Hermosa, Heuston, Homeward Bound *, Houston Hill, Idaho †, Idaho-Maryland group $70 million, Independence, Inkmarque, Kate Hayes *, Larimer, Le Duc, Lone Jack, Magenta, Mary Ann, Maryland †, Massachusetts, Massachusetts Hill *, New Brunswick †, New Eureka †, New Homeward Bound, New Ophir *, New York Hill, Normandy-

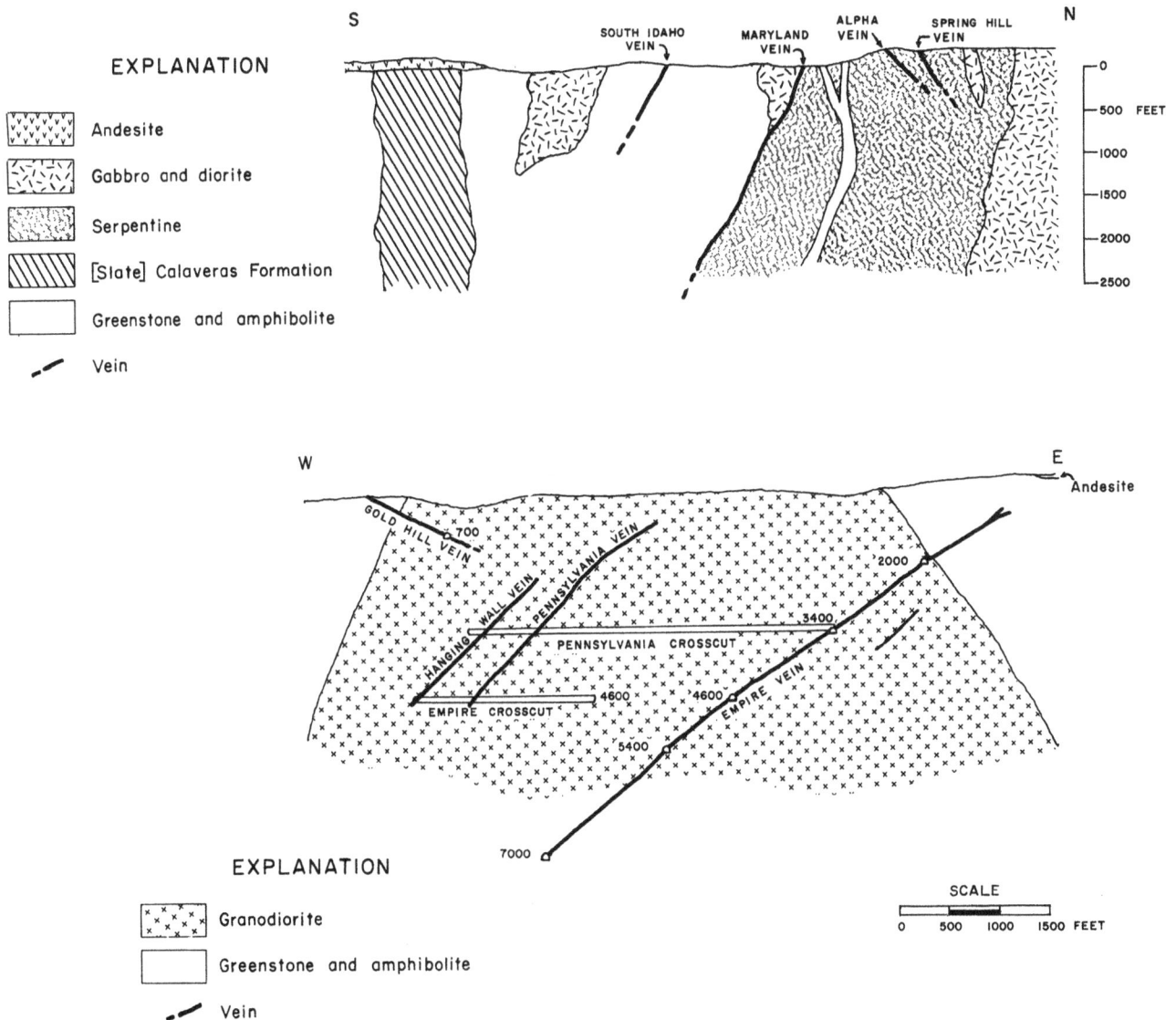

Figure 9 (top). Section through Idaho-Maryland Mine. After Johnston, 1940, figure 65.

Figure 10 (bottom). Section through Empire and Pennsylvania Mines. After Johnston, 1940, figure 62.

Dulmaine, Northern Bell, North Star *, Norumbagua $1 million+, Oakland $100,000, Old Brunswick †, Old Eureka †, Old Homeward Bound, Omaha, Orleans, Osborne Hill *, Peabody, Pennsylvania *, Phoenix, Polar Star, Prescott Hill, Prudential $100,000, Republic, Reward, Rich Hill, Rocky Bar, Rose Hill $100,000+, Scotia, Sebastopol, South Idaho †, Spring Hill $300,000+, Stockton Hill, St. John, Sultana *, Syndicate, Telegraph *, Union Hill † $750,000, Wisconsin, Wyoming, W.Y.O.D.

Bibliography

Crawford, J. J., 1894, Gold—Nevada County: California Min. Bur. Rept. 12, pp. 185–203.

Crawford, J. J., 1896, Gold—Nevada County: California Min. Bur. Rept. 13, pp. 234–271.

Farmin, Rollin, 1938, Dislocated inclusions in gold quartz veins at Grass Valley, California: Econ. Geology, vol. 33, pp. 579–599.

Farmin, Rollin, 1941, Occurrence of scheelite in Idaho-Maryland Mines at Grass Valley, California: California Div. Mines Rept. 37, p. 224.

Hanks, H. G., 1886, Nevada County: California Min. Bur. Rept. 6, pp. 44–49.

Hobson, J. B., 1890, Grass Valley district: California Min. Bur. Rept. 10, pp. 370–384.

Hobson, J. B., and Wiltsee, E. A., 1893, Grass Valley mining district: California Min. Bur. Rept. 11, pp. 267–285.

Hoover, H. C., 1896, Some notes on crossings: Min. and Sci. Press, vol. 72, pp. 166–167.

Howe, Ernest, 1924, The gold ores of Grass Valley, California: Econ. Geology, vol. 19, pp. 595–621.

Irelan, William, 1888, Grass Valley district: California Min. Bur. Rept. 8, pp. 425–435.

Johnston, W. D., Jr., 1932, Geothermal gradient at Grass Valley, Calif.: Washington Acad. Sci. Jour., vol. 22, pp. 267–271.

Johnston, W. D., Jr., 1940, The gold-quartz veins at Grass Valley, California: U. S. Geol. Survey Prof. Paper 194, 101 pp.

Johnston, W. D., Jr., and Closs, Ernst, 1934, Structural history of the fracture systems at Grass Valley, Calif.: Econ. Geology, vol. 29, pp. 39–54.

Knaebel, J. B., 1931, The veins and crossings of the Grass Valley district: Econ. Geology, vol. 26, pp. 375–398.

Lindgren, Waldemar, 1895, Smartsville folio, California: U. S. Geol. Survey Geol. Atlas of the U. S., folio 18, 6 pp.

Lindgren, Waldemar, 1896, Nevada City special folio, California: U. S. Geol. Survey Geol. Atlas of the U. S., folio 29, pp.

Lindgren, Waldemar, 1896, Gold-quartz veins of Nevada City and Grass Valley districts: U. S. Geol. Survey, 17th Ann. Rept. pt. 2, pp. 1–262.

Logan, C. A., 1930, Nevada County, Grass Valley district: California Div. Mines Rept. 26, pp. 96–99.

Logan, C. A., 1941, Mineral resources of Nevada County—gold quartz mining: California Div. Mines Rept. 37, pp. 380–431.

MacBoyle, Errol, 1919, Nevada County, Grass Valley district: California Min. Bur. Rept. 16, pp. 14–30.

In addition to the above references, many reports on mining methods, equipment, and mining activities in the district have appeared in various periodicals. Very many private reports also have been made on various mines.

Gravel Range

This district is in south-central Tuolumne County and north-central Mariposa County about 15 miles east of Groveland. The principal sources of gold have been several bodies of quartzitic Tertiary channel gravels that are part of the Tertiary Tuolumne River. The deposits are in the Gravel Range, at Dorseys and north of Smith Station. They were mined chiefly by hydraulicking. Bedrock consists of granodiorite in the east and slate and schist in the west. In places the gravels are capped by andesite.

* Part of the Empire-Star group.
† Part of the Idaho-Maryland group.

Bibliography

Lindgren, Waldemar, 1911, Tertiary gravels of the Sierra Nevada: U. S. Geol. Survey Prof. Paper 73, pp. 217–218.

Turner, H. W., and Ransome, F. L., 1897, Sonora folio: U. S. Geol. Survey Geol. Atlas of the U. S., folio 41, 7 pp.

Greenhorn Mountain

Location and History. This district is in Kern County about 28 miles northeast of Bakersfield. The first discovery of gold in Kern County was made in Greenhorn Creek in 1851 by a member of General John C. Fremont's party. A rush soon followed, and the town of Petersburg was established. Gold-mining activity declined before 1890, but there has been minor prospecting since. Most of the output has been from placer mining.

Geology. Much of the area is underlain by quartz diorite. There are a few bodies of metamorphic rocks and also some pegmatite dikes. The chief placer deposits were in Greenhorn, Fremont, Bradshaw, and Black Gulch Creeks. There are numerous small, poorly mineralized quartz veins, most of which are a few miles east of David Guard Station. The gold is in the free state and there is very little sulfide mineralization. Uranium-bearing peat bog was discovered in 1955 in the northwest part of the district.

Bibliography

Brown, G. C., 1916, Green Horn Mountain district: California Min. Bur. Rept. 14, p. 482.

Troxel, B. W., and Morton, P. K., 1962, Kern County, Greenhorn Mountain district: California Div. Mines and Geology, County Report 1, pp. 34–35.

Greenwood

Location. This district is in northwestern El Dorado County. It consists of the northwest segment of the Mother Lode gold belt, which splits at Garden Valley (the east segment continues through Georgetown). This segment of the belt is several miles wide and extends from the point of the split northwest through Greenwood and Spanish Dry Diggings to the Middle Fork of the American River, a distance of about eight miles.

History. Placer mining began in this area shortly after the beginning of the gold rush. The town was named for Caleb and John Greenwood, who established a trading post here in 1850. The district flourished during the 1850s, when the American River was mined and the seam deposits hydraulicked on a large scale. The river was mined by diverting the main stream with a series of flumes, tunnels, and wingdams. The gold-bearing gravels were removed from the bedrock and sent through sluices or long toms. Major mining activity continued through the early 1900s, much of the later placer mining done by Chinese. This district was quite productive again during the 1930s, when the Sliger, Taylor, and Grit lode mines were active. Since about 1955, numerous skin divers have been mining the Middle Fork of the American River by small-scale methods.

Geology. There are two northwest-trending belts of slate of the Mariposa Formation (Upper Jurassic) ½ to one mile apart. Chert, impure quartzite, and slate

Photo 27. Mining Operations, Grizzly Flat District. The photo was taken in the 1850s. *Photo courtesy of Bancroft Library.*

lie to the west, and greenstone and amphibolite schist lie in the center and to the east. A number of small lenticular bodies of serpentine and talcose schist are enclosed in the bedrock, which is deeply weathered in places.

Ore Deposits. There are several wide and sometimes extensive zones of quartz veins and veinlets and mineralized schist containing free gold and auriferous pyrite. Where deeply weathered, the bedrock was eroded, and the gold in the seams and veinlets remained and became concentrated. Such deposits are known as seam deposits or "seam diggings". The upper portions were mined by hydraulicking, and later the unweathered veins at depth were mined by conventional underground methods. Considerable specimen material has been recovered from this district, including crystallized gold. The famous Fricot nugget of crystallized gold (201 ounces), which was taken from the Grit mine in 1865, is on display in the Division of Mines and Geology exhibit in the Ferry Building, San Francisco. Milling-grade ore bodies commonly averaged 1/5 to more than 1/2 ounce of gold per ton. Some of the veins were mined to inclined depths of 2000 feet.

Mines. Admiral Schley, Argonaut $100,000+, Bazocoo, Cedarburg, Centennial, Esperanza $100,-000+, Eagle, French Hill $100,000+?, Greenwood, Grit, Hines-Gilbert $100,000+, Homestake, Maltby, Nancy Lee, Oakland Cons., Railroad Hill, Red Mount,

Revenge, Rosecranz $100,000+, San Martin, Sebastopol, Sliger $2.85 million, Taylor $1 million.

Bibliography

Clark, W. B., and Carlson, D. W., 1956, El Dorado County, Grit, Rosecranz, and Sliger mines: California Jour. Mines and Geology, vol. 52, pp. 415–416, 423–424, and 425–426.

Fairbanks, H. W., 1890, Geology of the Mother Lode region: California Min. Bur. Rept. 10, pp. 81–82.

Lindgren, Waldemar, and Turner, H. W., 1894, Placerville folio, California: U. S. Geol. Survey, Geol. Atlas of the U. S., folio 3, pp.

Logan, C. A., 1935, Mother Lode gold belt of California, Seam mines: California Div. Mines Bull. 108, pp. 43–47.

Preston, E. B., 1893, Taylor mines: California Min. Bur. Rept. 11, p. 205.

Ransome, F. L., 1900. Mother Lode district folio, California: U. S. Geol. Survey, Geol. Atlas of the U. S., folio 63, 11 pp.

Grizzly Flat

Location. The Grizzly Flat district is in south-central El Dorado County about 25 miles east of Placerville. It is in the Sierra Nevada east gold belt and includes the Hazel Valley and Baltic Peak areas. It is both a lode and placer gold-mining district.

History. The streams were originally mined during the gold rush. The camp was established in 1850 and named for a grizzly bear that surprised a group of miners during an evening meal. The Mt. Pleasant mine was discovered in 1851. There was much activity in both the lode and placer mines from the 1870s through the early 1900s. There was some mining activity again during the 1930s. The Hazel Creek mine was discovered in 1948 and was worked on a fair-sized scale until 1958.

Geology. The southern part of the district is underlain by granodiorite, which extends west from the mass of the Sierra Nevada batholith. To the north is slate, phyllite, and graphitic and mica schist of the Calaveras Formation (Carboniferous to Permian), which in places contains small tactite bodies. There are several patches of Tertiary gravels overlain by andesite that are part of the south-extending Tertiary channel of the Mokelumne River.

Ore Deposits. A number of north-trending quartz veins and stringers are found in both the granodiorite and metamorphic rocks. The ore contains free gold and abundant sulfides, especially galena. The ore averages ½ to more than one ounce of gold per ton, but few ore shoots were more than 100 feet long or attained much depth. The Tertiary channel gravels are thin and the channels narrow, but in some places they were extremely rich.

Mines. Lode: Blue Gouge, Cosumnes (Melton) $100,000+, Daily and Bishop, Eagle, Eagle King, Hazel Creek $1 million+, Morey, Mt. Hope, Mt. Pleasant $1 million+, Sunday. Placer: Grizzly Flat drift, Payne drift.

Bibliography

Clark, W. B., and Carlson, D. W., 1956, El Dorado County, Blue Gouge, Cosumnes, and Hazel Creek mines: California Jour. Mines, vol. 52, pp. 411, 413, and 416–418.

Irelan, William, Jr., 1888, Grizzly Flat district: California Min. Bur. Rept. 8, pp. 177–180.

Lindgren, Waldemar, and Turner, H. W., 1894, Placerville folio: U. S. Geol. Survey Geol. Atlas of the U. S., folio 3, 3 pp.

Lindgren, Waldemar, 1911, Tertiary gravels of the Sierra Nevada: U. S. Geol. Survey Prof. Paper 73, pp. 180–181.

Logan, C. A., 1938, El Dorado County, Blue Gouge, Melton, and Mt. Pleasant mines: California Div. Mines Rept. 34, pp. 224–225, 238, and 241–242.

Tucker, W. B., 1919, El Dorado County, Eagle King and Mt. Pleasant mines: California Min. Bur. Rept. 15, pp. 285 and 292.

Grub Gulch

Location and History. This district is in east-central Madera County at the site of the old town of Grub Gulch, seven miles north of Coarsegold and 35 miles northeast of Madera. The site is at the northwest end of a 20-mile-long belt that extends from here southeast through the Coarsegold and Fine Gold districts. This district has been the most productive portion of this belt. It includes part of the area that also was known as the Potter Ridge district. The camp was established shortly after the discovery of Coarsegold in 1849. There was much activity from the 1880s through the early 1900s but very little mining has been done since. A few small suction dredges were active in the 1940s and 1950s downstream in the Chowchilla River.

Geology and Ore Deposits. The district is underlain by a northwest-trending belt of mica schist and quartzite with granodiorite to the west. A series of gold-quartz veins are variously oriented. The veins range from one to 10 feet in thickness. The ore contains free gold and varying amounts of sulfides. The

milling-grade ore was reported to have yielded up to one ounce of gold per ton. The greatest depth of development is 800 feet.

Mines. Bullion, Butterfly, Conary, Crystal Spring, Enterprise $100,000, Gambetta $500,000, Hoboken, Josephine $360,000, Lucky Bill, Savannah, Starlight, Woodland.

Bibliography

Irelan, William, Jr., 1890, Potter Ridge mining district: California Min. Bur. Rept. 10, pp. 197–204.

McLaughlin, R. P., and Bradley, W. W., 1916, Madera County, gold: California Min. Bur. Rept. 14, pp. 539–553.

Hammonton

Location. The Hammonton district is in south-central Yuba County along the lower Yuba River about 10 miles east of Marysville. It is a major dredge field that extends along the river about eight miles. It also is known as the Yuba River district.

History. The river and streams here were first worked during the gold rush by small-scale placer methods. However, this soon ceased because the river level was raised by a large influx of hydraulic mine tailings. Bucket-line dredging began in the district in 1903 under the direction of W. P. Hammon. In 1905 his interests were taken over by Yuba Consolidated Gold Fields, which had just been organized. This concern perfected large-scale bucket-line dredging here into one of the most efficient methods for mining placer gold. Yuba Dredge No. 20 was one of the largest gold dredges in existence. The district was dredged almost continuously from 1903 to 1968 and was the principal source of gold in California for some time. The estimated total output from dredging was estimated in 1964 at 4.8 million ounces.

However, operations have been gradually curtailed; in 1967 only two dredges were operating. On October 1, 1968 the last dredge was shut down, thus ending a major industry that had existed for nearly 70 years. More than a billion cubic yards of gold-bearing gravels were dredged. The extensive piles of gravel have become increasingly important as sources of aggregate.

Gold-Bearing Gravels. The gold-bearing gravels are in and south of the Yuba River, which flows west-southwest through the area. Digging depths range from 60 to 80 feet on the upper end to 100 to 125 feet in the vicinity of the town of Hammonton. As much as 45 feet of the upper gravels are hydraulic mine-tailings. Bedrock in the upper eastern end of the field consists of metamorphic rocks, while, in the central and western portions, the gravels are underlain by clay. The gold recoveries have been as follows: 12¢/yd in 1915–16; 14¢–15¢/yd in 1920–22; 8¢–9¢/yd in 1928–29; 12¢/yd in 1948–49, and 16.56¢/yd in 1959. The hydraulic tailings were reported to have averaged 6¢/yd at the old price. The gravels are medium to fine and are free-washing. Minor amounts of platinum were recovered.

According to the April, 1960 issue of *Mining World*, in 1959 Yuba Cons. reported four dredges treated 16,642,265 cu. yds. with an average content of

Photo 28. Yuba Consolidated Dredge, Hammonton District. Dredge No. 17 operated in the district in Yuba County until 1966. This photo was taken a decade earlier.

16.56¢ per yard. Reserves in the area were estimated to be about 93 million cu. yds. Estimates are that about 235 million yds. of gold-bearing gravels are in the field but beyond depths of existing equipment.

Operations. Hammon and Evans, 1903–05 (bought by Yuba Cons.), two dredges; Marysville Dredging Co., 1906–25 (bought by Yuba Cons.), five dredges; Pacific Gold Dredging Co., 1916–23, one dredge; Yuba Cons. Goldfields, 1905–1968, 21 dredges, not all worked at the same time.

Bibliography

Doolittle, J. E., 1908, Yuba district: California Min. Bur. Bull. 36, pp. 88–91.

Lindgren, 1895, Smartsville folio, California: U. S. Geol. Survey Geol. Atlas of the U. S., folio 18, 6 pp.

Lindgren, Waldemar, 1895, Marysville folio, California: U. S. Geol. Survey Geol. Atlas of the U. S., folio 17, 2 pp.

Lindgren, Waldemar, 1911, Tertiary gravels of the Sierra Nevada, Yuba dredge field: U. S. Geol. Survey Prof. Paper 73, p. 221.

Logan, C. A., 1931, Yuba County, gold dredging: California Div. Mines Rept. 27, pp. 253–257.

O'Brien, J. C., 1952, Yuba County, Yuba Consolidated Goldfields: California Jour. Mines and Geology, vol. 48, pp. 150–151.

Sawin, Herbert, 1946, Placer mining for gold in California, Deep gravels dredged successfully: California Div. Mines Bull. 135, pp. 316–322.

Waring, C. A., 1919, Yuba County, gold dredgers: California Min. Bur. Rept. 15, pp. 425–437.

Winston, W. B., 1910, Gold dredging in California, Yuba County: Calif. Min. Bur. Bull. 57, pp. 165–174.

Hardin Flat

Location. This is a small Sierra Nevada east gold belt district in south-central Tuolumne County. It is west of Yosemite National Park on the Big Oak Flat road and two miles east of the town of Hardin Flat. The town was named for "Little Johnny Hardin", an eccentric Englishman who once owned a sawmill here.

There are a number of small prospects that are intermittently worked, mostly by weekend prospectors.

Geology. The principal rock in the area is granodiorite that is cut by narrow aplitic dikes. There are a number of narrow quartz veins that in places have yielded small but rich pockets of gold near the surface. Some sulfides are present.

Mines. Five Star, Golden Arrow, Huff, Mayflower, New Hope, Santa Maria.

Hildreth

Location and History. This district is in east-central Madera County at the site of the town of Hildreth about 35 miles east of Madera. It is on the northwest end of an indistinct belt of gold mineralization that extends southeast through the Temperance Flat and Big Dry Creek districts in Fresno County. Apparently the chief period of mining in the area was from about 1860 to 1890, with possibly some prospecting and development again during the 1920s and 1930s. The district was named for the Hildreth brothers, farmers who settled here about 1870.

Geology. The principal rock in the district is medium- to coarse-grained granodiorite with several narrow northwest-trending beds of slate and schist. There are a number of north-striking quartz veins containing free gold and often abundant sulfides. The veins are as much as 20 feet thick, and several have been developed to inclined depths of about 600 feet.

Mines. Abbey $100,000+?, Golconda, Hanover, Hildreth, Morrow (Moro, Bazinet), Mud Springs $250,000, Volcano No. 1 $100,000.

Bibliography

Goldstone, L. P., 1890, Hildreth mining district: California Min. Bur. Rept. 10, pp. 194–197.

Irelan, William, Jr., 1888, Hildreth mining district: California Min. Bur. Rept. 8, pp. 202–205.

McLaughlin, R. P. and Bradley, W. W., 1916, Madera County, gold: California Min. Bur. Rept. 14, pp. 539–553.

Hite Cove

Location and History. Hite Cove is in central Mariposa County on the South Fork of the Merced River. Placer mining began in the area shortly after the beginning of the gold rush, and the Hite mine was discovered in 1862 by John R. Hite. He operated the property for 17 years and became quite rich. The mine was active again during the early 1900s. There has been some prospecting in the area in recent years.

Geology and Ore Deposits. The district is underlain by graphitic schist and slate, quartzite, and hornfels. These rocks are cut by a variety of aplitic and granitic dikes, some of which are associated with gold-quartz veins. There are a number of northwest-striking quartz veins up to 12 feet thick. The ore contains native gold and often abundant sulfides. The greatest depth of development is about 800 feet.

Mines. Brown Bear, Bunker Hill, Confidence, Emma, Eureka, Georgia Point, Hite $3 million, Hite Central, Kaderitas, Mexican, Williams.

Bibliography

Bowen, O. E., Jr., 1957, Mariposa County, lode mines: California Jour. Mines and Geology, vol. 53, pp. 72–187.

Castello, W. O., 1921, Mariposa County, Hite Cove district: California Min. Bur. Rept. 17, p. 94.

Lowell, F. L., 1916, Mariposa County, Hite mine: California Min. Bur. Rept. 14, pp. 583–584.

Hodson

Location. This district is in the Sierra Nevada west gold belt in southwestern Calaveras County a few miles west of Copperopolis. A belt of lode-gold deposits extends from the site of the old town of Hodson northwest through Salt Springs Valley, a distance of about 10 miles. At one time this district was also known as the Felix district.

History. Small-scale placer mining probably was done here during the gold rush, and the mining of rich surface pockets soon followed. The district was highly productive during the 1890s and early 1900s when the Royal mine and other properties were worked on a large scale. The 120-stamp mill at the Royal mine, erected in 1903, was one of the largest mills in California. The mines were active again during the 1930s and early 1940s. Copper ore from Copperopolis was treated at the Mountain King Mill during World War II. More recent exploration work, including diamond drilling, has been done in the district, but very little of it has been gold mining.

Geology. On the west side of the mineralized belt are northwest-trending beds of slate of the Mariposa Formation (Upper Jurassic). Metavolcanic rocks, chiefly massive greenstones and amphibolite of the Logtown Ridge Formation (Upper Jurassic), are on the east. The central portion has been intruded by numerous serpentinized bodies in or adjacent to the northwest-trending Hodson fault zone.

Ore Deposits. The deposits consist of large low-grade bodies of mineralized schist and greenstone known as "gray ore," which contain some disseminated free gold, auriferous pyrite and minor amounts of other sulfides. The deposits are associated with the Hodson fault. One of the larger gray ore bodies is several thousand feet long, 500 feet wide, and has been mined to an inclined depth of several thousand feet. Some high-grade pockets have been taken from quartz veins and stringers containing free gold and sulfides.

Mines. Butcher Shop, Empire, Gold Knoll, Gold Metal, Mountain King $1 million, Pine Log, Ranch, Royal $5 million+, Wilbur Womble.

Bibliography

Clark, W. B., and Lydon, P. A., 1962, Calaveras County, gold: California Div. Mines and Geology County Report 2, pp. 32–93.

Knopf, Adolph, 1929, Royal mine: U. S. Geol. Survey Prof. Paper 157, p. 72.

Logan, C. A., 1936, Calaveras County, Royal mine: California Div. Mines Rept. 32, pp. 285–287.

Storms, W. H., 1900, Royal Consolidated gold mine: California Min. Bur. Bull. 18, pp. 126–127.

Taliaferro, N. L., and Solari, A. J., 1948, Geology of the Copperopolis quadrangle: California Div. Mines Bull. 145, pl. 1.

Tucker, W. B., 1919, Calaveras County, Royal Consolidated and Wilbur Womble mine: California Min. Bur. Rept. 15, pp. 103, 113.

Homer

Location. The Homer district is on the east slope of the Sierra Nevada in west-central Mono County in the vicinity of Lundy Lake, about six miles west of Mono Lake. The district has also been known as the May Lundy or Lundy district, because the May Lundy mine was the principal source of gold here.

History. Although this area was prospected during the Comstock silver rush of the 1860s, the lode deposits were not discovered until 1877. The district was organized in 1879. The May Lundy mine was named for the daughter of W. J. Lundy, who operated a sawmill here in the 1870s. This mine was worked on a major scale until 1911. Accumulated tailings were treated during the late 1930s, but there has been only minor prospecting since. The mine has a total production of $3 million.

Geology. The principal geologic feature is a two- to four-mile-wide belt or roof pendant of metamorphic rocks that extends northwest along the Sierran crest for many miles. These rocks consist of schist, slate, and hornfels of Triassic and Jurassic age. Granodiorite lies on both sides of this belt. The Sierra Nevada Mountains here have been prominently shaped by Pleistocene glaciation.

Ore Deposits. A series of northwest-striking and southwest-dipping quartz veins are found at or near the metamorphic-granitic contacts. The veins usually average two to three feet in thickness. The ore contains free gold, pyrite, and smaller amounts of other sulfides. Milling-grade gold ore yielded as much as one ounce per ton with a high content of silver. Several ore shoots at the May Lundy mine had stoping lengths of up to 300 feet.

Photo 29. Doss (Ginaca) Mine, Hornitos District. This 1934 view of the Mariposa County mine shows a Hadsel mill, at left. *Photo by Ralph Baverstock, from collection of Dr. Horace Parker.*

Bibliography

Bowen, O. E., Jr., 1962, Mines near Yosemite: California Div. Mines and Geology, Mineral Information Service, vol. 15, no. 3, pp. 1–4.

DeGroot, Henry, 1890, Homer district: California Min. Bur. Rept. 10, p. 342.

Eakle, A. S., and McLaughlin, R. P., 1919, Mono County, May Lundy mine: California Min. Bur. Rept. 15, pp. 166–167.

Sampson, R. J., and Tucker, W. B., 1940, Mono County, May Lundy and Parrot mines: California Div. Mines Rept. 36, pp. 128–129 and 130–131.

Whiting, H. A., 1888, Homer mining district: California Min. Bur. Rept. 8, pp. 367–371.

Honcut

Location and History. This is a gold-dredging district in southwest Butte County along Honcut and Wilson Creeks northeast of the town of Honcut. The name comes from Hoankut, an Indian village once situated on the Yuba River just below the mouth of Honcut Creek. The Bangor district is just to the east. The creeks were first worked by hand methods in the early days. Bucket-line dredging began in 1909 and continued until around 1920. There was some dragline dredging in the district during the 1930s.

Geology. Pleistocene gravels and Recent creek gravels overlie bedrock of greenstone and green schist. Digging depths averaged about 20 feet. The dredged area covers about 1000 acres.

Bibliography

Waring, C. A., 1919, Butte County, gold dredging: California Min. Bur. Rept. 15, pp. 187–198.

Winston, W. B., 1910, Honcut Creek dredging district: California Min. Bur. Bull. 57, pp. 158–159.

Honey Lake

There are a few small lode-gold mines and prospects several miles south of Honey Lake in southeast Lassen County and eastern Plumas County. These include the Plinco and Honey Lake mines. The Honey Lake or Badger mine was discovered in 1900 and worked dur- ing the 1920s and 1930s. The deposits consist of narrow and shallow quartz veins in granitic rock, which in places contain free gold and pyrite.

Bibliography

Averill, C. V., 1936, Lassen County, Honey Lake gold mines: California Div. Mines Rept. 32, pp. 435–436.

Hope Valley

This is a small gold- and tungsten-mining district in northwestern Alpine County about 10 miles west of Markleeville. The area was first prospected during the early 1860s, followed by minor intermittent prospecting and development since. Some tungsten was produced during World War II and the Korean War.

The ore deposits are associated with two north-trending roof pendants of hornfels, quartzite, and schist that are surrounded by granodiorite. The deposits consist of narrow gold-quartz veins and pyrite and tungsten-bearing garnetiferous tactite. Small amounts of copper also are present.

Hornitos

Location. The Hornitos district is in the Sierra Nevada west gold belt in western Mariposa County about 15 miles west of the town of Mariposa (see fig. 4). The district contains a several-mile wide belt of lode-gold mines that extends from the vicinity of the Exchequer Reservoir south-southeast through Hornitos to the Indian Gulch area.

History. The streams in the area were first worked in 1849, and lode mining began in 1850 at the Washington mine. The town was first settled by Mexicans who had been driven out of nearby Quartzburg. The name Hornitos is a diminutive of "horno" or small bake oven, from the Spanish. Mining activity was great from the 1860s through the 1880s, lesser from the 1890s to the 1920s. The Mt. Gaines mine was worked on a major scale during the 1930s. Since World War II the area has been prospected, but there has been very little recorded production. Historically, this is the most productive district of the Sierra Nevada west gold belt.

Geology. The district is underlain by greenstone and green schist in the west portion and slate in the east. Also present are smaller amounts of amphibolite, mica schist, and hornfels. A number of small granodiorite intrusions are exposed, along the margins of which chiastolite-mica schist has developed.

Ore Deposits. A number of north-trending quartz veins and stringers containing free gold and varying amounts of sulfides, chiefly pyrite, are present, as well as several large bodies of mineralized greenstone or "gray ore". Some of the veins have very flat dips. The ore shoots vary considerably in size with stoping lengths ranging from a few to as much as 400 feet. Some of the veins have been mined to inclined depths of more than 1500 feet. Milling-grade ore commonly contains from $\frac{1}{5}$ to more than $\frac{1}{2}$ ounce per ton in gold.

Mines. Badger $80,000+, Doss $100,000, Duncan, Lost Douglas, Martinez, Mt. Gaines $3.59 million, Numbers 1, 5, 8, and 9, Ruth Pierce $600,000, Washington $2,377,000.

Bibliography

Bowen, O. E., 1957, Mariposa County, gold: California Jour. Mines and Geology, vol. 53, pp. 69–187.

Castello, W. O., 1921, Mariposa County, Hornitos district: California Min. Bur. Rept. 17, p. 94.

Lowell, F. L., 1916, Mariposa County, gold: California Min. Bur. Rept. 14, pp. 575–600.

Hunter Valley

Location and History. This district is in the northwest corner of Mariposa County, in the general area of Hunter Valley, the Don Pedro Reservoir and Lake McClure. It was named for William W. Hunter, a well-known engineer. There was extensive placer gold-mining here during the 1850s and some copper mining in the 1860s. The lode gold mines were active until the early 1900s. Some mining was done again during the 1930s, and the Pyramid mine has been prospected recently.

Geology. The district is underlain by northwest-trending belts of slate of the Mariposa Formation (Upper Jurassic) and greenstones, chert and slate of the Amador Group (Middle to Upper Jurassic). Several small diorite and granodiorite intrusions are mapped.

Ore Deposits. A number of northwest-striking systems of gold-quartz veins are in the slate, chert and greenstone. Numerous stringers and cross veins are present. A number of high-grade pockets have been found. The ore contains free gold and often abundant sulfides, and milling ore commonly averaged one ounce per ton in gold. None of the veins have been mined to depths of more than a few hundred feet.

Mines. Blue Cloud, Cotton Creek, Iron Duke, Morning Star, Oak and Reese $500,000–$600,000, Or-

ange Blossom, Pyramid $200,000, Schoolhouse, Yellowstone.

Bibliography

Bowen, O. E., 1957, Mariposa County, Oaks and Reese and Pyramid mines: California Jour. Mines and Geology, vol. 53, pp. 145–147 and 158.

Castello, W. O., 1921, Mariposa County, Hunter Valley district: California Min. Bur. Rept. 17, pp. 94–95.

Turner, H. W., and Ransome, F. L., 1897, Sonora folio: U. S. Geol. Survey Geol. Altas of the U. S., folio 41, 7 pp.

Indian Diggings

Location. This district is in south-central El Dorado County about 30 miles southeast of Placerville. It includes the Indian Diggings, Henry Diggings, Omo Ranch and Brownsville areas. Indian Diggings is best known as a placer-mining district, but there are a number of lode deposits. The Fairplay district lies to the west.

Geology. The district is underlain by quartz-mica schist, graphitic slate, green schist, quartzite, and limestone. The central part of the area has been intruded by a round quartz-diorite stock. Portions of the bedrock are overlain by patches of Tertiary auriferous gravel and extensive andesite flows.

Ore Deposits. The channel gravels are part of the Tertiary Mokelumne River, which extends south into this area from the Grizzly Flat district. The channel then extended west and southwest toward Fiddletown in Amador County. Indian Diggings was on a branch of this channel. Large amounts of gold came from these channel deposits, especially those at Indian Diggings, where the bedrock is limestone that contained numerous rich potholes. Mining was done by both hydraulicking and drifting. A number of narrow north-striking quartz veins in quartz diorite contain small but often rich ore shoots. The ore contains free gold and abundant pyrite, galena, and smaller amounts of other sulfides.

Mines. Placer: April Fool, Armstrong, Carrie Hale, Chic, Christion, Deep Channel, Dorsey, Drusy, Hayward, Hidden Treasure, Irish Slide, Last Chance, Little Bill, Lucky Jack, Old Chink, Omo, Patterson, Payne, Peacock, Richmond, Syracuse, Telegraph, Tomcat, Yellow Aster, Yellowjacket. Lode: Black Oak, Gold Note, Independence, Polar Bear, Potosi, Stillwagon.

Bibliography

Clark, W. B., and Carlson, D. W., 1956, El Dorado County, Placer deposits: California Jour. Mines and Geology, vol. 52, pp. 429–435.

Lindgren, Waldemar, 1911, Tertiary gravels of the Sierra Nevada: U. S. Geol. Survey Prof. Paper 73, pp. 180–181.

Lindgren, Waldemar, and Turner, H. W., 1894, Placerville folio, California: U. S. Geol. Survey Geol. Atlas of the U. S., folio 3, 3 pp.

Indian Hill

Indian Hill is in western Sierra County about 10 miles west of Downieville and just south of the Brandy City district. Much of the production here has come from the Indian Hill and Depot Hill hydraulic mines. These mines were extensively worked from the 1850s to the 1880s, and intermittent development work and mining continued through the 1930s. The Depot Hill mine has been prospected recently. The deposits are on the LaPorte-Brandy City Tertiary channel. The lower gravels contain abundant quartz and are as much as 100 feet thick. They are overlain in places by intervolcanic gravels and andesite. Bedrock is granite with amphibolite to the west and slate, schist, and serpentine to the east. There are also a few gold-quartz veins in the district.

Bibliography

Logan, C. A., 1929, Sierra County, Indian Hill mine: California Div. Mines Rept. 25, pp. 192–193.

MacBoyle, Errol, 1920, Sierra County, Indian Hill mining district: California Min. Bur. Rept. 16, pp. 13–14.

Turner, H. W., 1897, Downieville folio: U. S. Geol. Survey Geol. Atlas of the U. S., folio 37, 8 pp.

Inskip

Location. Inskip is in northeastern Butte County about seven miles north of Sterling City. The Kimshew district lies to the east and the Magalia district to the south. The area was active before and during the early 1900s.

Geology. There are a number of narrow gold-quartz veins in slate, amphibolite and greenstone. The ore bodies usually are small but contain abundant sulfides.

Mines. Bluebird, Cain, Excelsior, Fitzpatrick, Inskip, Lost Treasure, Midas, Rawhide, Walker, Wild Yankee.

Bibliography

Waring, C. A., 1919, Butte County, gold-quartz mines: California Min. Bur. Rept. 15, pp. 211–224.

Iowa Hill

Location. The Iowa Hill district is in central Placer County in the vicinity of the old mining town of that name. It is an extensive placer-mining district that includes the Roach Hill, Monona Flat, Strawberry Flat, Succor Flat, Grizzly Flat, Shirttail Canyon, and Kings Hill areas.

History. Placer mining began along the American River and its tributaries soon after the beginning of the gold rush. Hydraulic and drift mining apparently began here in 1853, and by 1856 the output was as high as $100,000 per week. By 1880 more than $20 million had been produced from the district. Drift mining continued through the early 1900s, and there was appreciable activity again in the 1930s. Most of the town was destroyed by fire in 1922. The Big Dipper, Occidental and a few other mines have been intermittently worked in recent years. Also snipers and skin divers have been active in the district.

Geology. A main Tertiary channel of the American River crosses the area. There are numerous branches and intervolcanic channels, including the Succor Flat intervolcanic channel, which comes in from the northeast, and the west-trending Morning Star and Grizzly Flat deep channels. The deep channel gravels are well-cemented and in places yielded ½ ounce of gold or more per yard. The lowest seven feet were the richest but there also were some rich benches. The bedrock is uneven, and consists of hard slate and phyllite of the Cape Horn Formation (Carboniferous) and amphibolite, which contains a number of deep and rich potholes. To the east the gravels are overlain by thick beds of andesite. There are a few gold-quartz veins in the district.

Mines. Big Dipper $1.2 million, Blue Wing Quartz, Brunn, Buckeye, Campbell, Canyon, Carey, Copper Bottom, Dewey Cons., Drummond, Elizabeth Hill, Excelsior, Fitzpatrick, Gleason $1 million+, Golden Star, Golden Streak, Goodwin, Haymes, H and H, Iowa Hill, Irish and Bryne, Jupiter, Keystone, King's Hill Point, King's Hill Quartz, Lebanon, Mohawk, Morning Star, Occidental, Old Jupiter, Penn

Photo 30.　Big Dipper Drift Mine, Iowa Hill District. This early view of the Placer County mine looks east. *Photo courtesy of Calif. State Library.*

N

SECRET RAVINE

AMERICAN RIVER

NORTH FORK

PENN VALLEY
STRAWBERRY
MONONA FLAT
ROACH HILL
SUCCOR FLAT
COPPER BOTTOM
RANDALL
STAR UNITED
LITTLE INDIAN CREEK
BRUNN
IOWA HILL
IOWA HILL
MORNING STAR
GOLDEN STREAK
BIG DIPPER
GRIZZLY FLAT
DRUMMOND
OCCIDENTAL
JUPITER
GOODWIN CANYON
ELIZABETH HILL
GLEASON
TWENTY ONE
OLD JUPITER
CAMPBELL
INDIAN CREEK
HAYMES
WELCOME
KINGS HILL QUARTZ
H & H
KINGS HILL
KINGS HILL POINT
SHIRTTAIL CANYON
BRUSHY CANYON
BUCKEYE
KEYSTONE
SHIRTTAIL CANYON

SCALE
0 1/2 1 2 MILES

EXPLANATION

Hydraulic mine
Adit
Shaft
Tertiary channel (approximate course)
Direction of flow

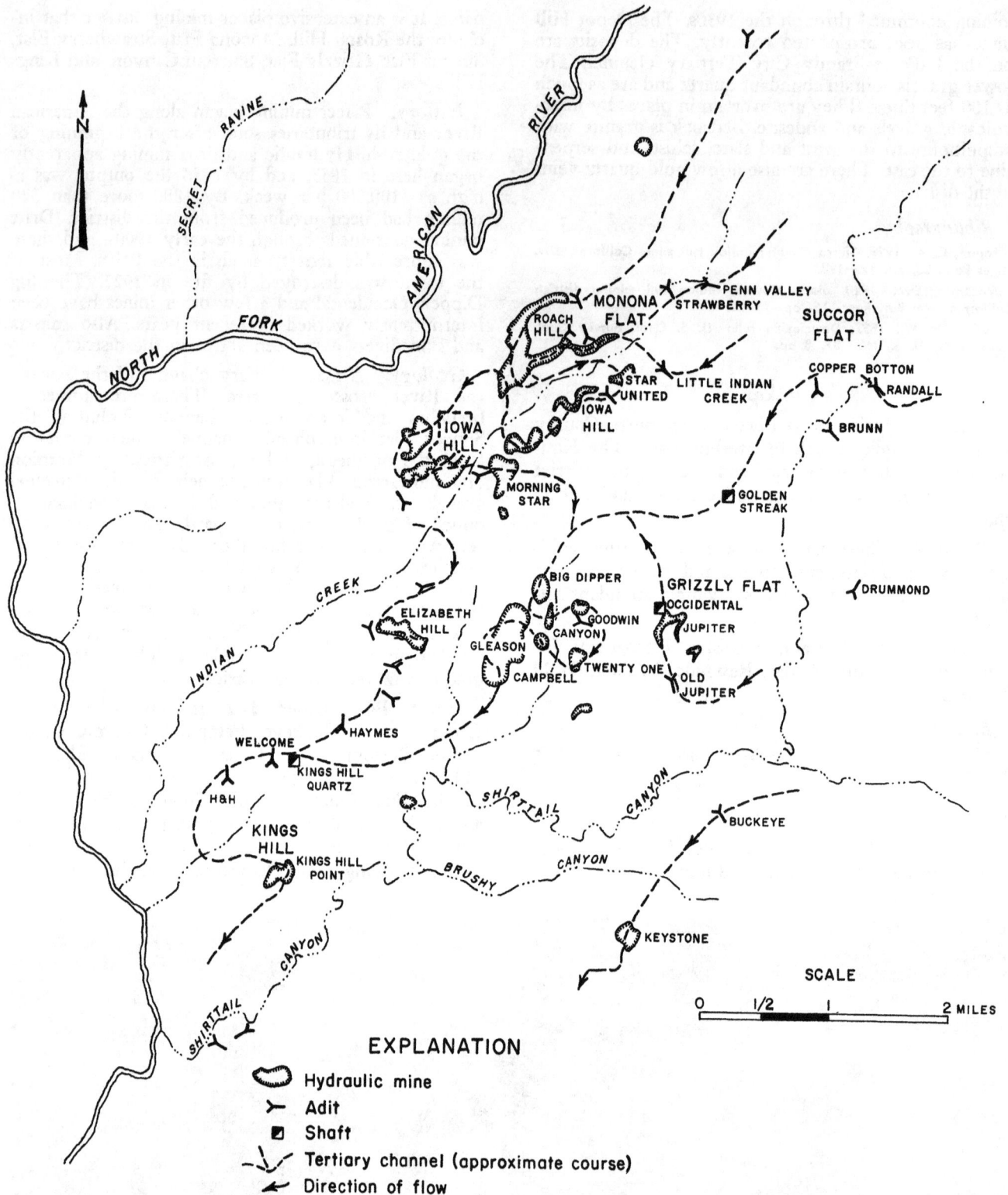

Figure 11. Map of Iowa Hill District, Placer County. The map shows mine locations and Tertiary channel courses. *After Chandra, 1961, Plate 3, and Hobson, 1890.*

Valley, Randall, Roach Hill, Star United, Strawberry, Twenty One, Welcome, Winchester, Wisconsin Hill, Union.

Bibliography

Chandra, D. K., 1961, Geology and mineral deposits of the Colfax and Foresthill quadrangles: California Div. Mines Spec. Rept. 67, 50 pp.

Photo 31. Amador-Star Mine, Jackson-Plymouth District. This 1952 view of the Amador County mine looks west. The dump is composed mainly of black slate. *Photo by D. W. Carlson.*

Hobson, J. B., 1890, Iowa Hill mining district: California Min. Bur. Rept. 10, pp. 419–425.

Jarman,´ Arthur, 1927, Iowa Hill: California Min. Bur. Rept. 23, pp. 86–87.

Lindgren, Waldemar, 1900, Colfax folio: U. S. Geol. Survey Geol. Atlas of the U. S., folio 66, 10 pp.

Lindgren, Waldemar, 1911, Tertiary gravels of the Sierra Nevada, Iowa Hill and Wisconsin Hill: U. S. Geol. Survey Prof. Paper 73, pp. 148–149.

Logan, C. A., 1936, Gold mines of Placer County, Big Dipper, Morning Star, and Succor Flat channel mines: California Div. Mines Rept. 32, pp. 52–54, 65–66, and 79.

Waring, C. A., 1919, Placer County, Iowa Hill district: California Min. Bur. Rept. 16, p. 318.

Irish Hill

Location. The Irish Hill district is in northwestern Amador County about five miles north of Ione. It includes the Muletown and the Forest Home areas.

Geology. Several extensive patches of Eocene quartz-rich channel gravels and younger gravels exist. They were first mined by ground sluicing and hydraulicking and later by dragline dredging. Bedrock consists of slate, phyllite, greenstone, and amphibolite. There are also several copper mines here.

Bibliography

Piper, A. M., Gale, H. S., Thomas, H. E., and Robinson, T. W., 1939, Geology and ground-water hydrology of the Mokelumne area: U. S. Geol. Survey Water Supply Paper 780, plate I.

Turner, H. W., 1894, Jackson folio: U. S. Geol. Survey Geol. Atlas of the U. S., folio 11, 6 pp.

Jackson-Plymouth

Location. A 20-mile-long belt of gold mineralization runs through western Amador County. On the belt, a portion of the Mother Lode, are the towns of Jackson, Sutter Creek, Amador City, Drytown, and Plymouth. Because of the uniform nature of the gold mineralization along this belt, the several districts and sub-districts have been grouped together here under a Jackson-Plymouth heading.

History. This entire belt was settled early in the gold rush when the streams were placer-mined. Jackson was settled by California Spanish at least as early as 1849. It was first known as Botilleas, but the name was soon changed in honor of Colonel Alden Jackson. Sutter Creek was named for Captain John A. Sutter who visited the region in 1846. Amador City was settled in 1851 and Plymouth in 1852. Drytown flourished from 1848 until 1857 when rich placer deposits were worked. Most of the important lode deposits were discovered during the 1850s. The Argonaut mine was first developed in 1850, the South Spring Hill and Lincoln in 1851, the Plymouth in 1852, the Original Amador and Keystone in 1853, the Central Eureka in 1855 and the Kennedy in 1856. Lode mining developed into a major industry that was to last 90 years.

By 1875 mines such as the Keystone, South Spring Hill, Oneida, Old Eureka, and Plymouth had become large and highly profitable operations. However, the Argonaut, Kennedy, Central Eureka, Bunker Hill, Fremont-Gover, and Lincoln Cons. (Lincoln, Wildman, and Mahoney), major gold sources of a later date, did not become important until the 1880s and 1890s. The properties constituting the Plymouth Cons. mine were consolidated in 1883, the Kennedy Mining and Milling Company was organized in 1885, and the Argonaut Mining Company in 1893. From the 1890s until 1942, this belt was one of the more important gold-mining districts in the nation. The value of pro-

Photo 32. Central Eureka Mine, Jackson-Plymouth District. This 1952 view of the Amador County mine looks northeast at the Old Eureka shaft. The mine was shut down a year after the photo was taken. *Photo by Jeffrey Schweitzer.*

duction ranged from $2 million to $4 million annually. Several thousand miners were employed, many of whom were of Italian, Austrian, and Serbian extraction.

There were two noted lawsuits between the Argonaut and Kennedy mines in 1894 and 1897 in which the former accused the latter of conducting mining operations in their ground. Several disastrous fires have occurred in the district, including one at the Argonaut mine in 1922 that caused the loss of 47 lives. This fire began on the 3350-foot level of the mine and trapped a whole shift of miners on the 4650-foot and 4800-foot levels.

Photo 33. Kennedy Mine and Mill, Jackson-Plymouth District. This view, looking north, shows the Amador County mine and 100-stamp mill in about 1936. At that time, the Kennedy was the deepest mine in the United States, with a vertical depth of 5,912 feet. *Photo courtesy of Calif. State Library.*

Photo 34. Kennedy Mine, Recent View. This view, to the north, shows the tailings wheels, headframe and remaining buildings at the Amador County mine. The structures are now a historical display. *Photo by Mary Hill.*

Photo 35. Argonaut Mine and Mill, Jackson-Plymouth District. This view of the Amador County mine, in about 1920, looks west. In 1922, an underground fire in this mine took 47 lives. *Photo courtesy of Calif. State Library.*

Photo 36. Plymouth Consolidated Mine, Jackson-Plymouth District. This view, to the southwest, shows the Empire shaft at the Amador County mine in 1952. *Photo by D. W. Carlson.*

As mining operations progressed to greater and greater depths, costs increased, especially because in some mines, the grade of ore decreased at depth and it became necessary to mine larger amounts. A number of immense mills were erected, including those at the Kennedy mine, which employed 100 "stamps" each, one at the South Eureka with 80 "stamps" and those at the Argonaut and Oneida, which had 60 "stamps" each. The ground became extremely heavy at depth and required much timbering. As costs continued to increase during the early 1900s and were accelerated during World War I, a number of mines were shut down. The South Spring Hill mine was shut down in 1902, the Lincoln Cons., in 1912, the Oneida and Zeila in 1914, the South Eureka in 1917, and the Bunker Hill and Treasure in 1922. However, the district continued to yield large amounts of gold as the Argonaut, Kennedy, Central Eureka and others increased the size of their operations. The Old Eureka and Central Eureka merged in 1924; the new operation was known as the Central Eureka, for a time as the Hetty Green, as it was controlled by that financier. The district's output increased after the 1934 rise in the price of gold. The veins continued to be devel-

oped at greater and greater depths until the Argonaut and Kennedy became the deepest mines in the country. Each has a vertical depth of more than 5900 feet. The Central Eureka, South Eureka, and Plymouth Cons. are more than 4000 feet deep.

All of the mines were shut down soon after the beginning of World War II. The Central Eureka mine was reopened in 1945, but because of greatly increased costs it was shut down again in 1953. This was the last active major gold mine on the Mother Lode.

Jackson-Plymouth was the most productive district of the Mother Lode belt, with a total output estimated by the author at about $180 million. If large-scale gold mining were ever to be done here again, it would be most desirable to consolidate the major mines and operate them as a unit. Some are connected underground and all produced considerable water.

The remaining surface plant of the Kennedy mine is now a museum. The large wooden tailing wheels and the superintendent's office at this mine, long noted landmarks, are historical displays. Several other distinctive old mine buildings have been preserved, including the Keystone mine office, which is a motel, and the Zeila office, which is a private home.

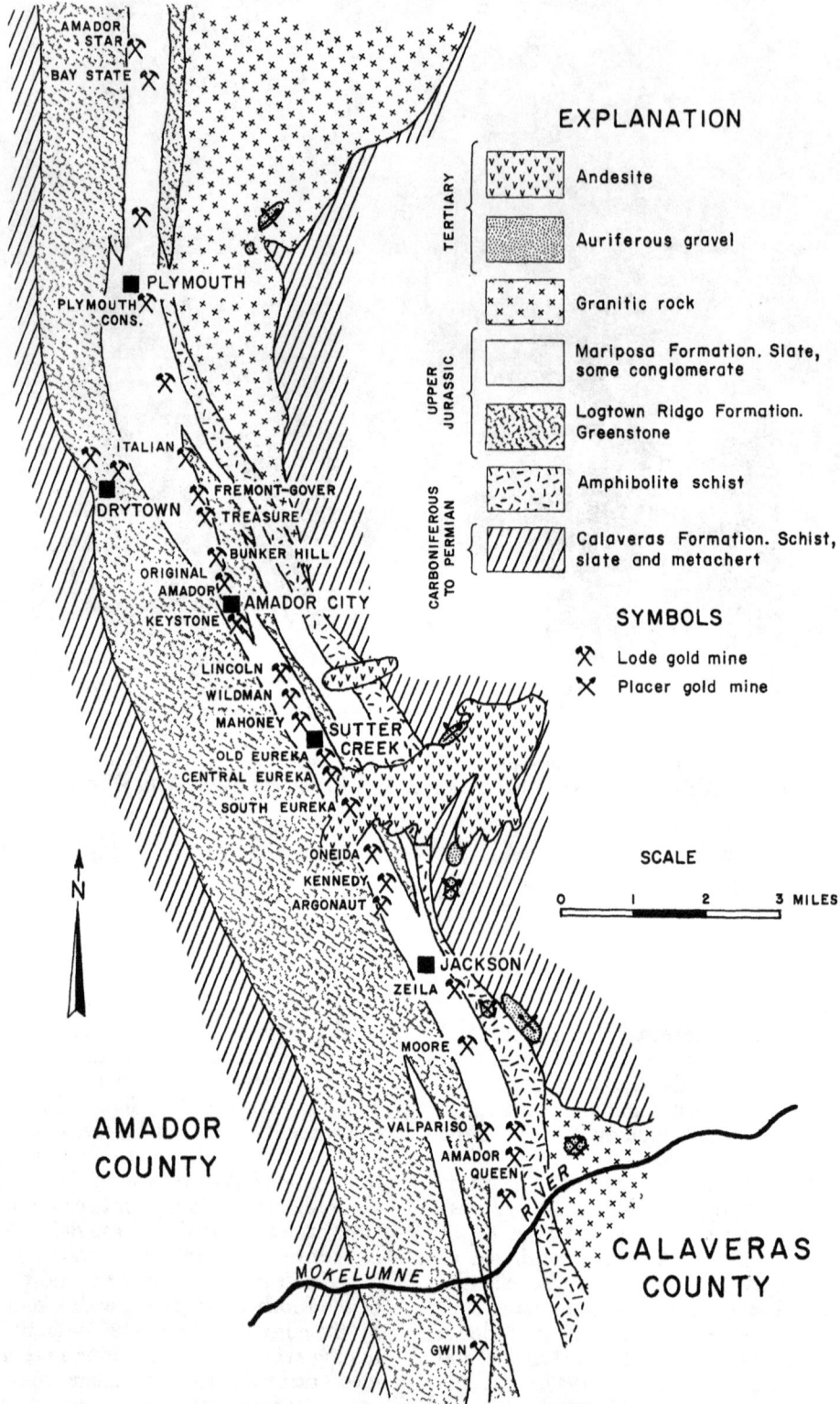

Figure 12. Geologic Map of Jackson-Plymouth District, Amador County. The locations of the mines are shown. *After Turner, 1894a; Lindgren and Turner, 1894; Knopf, 1929, and Carlson and Clark, 1954.*

Geology. The gold deposits are in a north- and northwest-trending mile-wide belt of gray to black slate of the Mariposa Formation (Upper Jurassic), with some interbedded coarse and occasionally sheared conglomerate and minor sandy and gritty layers (fig. 12). Massive greenstone of the Logtown Ridge Formation (Upper Jurassic) lies west of the belt of Mariposa Formation slate. Metasedimentary rocks, chiefly

W.

E.

4800' LEVEL

W.

E.

4200' LEVEL

W.

E.

2600' LEVEL

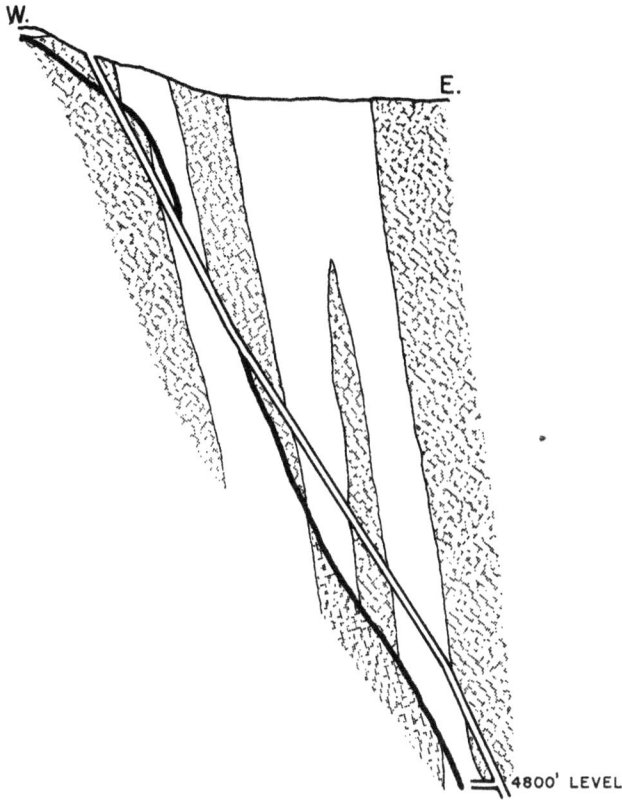

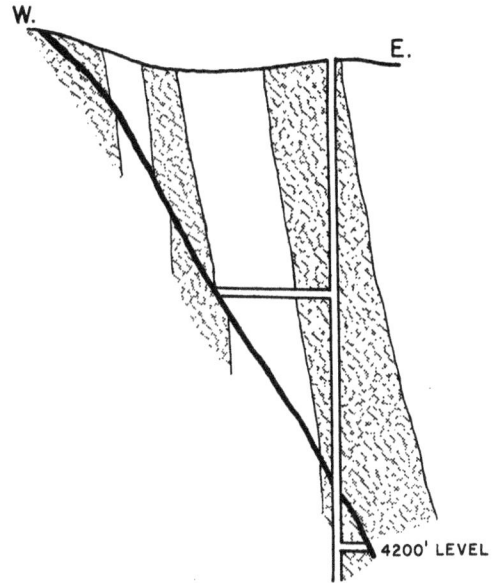

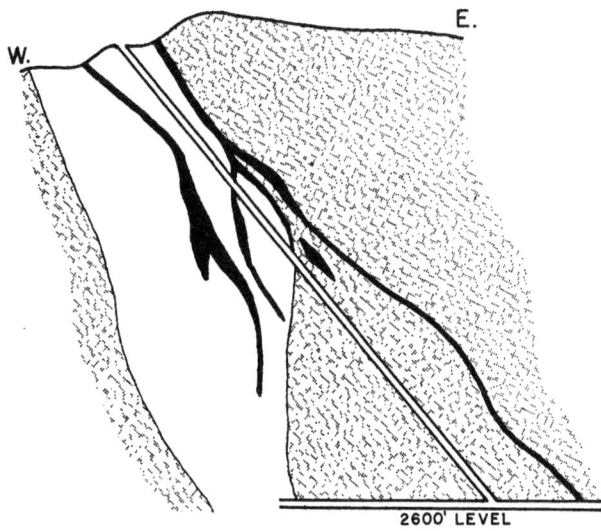

EXPLANATION
(For figures 13, 14, and 15)

Slate of Mariposa Formation

Greenstone of Logtown Ridge Formation

Gold–quartz vein

Figure 13 (topleft).   Section through Argonaut Mine. *After Knopf, 1929.*

Figure 14 (top right).   Section through Kennedy Mine. *After Knopf, 1929.*

Figure 15 (bottom).   Section through Keystone mine. *After Logan, 1935.*

graphitic schist, metachert and amphibolite schist of the Calaveras Formation (Carboniferous to Permian) are to the east. Several deposits of Tertiary auriferous channel gravels are exposed south of Jackson.

Ore Deposits. The ore bodies occur in massive and sheared quartz veins often with abundant fault gouge. The veins are mainly in slate of the Mariposa Formation. The veins sometimes are tens of feet thick; in places the Keystone vein is as much as 200 feet thick. Usually there are many stringers. The ore bodies contain disseminated fine free gold, pyrite, and minor amounts of other sulfides. The sulfides usually average one to two percent of the ore. In addition, greenstone bodies with disseminated auriferous pyrite known as "gray ore" sometimes are adjacent to the quartz veins at depth. The milling ore usually is low to moderate in grade ($\frac{1}{7}$ to $\frac{1}{3}$ ounce of gold per ton), but a number of the veins have been mined to inclined depths of 4000 to 6000 feet. The ore shoots usually had stope lengths of 200 to 500 feet, but pitch lengths were much greater, and often nearly vertical. A number of high-grade pockets were found. The ground was nearly always heavy and required much timbering. During mining operations, it was usually necessary to fill stoped-out areas with waste.

Mines. Alma, Alpine, Amador Gold $100,000+, Amador King $100,000+, Amador Queen No. 1 $100,000+, Amador Queen No. 2 $100,000+, Ama-

dor Star $100,000+, Anita, Argonaut $25.2 million, Ballard, Bay State $100,000+, Bellwether, Bunker Hill $5.1 million, Central Eureka Group $36 million, Crown Point, Detert Group, Fremont-Gover $5 million, Good Hope, Italian $140,000+, Hardenbergh $100,000+, Kennedy $34.28 million, Keystone $24 million, Lincoln $2.2 million, Mammoth, Mayflower, Mineral Point, Moore $564,000+, New London, North Star et al., Oneida $2.5 million+, Original Amador $3.5 million, Plymouth Cons. $13.5 million, Potosi, South Eureka $5.3 million, South Jackson, South Spring Hill $1.1 million, Treasure $1 million, Valpariso $100,000+, Wildman-Mahoney $5 million, Zeila $5 million+.

Bibliography

Brown, J. A., 1890, Amador County: California Min. Bur. Rept. 10, pp. 98–123.

Carlson, D. W., and Clark, W. B., 1954, Mines and mineral resources, Amador County, lode gold mines: California Jour. Mines and Geology, vol. 50, pp. 167–195.

Fairbanks, H. W., 1890, Geology of the Mother Lode region: California Min. Bur. Rept. 10, pp. 67–78.

Irelan, William, Jr., 1888, Amador County, Mother Lode mines: California Min. Bur. Rept. 8, pp. 41–96.

Knopf, Adolph, 1929, The Mother Lode system of California: U. S. Geol. Survey Prof. Paper 157, pp. 49–70.

Logan, C. A., 1927, Amador County, gold quartz mines: California Min. Bur. Rept. 23, pp. 149–185.

Logan, C. A., 1935, Mother Lode gold belt of California: California Div. Mines Bull. 108, pp. 55–124 and 141–142.

Preston, E. B., 1893, Amador County, quartz mines: California Min. Bur. Rept. 11, pp. 139–146.

Photo 37. Eagle-Shawmut Mine, Jacksonville District. This 1914 view, looking northeast, shows the 100-stamp mill, the tramway and the tailings pond at the Tuolumne County mine. The Tarantula mine is at left. *Photo courtesy of Tuolumne County Museum.*

Photo 38. Crystalline Mine, Jamestown District. This is an early view of the Tuolumne County mine. *Photo courtesy of Tuolumne County Museum.*

Ransome, F. L., 1900, Mother Lode district folio, California: U. S. Geol. Survey Geol. Atlas of the U. S., folio 63, 11 pp.

Storms, W. H., 1900, The Mother Lode region-Amador County: California Min. Bur. Bull. 18, pp. 43–87.

Turner, H. W., 1894, Jackson folio, California: U. S. Geol. Survey Geol. Atlas of the U. S., folio 11, 6 pp.

Tucker, W. B., 1916, Amador County, gold: California Min. Bur. Rept. 14, pp. 14–52.

Jacksonville

Location. This district is in southwestern Tuolumne County. It is in that portion of the Mother Lode belt that extends through the vicinity of southeast Jacksonville to the vicinity of Moccasin Creek. The Jamestown district is just to the north, the Big Oak Flat-Groveland district is to the east, and the Coulterville district is to the southeast.

History. Jacksonville, named for Colonel Alden Jackson—as was the town of Jackson in Amador County—was founded as a supply center in 1848. The placer deposits here were extremely rich, credited with a production of $9 million. Lode mining began in the district in the late 1850s. The Eagle-Shawmut mine was operated on a large scale from 1897 until 1942. The district was served for some years by the Hetch-Hetchy Railroad.

Geology and Ore Deposits. The deposits occur near or at the contact between serpentine on the west and slate and schist on the east interlayered with a number of narrow bands of greenstone. The ore deposits consist chiefly of large but low-grade bodies of pyrite ankerite-quartz and mariposite-ankerite-quartz rock and numerous pyrite-quartz stringers. Most of the gold values are in the sulfides; there is not much free gold. The ore zones were as much as 180 feet in thickness. There are several adjacent but relatively barren massive bull quartz veins. Mining extended to depths of 3000 feet. The ore usually contained 1/7 ounce or less gold per ton; for years the mill heads at the Eagle-Shawmut mine averaged about $2.75 per ton.

Mines. Clio $100,000+, Eagle-Shawmut $7.4 million, Harriman $100,000+, Mammoth $100,000, Moccasin, Orcutt, Republican, Tarantula $100,000+, Wheeler.

Bibliography

Knopf, Adolph, 1929, The Mother Lode system of California, Eagle-Shawmut and Clio mines: U. S. Geol. Survey Prof. Paper 157, pp. 79–83.

Logan, C. A., 1935, Mother Lode gold belt of California: Clio, Eagle-Shawmut and Harriman mines: California Div. Mines Bull. 108, pp. 159–160 and 162–165.

Ransome, F. L., 1900, Mother Lode district folio: U. S. Geol. Survey Geol. Atlas of the U. S., folio 63, 11 pp.

Storms, W. H., 1900, Eagle-Shawmut mine: California Min. Bur. Bull. 18, pp. 132–133.

Tucker, W. B., 1916, Tuolumne County, Eagle-Shawmut mine: California Min. Bur. Rept. 14, pp. 146–147.

Turner, H. W., and Ransome, F. L., 1897, Sonora folio California: U. S. Geol. Survey Geol. Atlas of the U. S., folio 41, 7 pp.

Jamestown

Location. The Jamestown district is in western Tuolumne County. It consists of that portion of the Mother Lode belt that extends from French Flat southeast through Rawhide, Jamestown, Quartz Mountain, and the town of Stent to the vicinity of the Belcher mine, a distance of about eight miles. It also has been called the "Jimtown" district.

History. The streams and rich surface ores were first worked in the gold rush. Jamestown was established in 1848 by Colonel George F. James, a lawyer. Hydraulic mining began at Stent soon afterward, and the lode mines were active from the 1860s on. The placers at nearby Campo Seco yielded $5.5 million and those at Jamestown $3 million. From around 1890 to World War I lode mining was a major industry; in 1906 more than 300 stamps were "dropping" in the various mills. There was some activity again during the 1920s and appreciable activity during the 1930s. There has been minor prospecting and development work in recent years at a few of the mines. The value

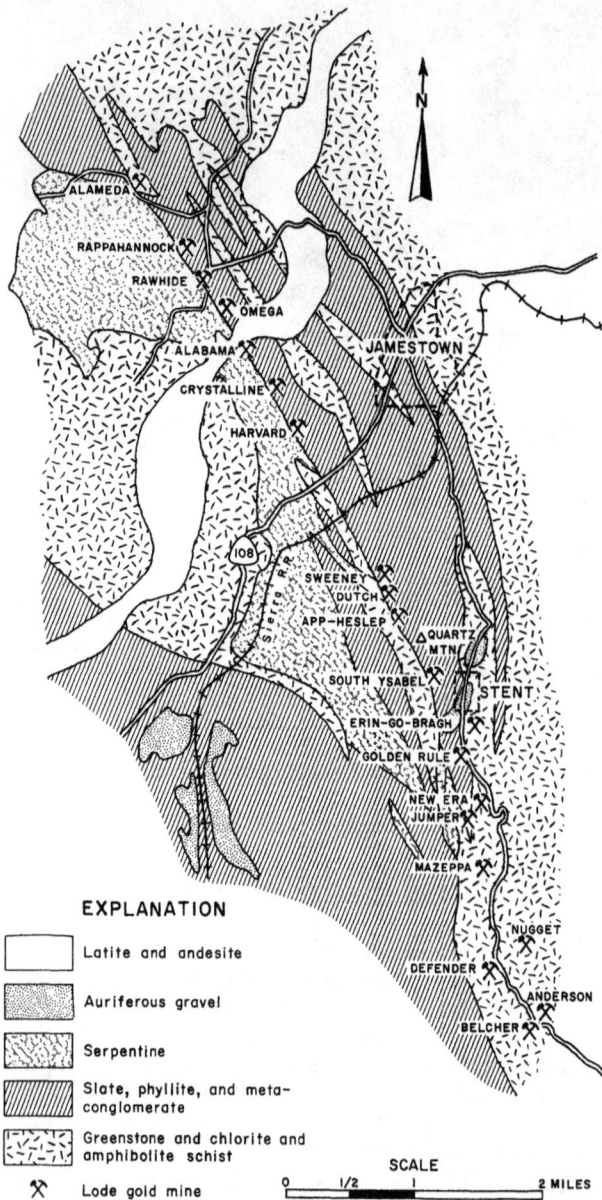

EXPLANATION

	Latite and andesite
	Auriferous gravel
	Serpentine
	Slate, phyllite, and meta-conglomerate
	Greenstone and chlorite and amphibolite schist
✗	Lode gold mine

SCALE

0   1/2   1   2 MILES

Figure 16. Geologic Map of Jamestown District, Tuolumne County. The map shows the locations of the mines. *Modified from Eric, Stromquist and Swinney, 1955.*

of the total output of this district is estimated at more than $30 million.

Geology. In the north portion of the district, the deposits occur along a northwest-striking contact with serpentine to the southwest and phyllite, slate, and metaconglomerate to the northeast (fig. 16). In the central and south portion, the deposits are at or near the contact between massive greenstones and slates on the west and chlorite and amphibolite schist to the east. Latite of Tuolumne Table Mountain crosses the belt north of Jamestown, and Tertiary gravel deposits underlie the latite in the vicinity of the town of Rawhide and to the southwest. At Quartz Mountain the Mother Lode belt swings from a northwest-southeast strike to almost due south.

Ore Deposits. Outcrops consist of massive quartz veins up to several tens of feet in thickness, adjacent bodies of ankerite-quartz-mariposite rock which sometimes are scores of feet thick, as well as bodies of mineralized schist and numerous parallel quartz stringers. These deposits often contain abundant disseminated sulfides (as much as eight to 10 percent of the total rock), which are mostly pyrite. The gold occurs in the native state or with pyrite. Milling-grade ore usually averaged $1/7$ to $1/3$ ounce gold per ton, but the ore shoots were large. The ore shoots had stoping lengths of as much as 400 feet or more, and several veins were mined to inclined depths of several thousand feet. A number of high-grade pockets have been found in this district. In places silver is abundant, and tellurides have been encountered.

Mines. Alabama $150,000, Alameda, Anderson, App-Heslep $6.5 million, Belcher, Crystalline $100,000, Defender, Dutch-Sweeney $3 million, Erin-go-bragh $282,000, Golden Rule, Harvard $2 million to $3 million, Hitchcock, Jumper $5 million, Mazeppa, New Era, Nugget, Omega, Rappahannock, Rawhide $6 million, Santa Ysabel $1.5 million.

Bibliography

Eric, J. H., Stromquist, A. A., and Swinney, C. M., 1955, Geology of the Angels Camp and Sonora quadrangles: California Div. Mines Spec. Rept. 41, 55 pp.

Fairbanks, H. W., 1890, Geology of the Mother Lode region: California Min. Bur. Rept. 10, pp. 50–56.

Irelan, William, Jr., 1888, App, Heslep, and Gem mines: California Min. Bur. Rept. 8, pp. 660–664.

Knopf, Adolph, 1929, The Mother Lode system of California: U. S. Geol. Survey Prof. Paper 157, 88 pp.

Photo 39. Harvard Mine, Jamestown District. This 1955 view shows a massive quartz vein at the Tuolumne County mine. *Photo by D. W. Carlson.*

Photo 40. Jumper Mine, Jamestown District. This view of the Tuolumne County mine was taken in about 1910. *Photo courtesy of Calif. State Library.*

Photo 41. Rawhide Mine, Jamestown District. This northward view, taken probably in the 1890s, shows the Tuolumne County mine and the town of Rawhide. The headframe and hoisting works are in the right center. *Photo courtesy of Tuolumne County Museum.*

Logan, C. A., 1928, Tuolumne County, quartz mines: California Min. Bur. Rept. 24, pp. 8–9.

Logan, C. A., 1935, App, Dutch, Harvard, Jumper, and Rawhide mines: California Div. Mines Bull. 108, pp. 156–158, 161–162, 165–168, and 171–172.

Ransome, F. L., 1900, Mother Lode district folio: U. S. Geol. Survey Geol. Atlas of the U. S., folio 63, 11 pp.

Storms, W. H., 1900, The Mother Lode region-Tuolumne County: California Min. Bur. Bull. 18, pp. 128–141.

Tucker, W. B., 1916, Tuolumne County, Dutch, Harvard, Jumper, and Rawhide mines: California Min. Bur. Rept. 14, pp. 145–146, 149–151, 152–153, and 159–160.

Turner, H. W., and Ransome, F. L., 1897, Sonora folio: U. S. Geol. Survey Geol. Atlas of the U. S., folio 41, 7 pp.

Jenny Lind

Location. This district is between the towns of Jenny Lind and Milton in western Calaveras County. It extends west into eastern San Joaquin and northeastern Stanislaus Counties. The area was first worked during the gold rush, and later hydraulicked. There was dredging here from 1903 until about 1940 and also small-scale lode mining. The district has yielded more than 100,000 ounces of gold.

Geology. The gold values are in river gravels and floodplain deposits in and adjacent to the Calaveras River. There are older terrace and shore gravels, some of which are overlain by hardpan. In places hydraulic mine tailings overlie the gravels. Dredging depths ranged from 20 to 40 feet, with the average nearer 20 feet. Recovered gold values ranged from 10¢ to 30¢ per yard and hydraulic tailings were around 10¢ per yard. There are a number of narrow gold-quartz veins in greenstone in the eastern portion of the district.

Dredging Operations. Butte Dredging Co., Calaveras Gold Dredging Co. 1903-16, El Oro Dredging Co., Isabel Dredging Co. 1908-25?, Milton Gold Dredging Co. 1935-?

Bibliography

Clark, W. B., and Lydon, P. A., 1962, Calaveras County, gold: California Div. Mines and Geology County Rept. 2, pp. 32–93.

Logan, C. A., 1919, Calaveras River area: California Min. Bur. Bull. 85, pp. 32–33.

Logan, C. A., 1936, Calaveras County, ancient shore-line deposits: California Div. Mines Rept. 32, pp. 324–325.

Tucker, W. B., 1916, Calaveras County, gold dredging: California Min. Bur. Rept. 14, pp. 124–127.

Turner, H. W., 1894, Jackson folio: U. S. Geol. Survey Geol. Atlas of the U. S., folio 11, 6 pp.

Winston, W. B., 1910, Gold dredging in California: California Min. Bur. Bull. 57, pp. 207–208.

Jerseydale

Location and History. The Jerseydale district is in west-central Mariposa County. It is about 10 miles northeast of the town of Mariposa and just east of Whispering Pines. It includes the Feliciana Mountain area. The streams were first mined during the gold rush, and the lode deposits were discovered soon afterward. Much activity continued from the 1870s until the early 1900s. Some mining was done again during the 1930s, with intermittent prospecting since.

Geology. The principal rocks in the area are slate, graphite schist and phyllite of the Calaveras Formation (Carboniferous to Permian) and granitic rocks. Also present are greenstone and dioritic dike rocks. Numerous narrow quartz veins contain small but rich high-grade shoots and "pockets". Sulfides are abundant.

Mines. Blue Bell, Buffalo, Comet, Early, Feliciana $159,000, King Solomon, Louisa, Monte Cristo, Roma $100,000.

Bibliography

Bowen, O. E., 1957, Mariposa County, Lode mines: California Jour. Mines and Geology, vol. 53, pp. 69–187.

Castello, W. O., 1921, Mariposa County, Jerseydale-Sweetwater district: California Min. Bur. Rept. 17, pp. 97–98.

Lowell, F. L., 1916, Mariposa County, Feliciana mine: California Min. Bur. Rept. 14, p. 582.

Photo 42. Alameda Mine, Jamestown District. This photo of the Tuolumne County mine was taken on July 4, 1890. Photo courtesy of Tuolumne County Museum.

Photo 43. Jamison Mine, Johnsville District. This northward view of the Plumas County mine dates back to about 1900. The Plumas-Eureka mine and the town of Johnsville are in the left background. *Photo courtesy of Calif. Division of Beaches and Parks.*

Johnsville

Location. The Johnsville district is in south-central Plumas County. Both a lode- and placer-gold district, it is at the north end of a major belt of gold mineralization that extends southward through the Sierra City district in Sierra County (see fig. 22, p. 116).

History. The river and stream gravels in the general area were first placer-mined in 1849 or 1850. The Eureka quartz vein, discovered in 1851, quickly brought many miners to the region, and most of the area was soon covered with claims. Considerable coarse gold was recovered from the creeks and considerable high-grade ore from the lode mines during those early years. Both the Plumas-Eureka and the Jamison mines were operated on a major scale until the early 1900s, when mining activity in the area declined. Johnsville was named for William Johns, manager of the Plumas-Eureka mine. There was intermittent activity in the district from the period of World War I until around 1943.

The area suffered from a number of disastrous fires. Part of the town and some of the mines became Plumas-Eureka State Park in 1959. The value of the total output of the district is unknown, and there have been a number of extravagant claims. The author estimates the production to be somewhere between $10 million and $20 million. This was a well-known early-day "snowshoe" or ski resort area.

Geology. A considerable variety of rocks crops out in this district, including north- and northwest-trending belts of slate, schist, quartzite, and limestone on the west; metadacite or quartz porphyry to the south; a gabbroic intrusion in the central portion; and greenstone to the east. Portions of the region are overlain by Tertiary andesite. Much of the central portion of the area is covered with glacial detritus. A number of patches of Tertiary gravels yielded gold in the early days. Massive bodies of magnetite are found to the west.

Ore Deposits. There are a number of north- and northwest-trending quartz veins and several wide complex systems of quartz veins. The individual veins usually are only a few feet thick. These contain free gold and often abundant pyrite and varying amounts of galena, chalcopyrite, and arsenopyrite. A number of high-grade pockets were taken from near the surface in the early days. Milling-grade ore contained from a few dollars to more than one ounce gold per ton. The sulfide concentrates sometimes held more than $150 in gold per ton. The ore shoots had horizontal stoping lengths of as much as several hundred feet.

Mines. Lode: Jamison $1.5 million+, Plumas-Eureka $8 million+, Plumas-Mohawk, Round Lake. Placer: Beckwith Cons. drift, Continental drift, Queen drift, Brown Bear hydraulic.

Bibliography

Averill, C. V., 1928, Mines and mineral resources of Plumas County: California Div. Mines Rept. 24, p. 261–316.

Averill, C. V., 1937, Mines and mineral resources of Plumas County: California Div. Mines Rept. 33, pp. 79–143.

Durrell, Cordell, 1959, Tertiary stratigraphy of the Blairsden quadrangle, Plumas County, California: Univ. of Calif., Pubs. in Geol. Sci., vol. 34, no. 3, pp. 161–192.

Irelan, William, 1888, Plumas-Eureka mine: California Min. Bur. Rept. 8, pp. 476–478.

Jackson, W. T., 1960, A history of mining in the Plumas-Eureka State Park area, 1851–1890: California Div. Beaches and Parks, 56 pp.

Jackson, W. T., 1961, A history of mining in the Plumas-Eureka State Park area, 1890–1943: Calif. Div. Beaches and Parks, 48 pp.

Lindgren, Waldemar, 1911, Tertiary gravels of the Sierra Nevada: U. S. Geol. Survey Prof. Paper 73, p. 112.

MacBoyle, Errol, 1920, Plumas County, Johnsville mining district: California Min. Bur. Rept. 16, pp. 21–27.

Turner, H. W., 1897, Downieville folio, California: U. S. Geol. Survey Geol. Atlas of the U. S., folio 37, 8 pp.

Jordan

Location and History. This district is in central Mono County just north of Mono Lake and about 15 miles south of Bridgeport. It is a lode and placer district that occupies an area on the east flank of the Sierra Nevada and the Mono Plains. It extends from Mono Lake north to the Keith district and includes the areas known as the Mono Diggings and Dogtown Diggings districts. Mono and Dogtown Diggings were first mined in 1857, and the district was organized in 1879. Work continued steadily to the early 1900s, but the greatest output was during the 1870s and 1880s.

Geology. The country rock in the area consists of hornfels, limestone, schist, and slate with granitic rocks in the Sierra Nevada. To the east are Tertiary andesites, which in places are overlain by sands and gravels derived from the Sierra Nevada. Auriferous gravel deposits occur in ill-defined channels that range from a few to more than 50 feet in thickness. The gold values varied considerably. Several quartz veins with abundant sulfides contain copper, silver and some manganese; there is some mineralized schist. The value of the total output of the district is unknown, but it is estimated to be several million dollars.

Photo 44. Plumas-Eureka Mine, Johnsville District. This photo of the Plumas County mine was taken in about 1900. Eureka Peak is in the background. The mine is now part of Plumas-Eureka State Park.

Bibliography

Whiting, H. A., 1888, Jordan mining district: California Min. Bur. Rept. 8, pp. 363–367.

Kearsarge

The Kearsarge district is on the east flank of the Sierra Nevada about eight miles west of Independence in Inyo County. The district was named in 1864 for the U.S.S. Kearsarge, a famous Union warship. The Rex Montis mine, the principal gold source, was worked on a substantial scale from 1875 to 1883, reportedly yielding 12,333 ounces of gold and silver in 1877. The Kearsarge mine also has yielded some values. The deposits consist of narrow quartz veins in quartz monzonite that contain native gold, sulfides, and reportedly native silver.

Bibliography

Moore, J. G., 1963, Geology of the Mount Pinchot quadrangle: U. S. Geol. Survey Bull. 1130, 152 pp.
Tucker, W. B., 1938, Inyo County, Rex Montis mine: California Div. Mines Rept. 34, pp. 415–416.

Keith

This district is on the east flank of the Sierra Nevada in western Mono County about 10 miles northwest of Mono Lake and just north of the Jordan district. The principal source of gold apparently has been the Dunderberg mine, which was first worked in the 1860s and has been intermittently prospected since. There are several quartz-barite veins with minor amounts of free gold· and abundant pyrite. Country rock is granite, quartzite, hornfels, and schist.

Bibliography

Eakle, A. S., and McLaughlin, R. P., 1919, Mono County, Dunderberg mine: California Min. Bur. Rept. 15, pp. 166–167.

Kelsey

Location. The Kelsey district is in northwestern El Dorado County. It is that portion of the Mother Lode gold belt that extends from the vicinity of the town of Kelsey northwest to Garden Valley.

History. This area was placer-mined soon after James Marshall's gold discovery in 1848 at Coloma, a few miles to the west. The camp was first settled by and named for Benjamin Kelsey. Marshall spent his last days at Kelsey, and a building on his property once housed a pioneer museum. There was much lode mining from the 1860s through the early 1900s and again in the 1930s. Some intermittent work has been done recently at the Black Oak mine.

Geology and Ore Deposits. A northwest-trending two-mile-wide belt of gray to black slate of the Mariposa Formation (Upper Jurassic) is in the central portion of the district, with greenstone, slate, graphite schist, and quartzite to the west. Amphibolite, slate, and schist lie to the east. Serpentine lenses are also present both to the east and west. The ore deposits occur in quartz veins with numerous stringers. The veins range from one to 10 feet in thickness. Nearly half of the known output of some of the mines has been from small but extremely rich ore shoots. None of the mines has been worked to depths of more than 600 feet.

Mines. Big Four, Big Sandy $100,000+, Black Oak $1.25 million, Gopher Hill, Gray Eagle, Hart, Ida Livingston, Kelsey $100,000+, Lady Emma, St. Clair, Veerkamp, War Eagle, Yuba.

Bibliography

Clark, W. B., and Carlson, D. W., 1956, El Dorado County, lode gold deposits: California Jour. Mines and Geol., vol. 52, pp. 401–429.
Fairbanks, H. W., 1890, Geology of the Mother Lode region: California Min. Bur. Rept. 10, p. 81.
Lindgren, Waldemar, and Turner, H. W., 1894, Placerville folio, California: U. S. Geol. Survey Geol. Atlas of the U. S., folio 3, 3 pp.
Logan, C. A., 1935, Mother Lode gold belt of California—Big Sandy, Black Oak, and Kelsey mines: California Div. Mines Bull. 108, pp. 19–21 and 29.

Kern River

The upper Kern River between Bakersfield and Bodfish was the scene of a rush soon after the discovery of gold at Greenhorn Creek in 1851. However, the deposits in the river are believed to have been worked out in a short time. Many lode-gold prospects are in the area, but the only one of any consequence is the Gem gold mine near Democrat Springs. Uranium was discovered at Miracle Hot Springs in 1954, and there was a "boom" that lasted for a few years. Practically the entire region is underlain by quartz diorite.

Bibliography

Troxel, B. W., and Morton, P. K., 1962, Kern County, Kern River Canyon district: California Div. Mines and Geology, County Rept. 1, p. 38.
Tucker, W. B., and Sampson, R. J., 1933, Kern County, Gem mine: California Div. Mines Rept. 29, p. 307.

Keyesville

Location and History. This district is in the southern Sierra Nevada in Kern County about 32 miles northeast of Bakersfield and two miles southwest of Isabella Dam. Gold was discovered here in 1852 by Richard M. Keyes, and for a time this was the largest community in Kern County. The chief periods of mining were the 1850s, 1860s, 1890s, and 1909–15: The area was prospected during the 1930s, but little has been done here since, and Keyesville has become a ghost town.

Geology and Ore Deposits. Virtually the entire district is underlain by quartz diorite. The gold deposits occur in a northeast-trending belt about three miles long. The veins consist of narrow quartz stringers with fault gouge that contain free gold and small amounts of pyrite, arsenopyrite, and pyrrhotite. There are some placer deposits, including one of possible Pleistocene age.

Mines. Bright Spot, High Grade, Homestake, Keyes $450,000, Keyesville, Keyesville Placer, Mammoth $500,000, Mooncastle, Nephi, Nob Hill, Opportunity, Sunrise, Virginia, Will Jean.

Bibliography

Brown, G. C., 1916, Kern County, Keyes district: California Min. Bur. Rept. 14, p. 483.
Troxel, B. W., and Morton, P. K., 1962, Kern County, Keyesville district, High Grade mine, and Mammoth mine: California Div. Mines and Geology, County Report 1, pp. 38–39, 111–112, and 115–117.
Tucker, W. B., and Sampson, R. J., 1933, Kern County, Keyes district: California Div. Mines Rept. 29, p. 283.

Kimshew

Location. The Kimshew mining district is in northeastern Butte County and northwestern Plumas County about 10 miles northeast of Stirling City. It includes the Golden Summit area to the north.

Geology. There are a number of moderate-sized deposits of Tertiary and Pleistocene gravel that have been mined both by hydraulicking and drifting and a few narrow gold-quartz veins. Bedrock consists of slate and amphibolite to the west and a granitic stock to the east.

Mines. Brown, Carr, Cash Entry, Gallagher and Perkins lode, Golden Summit, Little Johnnie, Ream and Burnside, Reese and Jones, Snow, Wescott.

Bibliography

Diller, J. S., 1895, Lassen Peak folio: U. S. Geol. Survey Geol. Atlas of the U. S., folio 15, 4 pp.

Turner, H. W., 1898, Bidwell Bar folio: U. S. Geol. Survey Geol. Atlas of the U. S., folio 43, 6 pp.

Kinsley

Location and History. This is an extensive district in the Sierra Nevada east gold belt in north-central Mariposa County. It is five miles east of Coulterville and about 25 miles north of Mariposa. The district includes several places that at various times have been classified as separate districts but are grouped together in this publication because they adjoin each other and are geologically similar. These are the Greeley Hill, Bull Creek, Gentry Gulch, Smith Ridge, and Dogtown areas. The Cat Town district lies just to the south and the Coulterville district to the west. The area was placer-mined during the gold rush. Much lode mining was conducted from the 1860s through 1900 and again in the 1930s. Several mines including the Hasloe and Horseshoe have been worked intermittently in recent years.

Geology. The district is underlain by slate, mica schist, quartzite, hornfels, phyllites and limestone lenses of the Calaveras Formation (Carboniferous to Permian). Present are several small granitic stocks and a number of diorite, quartz-diorite and aplite dikes. In places these dikes are associated with the gold-quartz veins and are important in the localization of the ore bodies.

Ore Deposits. Numerous north- and west-striking quartz veins range from one to five feet in thickness. The ore shoots generally are small, but commonly rich. The ore contains free gold and often abundant sulfides including galena and tetrahedrite, which are associated with rich ore. Molybdenite is present in a few places.

Mines. Argo $18,000, Bandarita $1.52 million, Bob McKee, Bondurant $390,000, Bunce, Carrie Todd, Contention, Cranberry, Garibaldi, Gold King, Hasloe $3 million, Horseshoe, Last Chance, Lovely Rogers, Louisiana, Marble Springs $200,000, Moonlight, Quail $400,000, Red Cloud, Red Mountain, Texas Hill $74,000+.

Bibliography

Bowen, O. E., 1957, Mariposa County, Bandarita, Bondurant, Hasloe, and Quail mines: California Journal of Mines and Geology, vol. 53, pp. 77–81, 106–108, and 159–161.

Castello, W. O., 1921, Mariposa County, Kinsley district: California Min. Bur. Rept. 17, pp. 95–96.

Turner, H. W., and Ransome, F. L., 1897, Sonora folio: U. S. Geol. Survey Geol. Atlas of the U. S., folio 41, 7 pp.

Knight's Ferry

Location. The Knight's Ferry district is in northeastern Stanislaus County and western Calaveras County. The town is located on the lower Stanislaus River about 12 miles east-northeast of Oakdale.

History. The district was placer-mined during the gold rush. The town, which was named for William Knight, was an important staging and supply center for the mines and camps of the southern Mother Lode region. It was the seat of Stanislaus County government from 1862 to 1872. The town also was once known as Dentville for the Dent brothers who were brothers-in-law of President U. S. Grant. The old wooden covered bridge that is still standing was reportedly designed by Grant in 1854. From the 1870s through the 1890s, numerous Chinese placer miners reworked the old tailings and small deposits overlooked by the Forty-Niners. There was dragline dredging in the district in the 1930s and early 1940s.

Geology. The east portion of the district is underlain by greenstone and quartz porphyry and the western portion by andesite. The gold was recovered from isolated patches of quartz-rich gravel of Eocene age, younger channel and terrace deposits buried under or adjacent to the Plio-Pleistocene latite of Tuolumne Table Mountain, and Recent gravels in and along the present channel of the Stanislaus River.

Bibliography

Taliaferro, N. L., and Solari, A. J., 1948, Geology of the Copperopolis quadrangle: California Div. of Mines Bull. 145, plates 1 and 2.

Watts, W. L., 1890, Stanislaus County, gold: California Min. Bur. Rept. 10, p. 681.

La Grange

Location. This district is in southeastern Stanislaus County. It is primarily a dredging field that extends westward from the town of La Grange along the Tuolumne River for nine miles. This district also includes a dredging field two miles to the south, on an older river channel, and surface "diggings" to the north. The town, originally known as French Bar, was founded in 1852, but the name was changed to La Grange in 1856 in honor of Lafayette's country home. For a time Bret Harte taught school here. The estimated output from dredging is $13 million worth of gold.

Geology. The stream gravels in and adjacent to the present Tuolumne River are medium to coarse, loosely consolidated, and average 30 to 35 feet in thickness. The gravels are underlain by tuff. Dredge recoveries during the 1920s averaged 11.6¢ per yard, but later recoveries are believed to have been less.

Minor amounts of platinum were recovered. The dredged area is about nine miles long and ½ mile wide. A Pleistocene river channel two miles to the south was dredged for a distance of 1½ miles and is ¼ mile wide. Here the gravels were compact with a thick volcanic overburden. Thin Eocene gravel patches with abundant quartz boulders and interbedded clay and sand are exposed to the north.

Concerns. La Grange Gold Dredging Co., 1907–42 and 1945–51, one dredge, the longest in the state; Tuolumne Gold Dredging Corp., 1938–43 and 1945–?, one dredge; Yuba Cons. Goldfields, 1941–42, one dredge.

Bibliography

Laizure, C. M., 1925, Stanislaus County, La Grange Gold Dredging Co.: California Min. Bur. Rept. 21, pp. 207–208.

Logan, C. A., 1919, Platinum and allied metals in California, Tuolumne River: California Min. Bur. Bull. 85, p. 33.

Logan, C. A., 1947, Gold—Stanislaus County: California Jour. Mines and Geology, vol. 43, pp. 92–94.

Lowell, F. L., 1916, Stanislaus County, La Grange Gold Dredging Company: California Min. Bur. Rept. 14, p. 629.

Winston, W. B., 1910, La Grange Gold Dredging Company: California Min. Bur. Bull. 57, pp. 210–211.

La Porte

Location. This district is in southwestern Plumas County in the general vicinity of the old mining town of La Porte, 25 miles south of Quincy and 50 miles northeast of Oroville. It was one of the great placer-mining districts of the state.

History. The streams were placer-mined early in the gold rush and were reported to have had very rich yields. The town, first known as Rabbit Creek, was renamed in 1857 after La Porte, Indiana. Hydraulic mining began in the middle 1850s and continued through the 1880s. During this time the district was enormously productive; the output from 1855 to 1871 alone was reported to have been at least $60 million. Appreciable drift mining and some lode mining were carried on. Some mining activity continued until the period of World War I. The district was prospected again during the 1930s, but apparently little mining has been done here since. La Porte was a noted early-day "snowshoe" or ski resort.

Geology. The main Tertiary channel of the North Fork of the Yuba River, known as the La Porte channel, extended south-southwest from Gibsonville into this district. From here the channel continued southwest and south again to be joined by a branch from the east from the St. Louis-Table Rock area. The main channel today continues on south to the Poverty Hill and Brandy City districts. At La Porte, the channel is 500 to 1500 feet wide and as much as 500 feet thick. The lower gravels are quartz-rich and up to 80 feet thick. Most of the gold was recovered from near bedrock. The gravels are capped by thick beds of sand and "pipe" clay. During the heyday of mining in the district, these lower gravels yielded from ⅒ to as much as one ounce of gold per cubic yard. To the east the channel deposits are capped by andesite as much as 800 feet thick. Also to the east, considerable fault-

ing has disturbed the channel gravels. Bedrock consists principally of amphibolite, with a one-mile wide belt of slate and quartzite of the Calaveras Formation (Carboniferous to Permian) in the central portion. There are some narrow gold-quartz veins in the district.

Bibliography

Lindgren, Waldemar, 1911, Tertiary gravels of the Sierra Nevada: U. S. Geol. Survey Prof. Paper 73, pp. 103–113.

Logan, C. A., 1928, Plumas County, La Porte mines: California Div. Mines and Mining Report 24, pp. 303–306.

MacBoyle, Errol, 1920, Plumas County, La Porte mining district: California Min. Bur. Rept. 16, pp. 27–31.

Turner, H. W., 1897, Downieville folio, California: U. S. Geol. Survey Geol. Atlas of the U. S., folio 37, 8 pp.

Last Chance

Location. This extensive placer-mining district is in eastern Placer County in the vicinity of the old mining camp of Last Chance, 10 miles northeast of Michigan Bluff and 15 miles northeast of Forest Hill. It includes the "diggings" here and at Star Town, Deadwood, and American Hill. Last Chance got its name when a starving miner used his last bullet to kill a deer. The mines in the district were operated almost steadily from the early 1850s until about 1920. There was some activity again in the 1930s, and the El Dorado and Last Chance mines have been intermittently worked in recent years.

Geology. Most of the gold has come from three southwest-trending Tertiary channels. The lowest but youngest is the El Dorado channel, which is steep, only about 100 feet wide, and contains coarse gold. The next youngest is the Sharp Stick channel, which contains clay and coarse boulders. The oldest is the Big Channel which is quartzitic, well-cemented, and up to 800 feet wide. The gravels are capped by andesite. Bedrock is slate and schist of the Blue Canyon Formation (Carboniferous), which encloses some narrow north-striking gold-quartz veins.

Mines. Beaman Ledge, Bear Wallow, Central, Darling, Deep Canyon, Double O, El Dorado, Elkhorn, Golden Riffle, Grizzly, Harkness, Home Ticket $200,-000+, Hornby, Last Chance, Little Hope, Missouri Flat, New Caledonia, Pacific, Pacific Slab, Peters, Rattlesnake, Rublin, Sharp Stick, Star Town.

Bibliography

Lindgren, Waldemar, 1900, Colfax folio: U. S. Geol. Survey Geol. Atlas of the U. S., folio 66, 10 pp.

Lindgren, Waldemar, 1911, Tertiary gravels of the Sierra Nevada: U. S. Geol. Survey Prof. Paper 73, p. 158.

Logan, C. A., 1936, Gold mines of Placer County, placer mines: California Div. Mines Rept. 32, pp. 49–96.

Waring, C. A., 1919, Placer County, drift and hydraulic mines: California Min. Bur. Rept. 15, pp. 352–379.

Light's Canyon

Location and History. This district is in northeastern Plumas County in the general area of Light's Canyon. It was named for Ephriam Light, a pioneer rancher. It includes the Moonlight Valley, Indian Valley, Engelmine, and Kettle Rock areas. It is also an important copper-mining district; the Engels and Superior mines are located here. Placer mining was originally done in the district during the 1850s and continued through the early 1900s. The Lucky S lode mine has been prospected in recent years. For some years the district was served by the Indian Valley Railroad.

Geology and Ore Deposits. The central portion of the district is underlain by granodiorite and quartz diorite. To the south are various metamorphic rocks of Paleozoic and Mesozoic age similar to those found in the Taylorsville district (see Taylorsville district). North and east are Tertiary gravels and volcanic rocks. Both the Recent and Tertiary gravels have yielded moderate amounts of gold. The gold-quartz veins are narrow, but the ore often is rich. The ore contains abundant sulfides including pyrite, galena, chalcopyrite, and sphalerite.

Bibliography
Diller, J. S., 1908, Geology of the Taylorsville region: U. S. Geol. Survey Bull. 353, 128 pp.
Lindgren, Waldemar, 1911, Tertiary gravels of the Sierra Nevada: U. S. Geol. Survey Prof. Paper 73, pp. 114–116.
MacBoyle, Errol, 1920, Plumas County, Light's Canyon mining district: California Min. Bur. Rept. 16, pp. 31–36.

Lincoln

Lincoln is in western Placer County, 15 miles west of Auburn and 11 miles north of Roseville. The town was named for Charles Lincoln Wilson, who built the California Railroad here in 1861. During the 1930s considerable amounts of gold were recovered by dragline dredging from gravels in lower Auburn Ravine, just east of Lincoln, and in Doty Ravine, a few miles to the north. This was probably the most profitable dragline dredge field in the state. Recoveries ranged from 15 to as high as 60 cents per yard. Digging depths ranged from 5 to 20 feet. The gravels are underlain by soft tuff. The total dredged area is about 1200 acres.

Bibliography
Lindgren, Waldemar, 1894, Sacramento folio: U. S. Geol. Survey Geol. Atlas of the U. S., folio 5, 3 pp.
Logan, C. A., 1936, Gold mines of Placer County, Lincoln Gold Dredging Company: California Div. Mines Rept. 32, pp. 82–83.

Long Tom

Location and History. The Long Tom district is in the southern Sierra Nevada in central Kern County. It is 23 miles northeast of Bakersfield and 10 miles south of Woody. The veins at the Long Tom mine, the chief source of gold in the district, were discovered prior to 1860 by prospectors looking for the source of placer gold in nearby creeks. The mine was considerably active during the 1880s and again from 1925 to 1939. It has an estimated total output of $800,000 to $900,000.

Geology. The country rock in the district is quartz diorite with small gabbroic inclusions. A number of fracture zones contain small gold-bearing quartz stringers with minor amounts of sulfides. The deposits do not extend to depths of more than a few hundred feet.

Bibliography
Brown, C. G., 1916, Kern County, Long Tom district and mine: California Min. Bur. Rept. 14, pp. 483 and 502.
Goodyear, W. A., 1888, Kern County, gold: California Min. Bur. Rept. 8, pp. 319–320.
Troxel, B. W., and Morton, P. K., 1962, Kern County, Long Tom mine: California Div. Mines and Geology, County Rept. 1, pp. 114–115.

Loraine

Location and History. This district is in the southern Sierra Nevada in central Kern County in the vicinity of the town of Paris-Loraine. It is about 35 miles east of Bakersfield and 12 miles north of Tehachapi. The area was first prospected in the 1850s, but the principal period of mining activity was from 1894 until around 1912. The district was active again in the 1920s and 1930s, and there has been intermittent prospecting since. It is also known as the Amalie district.

Geology. The district is underlain by a large roof pendant of slate and mica schist of the Kernville Series (Paleozoic?) in quartz diorite and granodiorite. There are a number of quartz veins ranging from one to 10 feet in thickness which contain free gold and abundant sulfides, especially silver sulfides. The veins occur in both the metamorphic and granitic rocks. Milling-grade ore commonly averages more than ½ ounce of gold and two ounces of silver per ton. Several ore shoots had stoping lengths of up to 300 feet, and several veins were mined to depths of 600 feet.

Mines. Amalie $600,000, Barbarossa, Cowboy $600,000, Deerhunter, Ella, Ferris, Golden Cross, Golden Peak, New Deal, Zenda 34,000 ounces+.

Bibliography
Brown, G. C., 1916, Kern County, Amalie district: California Min. Bur. Rept. 14, p. 482.
Crawford, J. J., 1894, Amalie mine: California Min. Bur. Rept. 12, p. 141.
Tucker, W. B., 1923, Kern County, Amalie mining district: California Min. Bur. Rept. 19, p. 156.
Tucker, W. B., and Sampson, R. J., 1933, Amalie district: California Div. Mines Rept. 29, pp. 280–281.

Lowell Hill

Location and History. This district is in south-central Nevada County about six miles northeast of Dutch Flat. It includes the Remington Hill, Negro Jack Hill, and Liberty Hill areas. The district was hydraulicked from the middle 1850s through the 1870s, and Liberty Hill was worked again from around 1896 to 1915.

The total output is unknown, but it exceeds $1 million. Lindgren, in 1911, estimated that two million yards had been removed and 16 million remained at Liberty Hill and that 1.75 million yards had been removed and 6 million remained at Remington Hill. Other estimates of remaining gravel at Liberty Hill range from six to 10 million yards.

Geology. The deposits are in a southwest-trending Tertiary channel that joins the Dutch Flat channel. There is a lower well-cemented blue gravel that contains gabbro and serpentine boulders and yielded 18 to 23 cents a yd. An upper quartz-rich gravel is in places covered by heavy clay. Bedrock is slate of the Blue Canyon Formation (Carboniferous) and some serpentine.

Bibliography

Lindgren, Waldemar, 1900, Colfax folio, California: U. S. Geol. Survey Geol. Atlas of the U. S., folio 66, 10 pp.

Lindgren, Waldemar, 1911, Tertiary gravels of the Sierra Nevada: U. S. Geol. Survey Prof. Paper 73, pp. 146–147.

MacBoyle, Errol, 1919, Nevada County, Lowell Hill mining district: California Min. Bur. Rept. 16, pp. 30–33.

Magalia

Location. The Magalia district is in north-central Butte County 15 miles northeast of Chico. It is bounded on the west by Doe Mill Ridge and on the east by the West Branch of the Feather River. It extends from Paradise on the south to several miles west of Powellton on the north. This district includes the placer deposits at Nimshaw, Forks of Butte, Mineral Slide and De Sabla and lode deposits at Toadtown. The Butte Creek dredging district adjoins this district at Centerville.

History. This region was extensively mined during and after the gold rush. The town was started in 1850 by E. B. Vinson and Charles Chamberlin. It was first known as Dogtown, renamed Magalia about 1862. The Magalia mine was discovered in 1855 and the Indian Springs mine in 1860. Large-scale mining continued until the 1890s; there was some activity from the early 1900s through the 1930s. There has been minor prospecting and development work since World War II. Some of the old mining properties have been made into housing subdivisions. The famous 54-lb. Willard, Dogtown, or Magalia nugget was found here in 1859. This is one of the more productive placer-mining districts in the state. Several local residents have estimated the total output to be $40 million, but that figure is too high (author). Much of the output has come from drift mines.

Geology. There are a number of south-southwest-trending steep, narrow, and rich channels. The longest channel is the Magalia or Mammoth channel that flowed along the east side of the district. Other productive channels include the Dix, Emma, Little Magalia, Pershbaker, and Nugget channels. In the south portion of the district there are shore gravels. The gold was extremely coarse, and a number of other large nuggets besides the Willard were taken here. Bedrock is slate and greenstone with smaller amounts of serpentine. The channels are faulted in places with the downstream side being thrown up. Water has always been a problem in the drift mines. A few gold-quartz veins in greenstone are associated with diorite dikes.

Mines. Drift: Bader, Black Diamond, Cole, Cory, Dix, Emma $1 million+, Ethel, Genii, Indian Springs, Kelly Hill, Lucky John, Lucretia, Magalia $1 million, Mammoth, Mineral Slide, Nuggett, Oro Fino, Parry, Pershbaker, Pete Wood, Pitts, Princess, Royal, Steifer, Willard. Hydraulic: Centerville, Kohl, Red Hill. Lode: Springer, Toadtown.

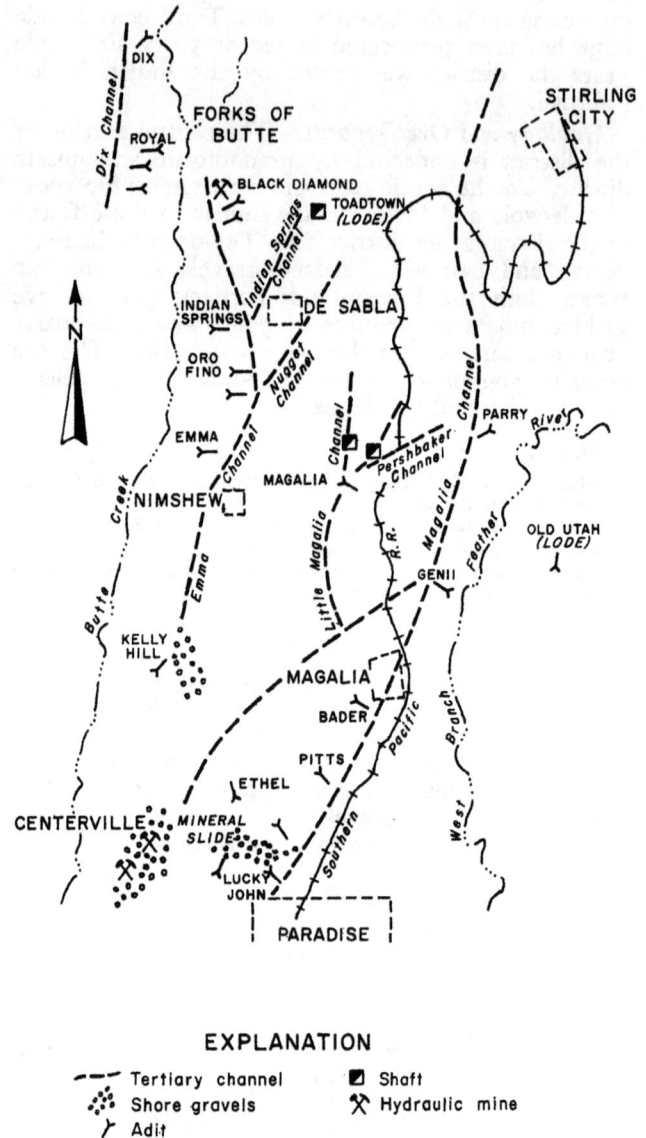

EXPLANATION

- - - Tertiary channel   ◾ Shaft
∴ Shore gravels   ⚒ Hydraulic mine
⌐ Adit

SCALE
0   1   2   3 MILES

Figure 17. Sketch Map of Magalia District, Butte County. The northern part of the Butte Creek district is also shown. The channels are not all of the same geologic age.

Bibliography

Irelan, Wm., 1888, Magalia Consolidated mine: California Min. Bur. Rept. 8, pp. 117–118.

Lindgren, Waldemar, 1911, Tertiary gravels of the Sierra Nevada: U. S. Geol. Survey Prof. Paper 73, pp. 84–86.

Logan, C. A., 1930, Butte County, placer mines: California Div. Mines Rept. 26, pp. 383–406.

Preston, E. B., 1893, Willard, Red Hill, and Indian Springs mines: California Min. Bur. Rept. 11, pp. 158–159.

Waring, C. A., 1919, Butte County, drift mines: California Min. Bur. Rept. 15, pp. 198–209.

Mammoth

Location. The Mammoth or Lake district is in southwestern Mono County about 50 miles south of Bridgeport. The district is on the east flank of the Sierra Nevada Mountains and is just east of Mammoth Lakes, a well-known resort area. The Devil's Postpile National Monument is about 10 miles to the west.

History. Gold and silver-bearing veins were discovered here in 1878, and a short-lived "rush" followed. Much of the production at that time was from the Mammoth mine, which yielded $200,000 in 1878–81. The district was organized in 1887. Several thousand people were in the area then, the principal settlements having been Mammoth City, Mill City, and Pine City. Some mining was done in the late 1890s, early 1900s, and again in the 1930s. The Beauregard mine was active from 1954 to 1958, and there has been minor prospecting and development work since. The value of the total production of the district is estimated at $1 million.

Geology and Ore Deposits. The district is underlain chiefly by northwest-trending beds of metamorphosed latite. Present in smaller amounts are schist, hornfels, marble, tactite, and quartzite. Granitic rocks are to the east and south. The ore deposits occur in a northwest-trending zone of alteration in the metamorphosed latite. This zone, 2½ miles long and ½ mile wide, contains disseminated pyrite. The pyrite has been oxidized so that the zone is stained a bright reddish brown.

The ore deposits consist of northwest-striking and steeply-dipping quartz veins and stringers that contain free gold, auriferous pyrite, pyrrhotite, and smaller amounts of other sulfides. Silver commonly is abundant. Milling-grade ore usually averages ¼ to ½ ounce of gold per ton. The veins range from a few to several tens of feet in thickness.

Mines. Argosy, Beauregard, Don Quixote, Lisbon, Mammoth $200,000, Mammoth Consolidated $100,000, Monte Cristo $100,000, Sierra Group.

Bibliography

DeGroot, Henry, 1890, Lake mining district: California Mining Bur. Rept. 10, pp. 340–342.

Mayo, E. B., 1934, Geology and mineral deposits of Laurel and Convict basins, southwestern Mono County: California Div. Mines Rept. 30, pp. 79–87.

Rinehart, C. D., and Ross, D. C., 1964, Geology and mineral deposits of the Mount Morrison quadrangle, Sierra Nevada, California: U. S. Geol. Survey Prof. Paper 385, pp. 97–100.

Sampson, R. J., and Tucker, W. B., 1940, Mineral resources of Mono County, gold: California Div. Mines Rept. 36, pp. 120–140.

Whiting, H. A., 1888, Lake mining district: California Min. Bur. Rept. 8, pp. 373–375.

Mariposa

Location and History. This district is in the vicinity of the town of Mariposa at the southeast end of the Mother Lode gold belt. The Mariposa mine was reported to have been discovered in 1849 by Kit Carson, and the first stamp mill in California was installed there that same year. Much of this district was part of the Las Mariposas Grant of General John C. Frémont. The old courthouse in Mariposa erected in 1854 is the oldest continuously used courthouse in California. The mines were worked until the early 1900s and again during the 1930s. The Mariposa mine has been prospected in recent years.

Geology and Ore Deposits. The district is underlain by northwest-trending belts of slate of the Mariposa Formation (Upper Jurassic), serpentine, and greenstone. There are several massive quartz veins in slate or greenstone. The ore contains free gold, pyrite, and arsenopyrite, which often is associated with high-grade ore. The Mariposa mine has been developed to an inclined depth of 1500 feet.

Mines. Evans II, Kane, Mariposa $2,395,000, Stockton Creek, Stockton Creek Tunnel.

Bibliography

Bowen, O. E., 1957, Mariposa County, Mariposa mine: California Jour. Mines and Geology, vol. 53, pp. 128–130.

Logan, C. A., 1935, Mother Lode gold belt—Mariposa County: California Div. Mines Bull. 108, pp. 180–190.

Ransome, F. L., 1900, Mother Lode district folio: U. S. Geol. Survey Geol. Atlas of the U. S., folio 63, 11 pp.

Storms, W. H., 1900, The Mariposa mine: California Min. Bur. Bull. 18, pp. 142–143.

Meadow Lake

Location. This is a small lode-gold district in eastern Nevada County just southwest of Meadow Lake and approximately seven miles northeast of Cisco. Gold was discovered here in 1863, and there was a "rush" to the area that lasted from 1865 to 1870. Minor prospecting was done afterward until the early 1900s and again during the 1920s and 1930s. The total output of the district has been estimated to be valued at $200,000.

Geology. The region is underlain chiefly by granodiorite with minor amounts of greenstone. There are a number of narrow quartz veins, which, in places, contain free gold and abundant sulfides. A number of high-grade pockets were encountered near the surface during the early days, but at depth the deposits pinch out or become very low in grade.

Mines. Baltimore group, Excelsior, Great Western, Hercules, Mammoth, New Hope, Of What, Philadelphia group.

Bibliography

Irelan, William, Jr., 1888, Meadow Lake district: California Min. Bur. Rept. 8, p. 454.

Lindgren, Waldemar, 1897, Truckee folio, California: U. S. Geol. Survey Geol. Atlas of the U. S., folio 39, 8 pp.

Lindgren, Waldemar, 1900, Colfax folio, California: U. S. Geol. Survey Geol. Atlas of the U. S., folio 66, 10 pp.

Logan, C. A., 1924, Nevada County, Meadow Lake district: California Min. Bur. Rept. 20, pp. 355–362.

MacBoyle, Errol, 1919, Nevada County, Meadow Lake mining district: California Min. Bur. Rept. 16, pp. 33–37.

Wisker, A. L., 1936, The gold-bearing veins of Meadow Lake district: California Div. Mines Rept. 32, pp. 189–204.

Meadow Valley

Location. This district is in west-central Plumas County about eight miles west of Quincy. It includes the Edmanton, Buck's Lake, Spanish Peak and Spanish Ranch areas. The Quincy district lies just to the east

Photo 45. Hydraulic Mining in the 1860s, Michigan Bar District. This is a view of hydraulic mining and ground sluicing in Sacramento County 100 years ago. The locomotive headlights at right made nighttime floodlights. *Photo courtesy of Calif. State Library.*

and the Granite Basin district to the southwest. The district was mined from the gold rush days through the early 1900s and has been prospected since.

Geology. The east portion of the district is underlain by quartzite, slate, schist, limestone, amphibolite, and serpentine. Meadow Valley is covered by Pleistocene lake beds. Granite lies to the west. There are a number of scattered patches of auriferous Tertiary gravel, in the vicinity of and northwest of Spanish Ranch, which were mined by hydraulicking and drifting. The Pleistocene lake gravels also yielded some gold. The only source of lode gold was the Diadem mine.

Bibliography

Lindgren, Waldemar, 1911, Tertiary gravels of the Sierra Nevada: U. S. Geol. Survey Prof. Paper 73, pp. 98–99.

MacBoyle, Errol, 1920, Plumas County, Edmanton and Spanish Ranch mining districts: California Min. Bur. Rept. 16, pp. 8–12 and 46–49.

Turner, H. W., 1898, Bidwell Bar folio, California: U. S. Geol. Survey Geol. Atlas of the U. S., folio 43, 6 pp.

Michigan Bar

Location and History. This is a placer-mining district in eastern Sacramento County in the vicinity of the old town of Michigan Bar. It extends to the west and south and includes the Sloughhouse area. Included are dredging fields in and near the Cosumnes River. The district was hydraulicked extensively in the 1850s and 1860s. Later it was worked by small-scale methods by many Chinese miners. The Cosumnes River and some of the older bench gravels were dredged in the 1930s and early 1940s. Also some drift mining was done. Fire clay is mined here now. The total gold production is unknown, but it has been estimated to be at least 1,700,000 ounces.

Geology. Eocene gravels interbedded with several layers of clay and thin sands are distributed over the eastern part of the district in the vicinity of Michigan Bar. The gravels are coarse, well rounded, have a sandy matrix, and usually are not too well-cemented. The dredged area is several miles long.

Bibliography

Carlson, D. W., 1955, Sacramento County, gold dredging: California Jour. Mines and Geology, vol. 51, pp. 135–142.

Lindgren, Waldemar, 1894, Sacramento folio: U. S. Geol. Survey Geol. Atlas of the U. S., folio 5, 3 pp.

Piper, A. M., Gale, H. S., Thomas, H. E., and Robinson, T. W., 1939, Geology and ground-water hydrology of the Mokelumne area: U. S. Geol. Survey Water-Supply Paper 780.

Michigan Bluff

Location. The Michigan Bluff district is in south-central Placer County. It is best known as a placer-mining district, and includes the Turkey Hill, Byrd's Valley, and Baker Ranch areas. The Damascus district is to the north and the Forest Hill district is to the west.

History. The town, first settled in 1850, was originally known as Michigan City. In 1858 the land began to slide into the river, so the town was moved higher up on the mountain side and became Michigan Bluff. Hydraulic mining began here in 1853, and the district soon became highly productive. During the middle and late 1850s, the gold output averaged $100,000 per month. Leland Stanford, Governor of California and one of the builders of the Central Pacific Railroad, operated a store here from 1853 to 1855. His old home still stands. Activity in the area declined during the 1870s, but some work continued intermittently through the early 1900s and again in the 1930s. Much of the region was devastated by fire in 1960.

Geology. This district is at the junction of two major Tertiary channels, one that comes in from the north from the Damascus district and the other comes in from the southeast from Ralston Divide. Just to the north at Baker Ranch there is an intervolcanic channel. The lower gravels at Michigan Bluff are nearly pure quartz with many large boulders. The gravels were extremely rich, the gold yield from six million yards reportedly having been $5 million. Much of the gold was coarse. Bedrock is slate and schist, and to the west there is serpentine. Some narrow gold-quartz veins are present.

Mines. Placer: Adams, Anna Sue, Baker Ranch, Beehive, Big Gun (Michigan Bluff) $1 million+, Bowen, Bower, Britt, Buckeye, Burnham, Burns, Burroughs, Drummond, Eastman, El Dorado Hill, Franklin, Golden Chief, Golden Gate, Gorman, Hazard, Hoffman, Imperial, Lightfoot, Manhattan, Mary Anna, Rainbow Land, Russel, Sage Hill, Turkey Hill, Washburn, Washington, Weeks, Weske. Lode: Bunker Hill and Nihill, Champion, Daniel Webster.

Bibliography

Jarman, Arthur, 1927, Michigan Bluff: California Min. Bur. Rept. 23, p. 90.

Lindgren, Waldemar, 1900, Colfax folio: U. S. Geol. Survey Geol. Atlas of the U. S., folio 66, 10 pp.

Lindgren, Waldemar, 1911, Tertiary gravels of the Sierra Nevada: U. S. Geol. Survey Prof. Paper 73, p. 152.

Logan, C. A., 1936, Gold Mines of Placer County, Michigan Bluff district: California Div. Mines Rept. 32, p. 70.

Waring, C. A., 1919, Placer County, Michigan Bluff district: California Min. Bur. Rept. 15, p. 318.

Mill Creek

Location. This is a small district in southeastern Fresno County in the vicinity of the town of Dunlap about 40 miles east of Fresno. Superficial placer mining was done in Mill Creek and other streams during the early days. Small-scale lode mining was done from the 1880s through the 1900s, but little or nothing has been done since.

Geology. The area is underlain by granodiorite and related rocks with narrow slate belts. Some limestone lenses and schist are present. A few shallow north-trending veins a few feet thick contain free gold and varying amounts of sulfides. The veins usually are at or near granite-schist contacts. Most of the output has come from the Dixie Queen and White Cross mines.

Bibliography

Bradley, Walter W., 1916, Fresno County, M. and M. Mining Company: California Min. Bur. Rept. 14, p. 447.

Irelan, Wm., Jr., 1888, Mill Creek district: California Min. Bur. Rept. 8, pp. 207–208.

Mineral King

Mineral King is in central Tulare County in the high Sierra Nevada, about 37 miles east of Lemon Cove near Sequoia National Park. It is a small mining district that has yielded minor amounts of gold, silver, copper, lead, and zinc. Gold- and silver-bearing ore was discovered here in August 1873, and a "rush" to the area was on during the following year or two. However, little mining has been done in the district since, and it is now a well-known resort area. The Mineral King deposits occur in a belt of contact metamorphism in calcareous slate, impure limestone, and granodiorite. The ore is complex and consists of quartz and epidote rock containing pyrite, chalcopyrite, sphalerite, galena, arsenopyrite and, in places, gold and silver.

Bibliography

Tucker, W. B., 1919, Tulare County, Mineral King mining district: California Min. Bur. Rept. 15, pp. 947–954.

U. S. Geological Survey Mineral Resources of the United States 1883–84, p. 642.

Mokelumne Hill

Location. Mokelumne Hill is in northwestern Calaveras County. It is both a placer- and lode-mining district and includes the Chili Gulch, Old Woman's Gulch, and Golden Gate Hill areas.

History. The streams in the area were placer-mined early in the gold rush. Mokelumne Hill, first known as Big Bar, developed as a mining camp in 1848. Chili Gulch was first known as Chilean Gulch, from the large number of Chilean miners who worked here. They were discriminated against and often were forced to leave the mining regions. There was also a number of French miners in the district who had disputes with the Americans. Later in the 1870s Chinese miners were active in this district in great numbers.

Large amounts of gold were recovered by hydraulicking and drifting, but output declined in the 1870s. The Quaker City, Boston, and other lode mines yielded substantial amounts of gold from the 1880s until about 1900. Mining was done in the district again in the 1930s, and there has been intermittent prospecting since. The gravels at Chili Gulch are now mined for aggregate. Many of the buildings in the old town of Mokelumne Hill are well preserved.

Geology and Ore Deposits. A complex system of Tertiary channels, extending south and southwest from Mokelumne Hill, included eight distinct channels and remnants of several others, which range from Eocene to Pliocene in age. The so-called Chili Gulch, Stockton Hill, and Deep Blue channels have been the most productive. These contain a high percentage of quartz pebbles and boulders. In places they also contain large clear quartz crystals, some of which are piezo-electric grade. Usually the channels are not more than a few hundred feet wide. Bedrock consists of slate, greenstone, and graphite schist.

The gold-quartz veins occur in the slate and greenstone and are up to 50 feet thick. The gold occurs in the native state and is associated with small amounts of pyrite. Several dacite volcanic domes crop out in the district, and to the northeast a granodiorite stock is exposed.

Mines. Placer: America, Chappellet, Concentrator, Coffee Mill, Duryea, French Hill, Gopher, Green Mountain, Happy Valley, Hexter, Mosher, Neilsen, North Star, South Diamond, What Cheer, Werle. Lode: Boston $1 million, Easy Bird $300,000?, Hamby, Lamphear $122,000, Nuner, Quaker City $1 million+.

Photo 46. Town of Monitor in the 1870s. This westward view shows the main street of Monitor, Alpine County, an area now traversed by the Monitor Pass highway. *Photo courtesy of Calif. State Library.*

Bibliography

Clark, W. B., and Lydon, P. A., 1962, Calaveras County, gold: California Division Mines and Geology County Rept. 2, pp. 32–93.

Haley, C. S., 1923, Gold placers of California: California Min. Bur. Bull. 92, pp. 147–148.

Knopf, A., 1929, The Mother Lode system of California: U. S. Geol. Survey Prof. Paper 157, 88 pp.

Lindgren, Waldemar, 1911, Tertiary gravels of the Sierra Nevada: U. S. Geol. Survey Prof. Paper 73, pp. 205–209.

Logan, C. A., and Franke, H., 1936, Calaveras County, placer mines: California Div. Mines Rept. 32, pp. 324–355.

Ransome, F. L., 1900, Mother Lode district folio: U. S. Geol. Survey Geol. Atlas of the U. S., folio 63, 11 pp.

Storms, W. H., 1894, Ancient channel system of Calaveras County: California Min. Bur. Rept. 12, pp. 482–492.

Turner, H. W., 1894, Jackson folio: U. S. Geol. Survey Geol. Atlas of the U. S., folio 11, 6 pp.

Monitor-Mogul

Location and History. Monitor and Mogul are in central Alpine County about six miles southeast of Markleeville. Mogul is in the north end, and Monitor is in the south end of the district.

The region was intensely prospected during the late 1850s and early 1860s. Substantial amounts of gold, silver, and copper were produced from such mines as the Alpine, Colorado, Curtz, and Morning Star. Operators had much difficulty in milling the sulfide-rich or "rebellious" ores, and shipped some ore all the way to Swansea, Wales, for treatment. Monitor, named for the famous iron clad warship of the Civil War, also was known as Loope.

Operations declined in the 1880s, but there was some mining activity in the early 1900s and 1930s. The Zaca mine was worked on a moderate scale during the 1960s. The value of the total gold and silver production is unknown, but some estimates have placed it between $3 million and $5 million.

Geology. The district is underlain by volcanic rocks of Tertiary age. Andesitic tuff breccia is most common, but also present are various types of flow rocks including obsidian. In places, particularly at Colorado Hill, which is in the middle of the district, the volcanic rocks have been intensely altered and silicified. These altered rocks have a bleached appearance, in places stained yellow, red, and brown by iron oxide. These bleached zones stand out prominently from the unaltered rock.

Ore Deposits. The ore deposits occur in the zones of alteration and silicification. The gold, silver, and copper are nearly always associated with various sulfide minerals, which include pyrite, chalcopyrite, enargite, sphalerite, galena, argentite, and arsenopyrite. A number of other ore minerals also are present. The ore occurs in disseminated form, in veins and seams, and occasionally in tabular sulfide masses. In places high-grade pockets have been found.

Mines. Alpine, Curtz, Georgiana, Globe, Lincoln, Morning Star, Orion, Red Gap, Silver Hill. Zaca mines: Advance, Colorado, Tarshish.

Bibliography

Eakle, A. S., 1919, Alpine County, Mogul and Monitor districts: California Min. Bur. Rept. 15, pp. 8–25.

Evans, J. R., et al., 1966, Guidebook along the east-central front of the Sierra Nevada-Zaca mine: Geol. Society Sacramento annual field trip, June 18 and 19, 1966.

Logan, C. A., 1921, Alpine County, copper: California Min. Bur. Rept. 17, pp. 402–404.

Logan, C. A., 1923, Alpine County mines: California Min. Bur. Rept. 18, pp. 358–361.

Moore's Flat

Location. The Moore's Flat district is in north-central Nevada County about 15 miles northeast of Nevada City. It is both a lode and placer district and includes the "diggings" at Moore's Flat, Oreleans Flat, Woolsey Flat, Snow Point, and Snow Tent. The Alleghany district adjoins the Moore's Flat district on the northeast. Moore's Flat was named for H. M. Moore who built a store there in 1851.

Geology. A number of gravel deposits were accumulated in a west-southwest-trending Tertiary channel of the Yuba River that continues west and southwest into the North Bloomfield district. At Moore's Flat, Lindgren (1911, p. 141) estimated, 26 million cubic yards were removed and 15 million remained. The gravels are quartz-rich and in places more than 100 feet thick. Hydraulic mining here during the 1880s had reported gold recoveries of 11 to 15 cents per yard. The gravels are capped by andesite on the south side of the district. Bedrock consists of amphibolite, slate, and serpentine. The gold-quartz veins usually are narrow and contain small but often rich pockets.

Bibliography

Lindgren, Waldemar, 1911, Tertiary gravels of the Sierra Nevada: U. S. Geol. Survey Prof. Paper 73, p. 141.

Lindgren, Waldemar, 1911, Colfax folio: U. S. Geol. Survey Geol. Atlas of the U. S., folio 66, 10 pp.

Mooreville Ridge

Mooreville Ridge is in southwest Plumas and southeast Butte Counties. It includes the Camel Peak and American House areas. A number of Tertiary channel gravel deposits in the area have been mined largely by hydraulicking. The South Fork of the Feather River, which flows through the district, was mined also in the early days and skin divers are active in the area now. Bedrock in the district is granite, slate, serpentine, and amphibolite. The ridges are capped by Tertiary andesite and basalt. A few gold-quartz veins occur in the district.

Mines. Butte County: Dodson hydraulic, Golden Trant, Ludlam hydraulic, Walters. Plumas County: American House hydraulic, Browns Hill, Davis Point hydraulic, Fall River, Sanborn, Walters.

Bibliography

Lindgren, Waldemar, 1911, Tertiary gravels of the Sierra Nevada: U. S. Geol. Survey Prof. Paper 73, pp. 99–100.

Turner, H. W., 1898, Bidwell Bar folio: U. S. Geol. Survey Geol. Atlas of the U. S., folio 43, 6 pp.

Mormon Bar

Mormon Bar is in south-central Mariposa County about three miles south of the town of Mariposa. The area was placer-mined during the 1850s and 1860s, and, by 1870, the easily worked placers were largely exhausted. The area was mined again in the 1930s by dragline dredges, and there has been minor prospecting since. The total gold production for the district is estimated at 75,000 ounces. The deposits are in and adjacent to Mariposa Creek. The average depth of the mined gravels was about six feet.

Morris Ravine

Location. This district is at the south side of Oroville Table Mountain three miles north of Oroville. It includes the Monte de Oro area. It is both a lode- and placer-mining district. Morris Ravine and other nearby ravines were first placer-mined during the gold rush. Drift and lode mining began soon afterward and continued until around World War I. There was some work again during the 1930s. The Morris Ravine mine has been intermittently worked in recent years.

Geology. The bedrock in the district consists of amphibolite with smaller amounts of slate and phyllite. The slate contains fossil ferns. The bedrock is overlain by sedimentary and volcanic rocks of Oroville Table Mountain, a mesa-like hill consisting of thick beds of sands, tuffs, clays and auriferous channel gravels capped by black basalt. Fossil leaves also have been found in the clays.

Ore Deposits. The channel gravels are quartz-rich, well-cemented, and interbedded with sands and clays. The gold ranges from fine to coarse. A few small diamonds have been recovered here. The veins usually are narrow but the Banner vein has been mined to a depth of 1000 feet. The ore bodies are large but low in grade ($\frac{1}{10}$ ounce of gold/ton). Small rich pockets were mined also. Sulfides are spotty but often rich.

Mines. Placer: Monte de Oro, Morris Ravine, Perkins and Goodall, Yuba. Lode: Banner $1 million, Bumble Bee $100,000+.

Bibliography

Creely, R. S., 1965, Geology of the Oroville Quadrangle: California Div. Mines and Geology Bull. 184, 86 pp.

Lindgren, Waldemar, 1911, Tertiary gravels of the Sierra Nevada: U. S. Geol. Survey Prof. Paper 73, pp. 86–89.

Logan, C. A., 1930, Butte County, Banner mine: California Div. Mines Rept. 26, pp. 369–370.

O'Brien, J. C., 1949, Butte County, gold: California Jour. Mines and Geology, vol. 45, pp. 426–433.

Waring, C. A., 1919, Butte County, gold-quartz mines: California Min. Bur. Rept. 15, pp. 211–224.

Mountain Meadows

Mountain Meadows is in southwestern Lassen County a few miles southeast of Westwood and eight miles north of Greenville. Years ago placer gold was recovered from Tertiary gravels in the southeast end of the valley. These deposits are on the northwest end of a Tertiary channel known as the Jura channel, which extends southeast to the Taylorsville and Genessee districts. Bedrock consists of greenstone and amphibolite.

Bibliography

Lindgren, Waldemar, 1911, Tertiary channels of the Sierra Nevada: U. S. Geol. Survey Prof. Paper 73, 116.

Mountain Ranch

Location. This is a small district in central Calaveras County in the vicinity of the town of Mountain Ranch. It includes the Cave City area.

Photo 47. Princeton Mine, Mount Bullion District. This photo of this highly productive Mariposa County mine was taken probably around 1900. Photo courtesy of Calif. State Library.

Geology. The district is largely underlain by graphite schist, quartzite, limestone, and thin bodies of talcose schist. Patches of Tertiary auriferous gravels overlie the bedrock. A few narrow gold-quartz veins, in schist, contain small but sometimes rich ore shoots. The ore contains free gold and abundant sulfides, especially galena.

Mines. Lode: Gaston Hill. Placer: Cotton Flat, Foley, Hidden Cave, Humboldt, Mountain Ranch, Rose Hill.

Bibliography

Clark, Lorin, 1954, Geology and mineral deposits of the Calaveritas quadrangle: California Div. Mines Spec. Rept. 40, 23 pp.

Clark, W. B., and Lydon, P. A., 1962, Calaveras County, gold: California Div. Mines and Geology County Rept. 2, pp. 32-93.

Turner, H. W., 1894, Jackson folio: U. S. Geol. Survey Geol. Atlas of the U. S., folio 11, 6 pp.

Mount Bullion

Location. The Mount Bullion district is in west-central Mariposa County about seven miles northwest of Mariposa. The district is in the southern end of the Mother Lode gold belt and extends northwest towards Bagby and Bear Valley (fig. 18). It includes the Agua Fria and Mount Ophir areas.

History. This region was first placer-mined in 1848, many of the miners having been of Spanish descent. Agua Fria Creek and other streams were highly productive (Agua fria means cold water in Spanish). Lode gold-mining began shortly afterward. Much of this district is in the Las Mariposa land grant, which originally belonged to General John C. Frémont. The mines in this grant were not located and surveyed in the same fashion as those on public lands, and to this day the land plats within this grant are difficult to coordinate with established survey lines. The grant later underwent lengthy litigation, and Frémont eventually

went bankrupt. He named nearby Mount Bullion for his father-in-law, Senator Thomas Hart Benton, who was sometimes known as "Old Bullion".

At Mount Ophir, which is now a ghost town, are the ruins of an early-day mint. From 1849 until 1854 private coinage subject to federal inspection was authorized in California. It is believed that some of the now extremely rare and valuable octagonal fifty-dollar gold slugs were minted here from locally mined gold.

Gold mining in the district continued fairly steadily from the 1850s through the 1870s. There was considerable activity from around 1900 to 1920 when the Princeton and other mines were worked. Some mining was done in the 1930s and early 1940s, and there have been a few intermittent small-scale operations since.

Geology. As shown in figure 18, the gold mineralization is confined chiefly to a northwest-trending belt of slate, phyllite, and metasandstone of the Mariposa Formation (Upper Jurassic). Within this formation are two belts of pyrite-bearing metarhyolite that may have possible future economic significance. Greenstone of the Peñon Blanco Formation (Upper Jurassic) crops out to the east and west. Also present are thin bands of serpentine and numerous aplite dikes.

Ore Deposits. Several north-northwest-striking systems of quartz veins occur principally in slate. The veins usually range from four to 10 feet in thickness, although there are some massive ones that are considerably thicker. The ore contains free gold and pyrite, which is abundant in places. Milling ore yielded from $\frac{1}{4}$ to $\frac{1}{2}$ ounce of gold per ton, and considerable high-grade ore was recovered close to the surface. Some of the ore shoots were extensive; several in the Princeton mine had stoping lengths of more than 500 feet. The greatest depth of development is 1600 feet on the incline.

A number of extensive vein systems have not been thoroughly explored. Also, there are several extensive deposits of pyritic metarhyolite in the Mariposa Formation that in places contain gold. These bodies are several miles long and 60 or more feet thick.

Mines. Greens Gulch $119,000+, King Midas, Louis, Mt. Ophir $250,000 to $300,000, Mountain View I, Nellie Kahoe, Ortega, Princeton $5 million, Sorrel.

Bibliography.

Bowen, O. E., 1957, Mariposa County, Mount Ophir and Princeton mines: California Jour. Mines and Geology, vol. 53, pp. 139–140 and 155–158.

Castello, W. O., 1921, Mariposa County, Mt. Bullion-Bear Valley district: California Min. Bur. Rept. 17, p. 98.

Knopf, Adolph, 1929, The Mother Lode belt of California, Princeton mine: U. S. Geol. Survey Prof. Paper 157, pp. 84–85.

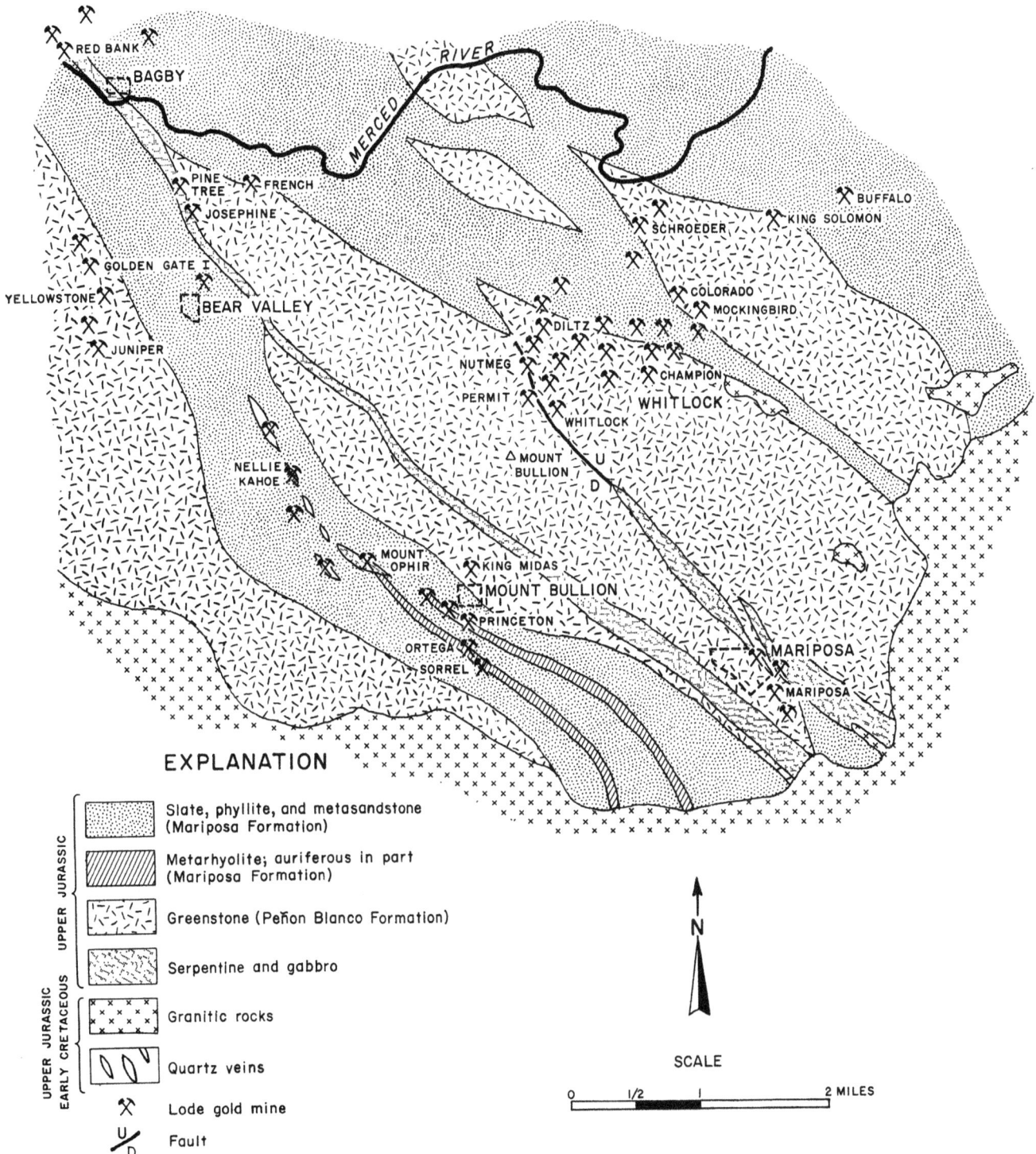

Figure 18. Geologic Map of Bagby, Mariposa, Mount Bullion and Whitlock Districts, Mariposa County. By O. E. Bowen and J. R. Evans, 1966.

Photo 48. Champion Mine, Nevada City District. This 1893 view of the Nevada County mine looks east.

Logan, C. A., 1935, Mother Lode belt, Mariposa County: California Div. Mines Bull. 108, pp. 180–190.

Ransome, F. L., 1900. Mother Lode district folio: U. S. Geol. Survey Geol. Atlas of the U. S., folio 63, 11 pp.

Storms, W. H., 1900, The Princeton mine: California Min. Bur. Bull. 18, p. 143.

Murphys

Location. The Murphys district is in south-central Calaveras County about seven miles northeast of Angels Camp. It extends west to include the Esmeralda and Fricot Ranch areas.

History. The streams were first mined during the gold rush. Murphys was established in 1848 or 1849 and named for John M. Murphy, a member of Captain Weber's Stockton Mining Company. Lode mining probably began shortly afterward and continued almost steadily until around World War I. Some of the mines have been prospected in recent years. The old town of Murphys, one of the best-preserved mining towns in the Sierra Nevada, is a popular tourist attraction.

Geology. The eastern part of the district is underlain by graphite schist, slate, and a large limestone lens of the Calaveras Formation (Carboniferous to Permian). The western part is underlain by slate, schist, green schist and numerous lenses of talcose rock derived from serpentine. A number of patches of auriferous Tertiary gravels overlie the bedrock.

Ore Deposits. Quartz, occurring in a great many west-trending veins—mostly in schist and slate—is glassy and white, rose, or occasionally black. Rose-colored quartz is characteristic of this district. The ore bodies usually are small and shallow, but often they are rich. The ore contains free gold and often abundant sulfides. Although there are no large mines, the district was quite productive because of the large number of veins.

Mines. Basco, Beatrice, Bence $200,000, Bonehard, Buckeye, Buckhorn, Crown Point, Cowbell, Dora Cons., Dragone, Economic, Esmeralda $300,000, Eureka, Fairplay, Falcon, Fricot Group, Great Divide, Gumboot, Hidden Treasure, K and J, Last Chance, Malteson, Manhatten, Miralda, Oro y Plata, Piety Hill, Rocky Bar, Total Wreck.

Bibliography.

Clark, L. D., 1954, Geology and mineral deposits of the Calaveritas quadrangle: California Div. Mines Spec. Rept. 40, 23 pp.

Clark, W. B., and Lydon, P. A., 1962, Calaveras County, gold: California Div. Mines and Geology County Report 2, pp. 32–93.

Logan, C. A., 1936, Calaveras County, gold quartz mines: California Div. Mines Rept. 32, pp. 235–323.

Turner, H. W., 1894, Jackson folio: U. S. Geol. Survey Geol. Atlas of the U. S., folio 11, 6 pp.

Tucker, W. B., 1916, Calaveras County, gold quartz mines: California Min. Bur. Rept. 14, pp. 66–114.

Turner, H. W., and Ransome, F. L., 1898, Big Trees folio: U. S. Geol. Survey Geol. Atlas of the U. S., folio 51, 8 pp.

Nashville

Location and History. The Nashville district is in southwestern El Dorado County about 15 miles south of Placerville. It is in the Mother Lode gold belt. The area was mined during the gold rush when considerable quantities of high-grade ore were taken from near the surface. Originally known as Quartzburg, the town of Nashville was renamed by miners who came from Tennessee. Activity was considerable here during the 1930s, when the Montezuma-Apex and Nashville mines were worked. There has been minor prospecting since.

Geology. A north-trending belt of gray to black slate of the Mariposa Formation (Upper Jurassic) is in the central portion of the district. Massive greenstone is to the west, and schist, amphibolite, quartzite, and granitic rocks are to the east.

Ore Deposits. Several long north-striking massive quartz veins in the slate are up to 25 feet thick. These veins contain large but low- to moderate-grade ore bodies (1/7 to 1/4 ounce of gold per ton). Stoping lengths were up to 500 feet, and the veins were mined to inclined depths of 2000 feet. The ore contains free gold and pyrite. Considerable fault gouge is present. The veins in greenstone to the west and amphibolite and schist to the east are usually only a few feet wide, but they have yielded appreciable amounts of high-grade ore.

Mines. Bonanza, Briarcliff $120,000, Balmaceda, Last Chance, Manhattan, Monarch-Sugar Loaf $100,-000, Montezuma-Apex $1 million, Nashville $2 million.

Bibliography.

Clark, W. B., and Carlson, D. W., 1956, El Dorado County, lode gold mines: California Jour. Mines and Geology, vol. 52, pp. 401–429.

Fairbanks, H. W., 1890, Geology of the Mother Lode region: California Min. Bur. Rept. 10, pp. 80–81.

Lindgren, Waldemar, and Turner, H. W., 1894, Placerville folio: U. S. Geological Survey Geol. Atlas of the U. S., folio 3, 4 pp.

Logan, C. A., 1935, Montezuma mine: California Div. Mines Bull. 108, pp. 30–34.

Nevada City

Location. The Nevada City district is in western Nevada County. The district covers an extensive area, from the vicinity of Indian Flat east through Nevada City, northeast to Willow Valley and southeast through Canada Hill and Banner Hill to the vicinity of the Lava Cap mine. It is both a lode- and placer-mining district and once was an important center of gold mining in California.

History. Gold was first mined in this district in Deer Creek, which, in 1849, was called Deer Creek Diggings. The name Nevada was adopted in May 1850 at a public meeting. The placers were rich, and the town grew fast. Hydraulic mining was first practiced in California at American Hill here in 1852, by E. G. Matteson (hydraulicking was also done that same year at Yankee Jims in Placer County). Hydraulic mining flourished until around 1880. Drift mining began in the 1850s, and the drift mines were continuously active until around 1900. Gold-quartz was discovered in 1850, when the Gold Tunnel vein was found. However, important production of lode gold did not commence until the early 1860s because of difficulties in milling the ore. By 1865, the output from lode-gold mining was averaging $500,000 per year and later ranged from $300,000 to $600,000. The Champion and Providence mines were the major producers during these years. Later these two mines were in litigation, and in 1902 the Champion owners bought the Providence. Large-scale lode-gold mining was resumed in the district again during the 1930s when the Lava Cap and Banner mines were operating. From 1933 to 1942, the Lava Cap yielded $12 million. There has been only minor activity since 1942. The old town of Nevada City, the county seat, is now a popular tourist center with numerous well-preserved old buildings. The Nevada County Narrow Gauge railroad served the area from 1877 to 1942. The total output of the district is unknown, but it is estimated by the author at more than $50 million and may have exceeded $70 million.

Photo 49. Lava Cap Mine, Nevada City District. This is a recent photo of a mine in Nevada County that yielded $12 million in gold in 1933–42.

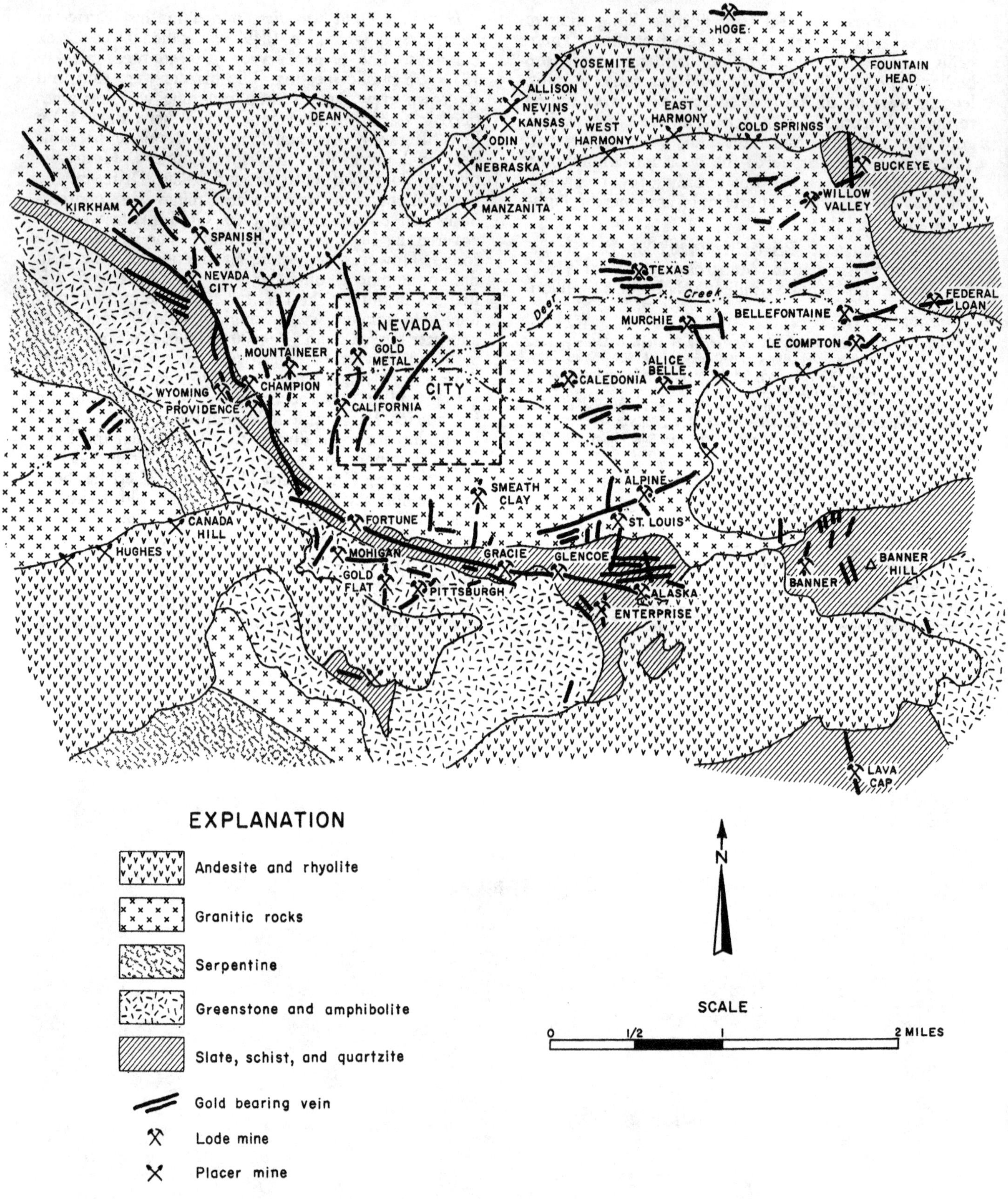

EXPLANATION

- Andesite and rhyolite
- Granitic rocks
- Serpentine
- Greenstone and amphibolite
- Slate, schist, and quartzite
- Gold bearing vein
- Lode mine
- Placer mine

SCALE

0   1/2   1   2 MILES

Figure 19.   Geologic Map of Nevada City District, Nevada County. The map shows vein systems and the principal mines. *After Hobson, 1890, and Lindgren, 1896a.*

Photo 50. Providence Mine, Nevada City District. This 1893 view of the mine, in Nevada County, looks southeast. The Champion mine is at left, Deer Creek in the foreground.

Geology. The central portion of the district is underlain by granitic rocks, chiefly granodiorite (fig. 19). Adjacent are beds of slate, mica schist, and quartzite, most of which are part of the Calaveras Formation (Carboniferous to Permian). To the west and southwest are fairly extensive beds of massive greenstone, amphibolite, and serpentine. There are a number of fine- to medium-grained dioritic and aplitic dikes, some of which are associated with the gold-quartz veins. In places these rocks are overlain by Tertiary channel gravels capped by rhyolite and andesite.

Ore Deposits. Several major gold-quartz vein systems traverse the district. In the west portion one system extends northwest along a granodiorite-metasedimentary rock contact. In the southern and eastern portion of the district the veins strike nearly west and dip either north or south. There are also a few northeast-striking and southeast-dipping veins. The veins usually are one to four feet thick, but in places a few are as much as 15 feet thick. The ore contains varying amounts of free gold, often abundant pyrite and smaller amounts of other sulfides. Some of the ore bodies are extensive; the ore body at the Providence mine persisted to an inclined depth of more than 2700 feet. Considerable high-grade ore has been recovered in the district.

Several important Tertiary channels were sources of ore-bearing gravels. One, the Harmony channel, which enters the district from the northeast, was extensively mined by drifting. The pay gravel in this channel was 150 to 200 feet wide, two to four feet deep, quartzitic, often sub-angular and well-cemented. These pay streaks yielded $1.55 to $2.50 in gold to the ton, at the old price. The Manzanita channel, which yielded $3 million, is just to the west. Northwest of town is the northwest-trending Cement Hill channel. In the southern part of the district is the Town Talk channel, which was narrow but rich in places. Much of the placer gold taken from the channel deposits in this district was coarse.

Mines. Lode: Alaska, Alice Belle, Alpine, Bagley, Banner $1 million+, Belle Fontaine, Buckeye, Caledonia, California Cons. $1 million, Canada Hill $1.13 million, Carter, Central South Yuba, Champion $3 million, Coan, Deadwood $300,000, Enterprise, Federal Loan $200,000, Fortune, Franklin, Glencoe, Gold Flat, Gold Metal, Gold Tunnel $300,000, Gracie, Hoge $600,000, Kirkham, Lava Cap $12 million, Le Compton, Massachusetts, Mayflower, Merrifield, Merrimac, Montana, Mohigan, Mountaineer $2 million to $3 million, Mt. Auburn, Murchie, National, Neversweat, Nevada City, Oustomah, Phoenix $200,000, Pittsburgh $1 million+, Sneath and Clay $180,000, Soggs, Spanish, St. Louis, Texas, Union, Willow Valley $130,000, Wyoming. Drift: Allison, Cold Springs, Coleman, Dean, East Harmony, Fountain Head, Grover, Hughes, Kansas, Knickerbacker, Live Oak, Manzanita, Nebraska, Nevins, Odin, Pennsylvania, Phoenix, West Harmony, Yosemite. Hydraulic: American Hill, Buckeye Hill, Canada Hill, Hirschmann.

Bibliography

Chandler, J. W., 1941, Mining methods and costs of the Lava Cap Gold Mining Corporation, Nevada City: Calif. Div. Mines Rept. 37, pp. 409–425.

Crawford, J. J., 1896, Champion, Harmony, Mayflower, and Providence mines: Calif. Min. Bur. Rept. 13, pp. 239, 247–248, 252–253, and 260.

Hobson, J. B., 1890, Nevada City district: Calif. Min. Bur. Rept. 10, pp. 384–389.

Hobson, J. B., and Wiltsee, E. M., 1893: Nevada City mining district: Calif. Min. Bur. Rept. 11, pp. 285–296.

Photo 51. Malakoff Mine, North Bloomfield District. This is a recent view of the mine, in Nevada County, once one of the largest hydraulic mines in the state. The banks are as high as 600 feet. *Photo by Mary Hill.*

Irelan, William, Jr., 1888, Nevada City district: Calif. Min. Bur. Rept. 8, pp. 418–425.

Lindgren, Waldemar, 1895, Smartsville folio, California: U. S. Geol. Survey Geol. Atlas of the U. S., folio 18, 6 pp.

Lindgren, Waldemar, 1896, Nevada City special folio, California: U. S. Geol. Survey Geol. Atlas of the U. S., folio 29, 7 pp.

Lindgren, Waldemar, 1896, Gold-quartz veins of the Nevada City and Grass Valley districts: U. S. Geol. Survey Ann. Rept. 17, pt. 2, pp. 1–262.

Lindgren, Waldemar, 1911, The Tertiary gravels of the Sierra Nevada: U. S. Geol. Survey Prof. Paper 73, pp. 125–132.

Logan, C. A., 1930, Nevada County, Geology of Grass Valley and Nevada City districts: Calif. Div. Mines Rept. 26, pp. 97–99.

Logan, C. A., 1941, Mineral resources of Nevada County: Calif. Jour. Mines and Geology, vol. 37, pp. 380–431.

MacBoyle, Errol, 1919, Nevada County, Nevada City district: Calif. Min. Bur. Rept. 16, pp. 37–44.

Newtown

Location. Newtown is in central El Dorado County 10 miles east of Placerville. It was originally known as Dogtown. Chiefly a placer-mining district, it includes the Camino and Pleasant Valley areas.

Geology. Numerous deposits of auriferous gravels were deposited in the west-trending Tertiary channel of the South Fork of the American River. They were mined during the early days by hydraulicking and drifting, and again by drifting through the 1930s. The lower gravels are quartz-rich and in places contain coarse gold. They are overlain by rhyolite and andesite. Bedrock is slate, schist, and quartzite with thin bands of serpentine. Granodiorite crops out to the north and south.

Bibliography

Clark, W. B., and Carlson, D. W., 1956, El Dorado County, placer deposits: California Jour. Mines and Geology, vol. 52, pp. 429–435.

Lindgren, Waldemar, and Turner, H. W., 1894, Placerville folio: U. S. Geol. Survey Geol. Atlas of the U. S., folio 3, 3 pp.

North Bloomfield

Location. The North Bloomfield mining district is in north-central Nevada County about 10 miles northeast of Nevada City. This district also includes the "diggings" at Lake City to the west, Derbec to the north, and Relief to the east.

History. Gold was discovered here originally in 1851. Hydraulic mining began about 1853 and, by 1855, had become a major industry. An extensive system of ditches and flumes supplied water to the mines from Bowman Lake and other reservoirs to the east in the high Sierra Nevada. The town of North Bloomfield was first known as Humbug City. Its name was changed to Bloomfield and then to North Bloomfield when someone discovered there was a Bloomfield in Sonoma County.

As more and more gold-bearing gravel was excavated, the hydraulic pits here became enormous. The pit at the famous Malakoff mine is more than 7000 feet long, 3000 feet wide, and up to 600 feet deep. The tailings from the hydraulic operations were allowed to flow into the rivers, a procedure that led to litigation with the farmers who lived downstream. In a famous court case in 1884 (Woodruff vs. North Bloomfield Gravel Mining Company (16 Fed. Rep. 25)), Judge Lorenzo Sawyer issued an injunction against the dumping of mine debris into the Sacramento and San Joaquin Rivers and their tributaries. Injunctions against other mines soon followed, and hydraulic mining in the Sierra Nevada has not been important since that date. The Malakoff diggings and part of the old town are now a state park. The U. S. Geological Survey, beginning in 1966, initiated an exploration program in this area to determine the extent of unmined gravels. Drilling and geophysical exploration have been done.

The total output of the Malakoff mine is about $3.5 million, and the Derbec mine has probably yielded $1 million to $2 million according to Lindgren (1911). He estimated that 30 million yards had been removed and 130 million remained at North Bloomfield; Jarman (1927) estimated that 40 million yards had been removed and more than 50 million remained.

Geology. The main channel of the Tertiary Yuba River entered this district from the northeast via Derbec. A branch joined this channel from Relief Hill to the southeast. At North Bloomfield the main channel curves west and north and then west again as it continued toward the North Columbia district. Although the gravels are as much as 600 feet thick, most of the values were obtained from the lower 130 feet of blue gravel. These gravels yielded from four to 10 cents of gold per cubic yard. Bedrock consists of slate, schist, and phyllite.

Bibliography

Hobson, J. B., and Wiltsee, E. A., 1893, North Bloomfield mining district: California Min. Bur. Rept. 11, pp. 311–312.

Irelan, William, Jr., 1888, North Bloomfield mine: California Min. Bur. Rept. 8, pp. 454–459.

Jarman, Arthur, 1927, Bloomfield hydraulic mine: California Min. Bur. Rept. 23, pp. 107–110.

Lindgren, Waldemar, 1900, Colfax folio, California: U. S. Geol. Survey Geol. Atlas of the U. S., folio 66, 10 pp.

Lindgren, Waldemar, 1911, Tertiary gravels of the Sierra Nevada: U. S. Geol. Survey Prof. Paper 73, pp. 139–141.

MacBoyle, Errol, 1919, Nevada County, North Bloomfield mining district: California Min. Bur. Rept. 16, pp. 45–51.

North Columbia

Location. This district is in north-central Nevada County about seven miles northeast of Nevada City. It includes placer deposits in the North Columbia and Columbia Hill areas and lode deposits in the Delhi mine area.

History. The district was first placer-mined during the gold rush and hydraulicked on a large scale from the middle 1850s to the early 1880s. It was named after Columbia Hill, and "North" was added to distinguish it from Columbia in Tuolumne County. The Delhi lode mine was active from the 1860s to the 1890s. Chinese carried on small-scale placer mining from the 1890s to the early 1900s. Some work was done again during the 1930s and the area has been prospected since. The total output of the district is unknown. The placer mines are estimated to have yielded $2 million to $3 million. Lindgren (1911) estimated that 25 million yards of gravel had been removed and 165

Photo 52. Main Hydraulic Pit, North Columbia District. The pit, in Nevada County, is nearly two miles long and a mile wide. The view is north.

million remained. The lode mines have yielded more than $1 million. Beginning in 1966, the U.S. Geological Survey and U.S. Bureau of Mines have studied this district as part of their "heavy metals" programs (see also sections on Badger Hill and North Bloomfield).

Geology. Extensive channel deposits lie at the junction of two major streams of the Tertiary Yuba River. A west-trending channel extends through the district from North Bloomfield and continues on to Badger Hill, and a branch enters the district from Blue Tent to the south. In places the gravels are as much as 500 feet thick. The lower gravels are coarse while upper bench gravels are fine. Much sand and clay are present. Bedrock is phyllite of the Delhi formation (Carboniferous). Several gold-quartz veins have yielded high-grade ore.

Mines (all lode). Delhi $1 million, Enterprise, Grizzly Ridge, St. Gothard.

Bibliography

Hobson, J. B., and Wiltsee, E. A., 1893, Columbia Hill district: California Min. Bur. Rept. 11, pp. 305–308.

Irelan, William, Jr., 1888, Columbia Hill district: California Min. Bur. Rept. 8, pp. 444–447.

Lindgren, Waldemar, 1900, Colfax folio, California: U. S. Geol. Survey Geol. Atlas of the U. S., folio 66, 10 pp.

Lindgren, Waldemar, 1911, Tertiary gravels of the Sierra Nevada: U. S. Geol. Survey Prof. Paper 73, p. 139.

MacBoyle, Errol, 1919, Nevada County, North Columbia mining district: California Min. Bur. Rept. 16, pp. 48–51.

North San Juan

Location. This district is in northwestern Nevada County nine miles northwest of Nevada City. It consists of the placer deposits that extend from the vicinity of North San Juan southwest for about 2½ miles to the Sweetland area. It was named for a nearby hill with the "North" added to distinguish it from San Juan (now San Juan Bautista) in San Benito County.

Geology. A main Tertiary channel of the Yuba River extends into the area from the Badger Hill and North Columbia districts from the east and another from Camptonville from the northwest. The channel then extends southwest through Sweetland into the French Corral district. The gravels are 150 to 400 feet thick and up to 1000 feet wide. The low gravels yielded up to 30¢ per yard in gold, mostly from hy-

draulicking. The bedrock is granodiorite with amphibolite to the west. There are several narrow gold-quartz veins in the amphibolite.

Bibliography

Lindgren, Waldemar, 1895, Smartsville folio, California: U. S. Geol. Survey Geol. Atlas of the U. S., folio 18, 6 pp.

Lindgren, Waldemar, 1911, Tertiary gravels of the Sierra Nevada: U. S. Geol. Survey Prof. Paper 73, pp. 121–125.

MacBoyle, Errol, 1919, Nevada County, North San Juan mining district: California Min. Bur. Rept. 16, pp. 51–54.

Ophir

Location. This district is in southwestern Placer County in the vicinity of the old town of Ophir. It extends east to Auburn and west to Gold Hill and includes the area known as the Duncan Hill mining district.

History. Gold-bearing surface gravels were discovered at Ophir and Auburn in 1848 and for several years yielded substantial amounts of the mineral. The quartz veins were then developed, and the Ophir and Duncan Hill districts were organized. Appreciable amounts of high-grade gold ore were recovered during the 1860s, 1870s, and 1880s, but mining activity in the district declined after that. The mines were active again from the early 1900s through the 1930s with substantial production, but little has been done since. The value of the total output of the district is estimated at more than $5 million.

Geology. The mineralized zone is on the northeast flank of a granodiorite and quartz-diorite stock that is intrusive into amphibolite schist (fig. 20). A series of west-northwest-striking and south-dipping quartz veins occur in the granitic rocks or along the granitic rock-amphibolite schist contact. A few veins are in the amphibolite. The ore contains free gold in places with often abundant pyrite, galena, and chalcopyrite; much of the ore is base. Milling ore ranged from ¼ to one ounce of gold per ton. Considerable high-grade ore was taken close to the surface during the early days. The veins range from one to five feet in thickness, and several were mined to depths of more than 1000 feet. The ore shoots had stoping lengths of up to 250 feet.

Mines. Belmont, Black Ledge, Centenial $150,000+, Conrad $50,000, Crater $750,000, Doig, Eclipse $100,-000+, Gold Blossom $216,000+, Grass Ravine, Green $150,000+, Green Emigrant $150,000+, Hathaway $336,000, Julian, Mina Rica $55,000, Moore $180,000, Oro Fino $500,000+, Pine Tree, Rock Creek $200,-000?, St. Lawrence, St. Patrick $148,000+, Three Stars $415,000.

Bibliography

Hobson, J. B., 1890, Ophir mining district: California Min. Bur. Rept. 10, pp. 427–433.

Irelan, William, Jr., 1888, Auburn district: California Min. Bur. Rept. 8, pp. 460–462.

Lindgren, Waldemar, 1892, Gold-silver veins at Ophir: U. S. Geol. Survey 14th Ann. Rept., pt. 2, pp. 249–284.

Lindgren, Waldemar, 1894, Sacramento folio: U. S. Geol. Survey Geol. Atlas of the U. S., folio 5, 3 pp.

Logan, C. A., 1936, Gold mines of Placer County, Ophir district: California Div. Mines Rept. 32, pp. 28–31.

Lydon, P. A., 1959, Geology along U. S. Highway 40: California Div. Mines Mineral Information Service, Vol. 12, no. 8. pp. 1–9.

Waring, C. A., 1919, Placer County, Ophir district: California Min. Bur. Rept. 15, p. 319.

Oroville

Location. This district is in southwestern Butte County. It is mainly a dredging field that extends from just west of the city of Oroville southwest along the Feather River to a point about five miles due east of Biggs. The field is one to two miles wide and nine miles long.

History. Shallow placers were mined here during the gold rush. The area was settled in 1849: Oroville originally was known as Ophir City, but the name was changed in 1855. Around 1895, W. P. Hammon and others tested the area to determine the feasibility of mining on a large scale. They introduced bucket-line dredging in 1898, the first in California. The field was highly productive from 1903 to 1916; in 1908 there were 35 dredges and 12 dredging companies active in

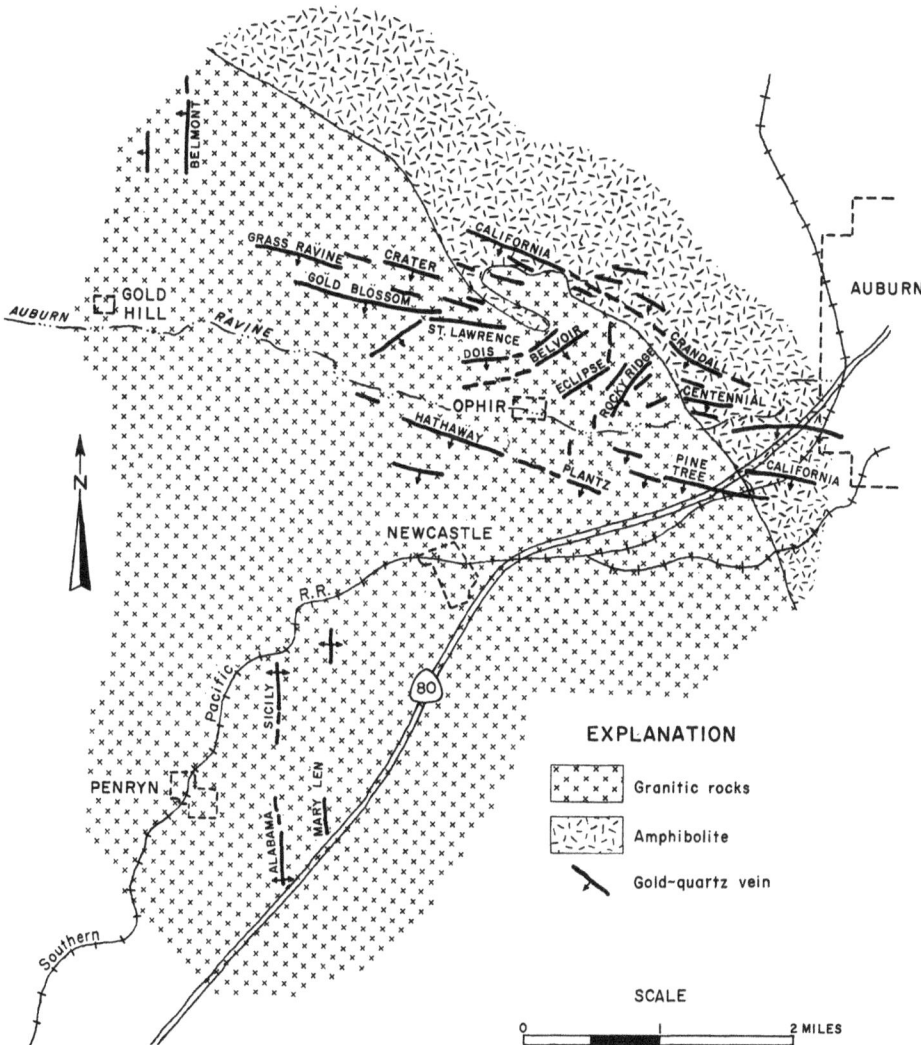

Figure 20. Geologic Map of Ophir and Penryn Districts, Placer County. The principal gold veins are shown. *After Lindgren, 1892 and 1894.*

Photo 53. Cherokee Mining Company Dredge, Oroville District. This 1904 photo shows one of the earliest bucket-line dredges in California, operating here in Butte County.

the field. Output later declined, but dredging was done again from 1936 to 1942 and 1945 to 1952. The dredge field is now an important source of sand and gravel. The total output from dredging is estimated to be about 1,964,000 ounces of gold.

Geology. The gold occurs in river gravels and adjacent terrace gravels on the flood plain. The gravels rest on a bedrock of soft but compact andesite and rhyolite tuff. Coarse boulders, which become finer downstream, are present along with alternating sand layers. Digging depths ranged from 25 feet upstream to as much as 55 feet downstream. The gold was fine and occurred chiefly in the gravels. It was 915 to 930 in fineness. Dredge recoveries ranged from 15 cents to 25 cents per yard of gold at the new price. Minor amounts of platinum also were recovered.

Concerns. Butte Dredging Co., El Oro Dredging Co., Feather River Development Co., Gold Hill Dredging Co. 1938–50, Gold Run Dredging Co. 1906, Indiana Gold Dredging Co. 1908, Kentucky Ranch Gold Dredging Co. 1909, Natomas Cons. 1909–17, Oroville Dredging Ltd. 1906–16, Oroville Gold Dredging Co. 1941–44, Oroville Union Gold Dredg. Co. 1914, Oro Water Light & Power Co. 1906–16, Pacific Gold Dredging Co. 1906–16?, Pennsylvania Dredging Co., Shasta-Butte Gold Dredg. Co. 1928–?, Viloro Syndicate Ltd. 1904–16?, Yuba Cons. 1935–52. Many other earlier concerns were consolidated around 1906.

Bibliography

Doolittle, J. E., 1908, Gold dredging in California, Oroville district: California Min. Bur. Bull. 36, pp. 68–88.

Lindgren, Waldemar, 1911, Tertiary gravels of the Sierra Nevada: U. S. Geol. Survey Prof. Paper 73, pp. 220–221.

Logan, C. A., 1930, Butte County, gold placer mines: California Div. Mines Rept. 26, pp. 383–384.

O'Brien, J. C., 1949, Butte County, Yuba Consolidated Goldfields dredges: California Jour. Mines and Geology, vol. 45, pp. 432–433.

Waring, C. A., 1919, Butte County, gold dredging: California Min. Bur. Rept. 15, pp. 187–198.

Winston, W. B., 1910, Gold dredging in California, Dredging in the Oroville district: California Min. Bur. Bull. 57, pp. 111–158.

Pacific

This is a placer-mining district in east-central El Dorado County, in the vicinity of Pacific House and about 20 miles east of Placerville. Several hydraulic and drift mines here were originally worked in 1850s or 1860s, with some work at the Pacific Channel drift mine in the early 1920s. The gravel deposits are on a Tertiary channel that extended west and southwest toward Placerville. The gravels are capped by andesite and rest on a granite bedrock.

Bibliography

Lindgren, Waldemar, and Turner, H. W., 1894, Placerville folio, California: U. S. Geol. Survey Geol. Atlas of the U. S., folio 3, 3 pp.

Logan, C. A., 1920, El Dorado County, Pacific channel mine: California Min. Bur. Rept. 17, pp. 428–429.

Paloma

This district is in northwestern Calaveras County in the vicinity of the old mining town of Paloma. Much of the gold produced in the district has come from the famous Gwin mine, which was operated on a large scale during the 1860s, 1870s, and from 1894 to 1908.

The mine, named for U.S. Senator William Gwin, had a total output valued at about $7 million.

The gold-bearing veins are in the same belt of slate of the Mariposa Formation (Upper Jurassic) on which the large Jackson district mines lie, in Amador County to the north (see fig. 12). The quartz veins strike north-northwest and dip steeply to the east. The ore contains free gold, pyrite, and arsenopyrite. Milling ore recovered from the Gwin mine averaged about ¼ ounce gold per ton. The great north ore shoot in this mine had a horizontal stoping length of up to 800 feet and a pitch length of 1500 feet.

Bibliography

Clark, W. B., and Lydon, P. A., 1962, Calaveras County, Gwin mine: California Div. Mines and Geology County Report 2, pp. 56–59.

Turner, H. W., 1894, Jackson folio, California: U. S. Geol. Survey Geol. Atlas of the U. S., folio 11, 6 pp.

Penryn

Location. This district in southwestern Placer County in the vicinity of the town of Penryn, also has been known as Stewart's Flat. The Ophir district adjoins it on the north and northeast (fig. 20). The town of Penryn was named by Griffith Griffiths, who opened granite quarries here in 1860, after Penrhyn, Wales. Appreciable mining activity here during the 1930s and early 1940s was highlighted by work at the Alabama, Chicago, and Sicily mines.

Geology. The principal country rocks are granodiorite and quartz diorite with some diorite and gabbro. A number of north- to northwest-striking quartz veins, cropping out with nearly vertical dips, range from one to five feet in thickness. The ore contains free gold with varying amounts of sulfides. Argentite and tellurides are occasionally present. The milling-grade ore usually averaged ⅓ to ½ ounce of gold per ton, but some was considerably richer.

Mines. Alabama $1 million+, Chicago $100,000+, Elizabeth, Highway 40, Jenny Lind, Mary Len, Penryn, Sicily $100,000+.

Bibliography

Lindgren, Waldemar, 1894, Sacramento folio: U. S. Geol. Survey Geol. Atlas of the U. S., folio 5, 3 pp.

Logan, C. A., 1936, Gold mines of Placer County, Alabama and Chicago mines: California Div. Mines Rept. 32, pp. 10–11 and 16–17.

Pike

Location and History. This district is in the southwest corner of Sierra County at the site of the old town of Pike or Pike City. Some of the early-day miners came from Pike, Missouri. The district, 12 miles northeast of North San Juan and 26 miles northeast of Nevada City, includes the Tippicanoe, Negro Tent, Snowden Hill, and Grizzly Gulch areas. The area was originally mined during the gold rush, and the Alaska mine was worked on a large scale from 1863 to 1916. The Pleasant View hydraulic mine was active in 1962 and 1963.

Geology and Ore Deposits. The district is underlain by metadiabase, serpentine, and irregular bodies of amphibolite. Fine-grained phyllite and quartzite of the Delhi Formation (Carboniferous) lie on the east side. Lenticular quartz veins with calcite occur chiefly in the metadiabase. The ore bodies contain free gold and often abundant sulfides, including galena. There are some extensive ore shoots and some high-grade pockets. The gold often is coarse. Several patches of Tertiary quartz-rich gravel are a few acres in extent, some of the gravel capped by andesite.

Mines. Lode: Alaska $1 million+, American Flat, Beame, Blue Grouse, Bowman, General Grant. Placer: Grizzly Gulch, Mt. Alta, Orient, Pleasant View, Tippicanoe, True Grit.

Bibliography

Averill, C. V., 1942, Sierra County, Alaska mine: California Div. Mines Rept. 38, pp. 17–18.

Lindgren, Waldemar, 1900, Colfax folio, California: U. S. Geol. Survey Geol. Atlas of the U. S., folio 66, 10 pp.

Lindgren, Waldemar, 1911, Tertiary gravels of the Sierra Nevada: U. S. Geol. Survey Prof. Paper 73, p. 138.

MacBoyle, Errol, 1920, Sierra County, Pike mining district: California Min. Bur. Rept. 16, pp. 14–15.

Pilot Hill

Location. This district is in northwestern El Dorado County in the vicinity of the town of Pilot Hill. The area was first mined during the gold rush. It was worked again in the 1930s with draglines and power shovels.

Geology. The principal deposit is a 20- to 30-acre remnant of the main Tertiary channel of the American River that extended northwest from Placerville. The gravels rest on greenstone and green schist. To the south are several bodies of gabbro and diorite. During the 1930s the yield from these gravels was 13 to 60 cents per yard in gold. Also, there are a few narrow gold-quartz veins.

Bibliography

Lindgren, Waldemar, and Turner, H. W., 1894, Placerville folio: U. S. Geol. Survey Geol. Atlas of the U. S., folio 3, 3 pp.

Lindgren, Waldemar, 1911, Tertiary gravels of the Sierra Nevada: U. S. Geol. Survey Prof. Paper 73, p. 164.

Pine Grove

Location. Pine Grove is in east-central Amador County about 10 miles east of Jackson. It includes the Irishtown and Clinton areas. It is a small Sierran east gold belt district that was first worked during the gold rush and has been intermittently prospected ever since.

Geology. The district is underlain chiefly by graphitic slate, schist, and metachert. Granodiorite lies to the east and west. A number of narrow quartz veins containing small ore shoots are rich in places. The ore commonly contains abundant sulfides, especially galena and chalcopyrite. A few small patches of Tertiary auriferous channel gravels overlie the bedrock.

Mines. Black Wonder, Contini, Mikado, Peterson, Pine Grove Unit, Rainbow, Red Hill.

Bibliography

Carlson, D. W., and Clark, W. B., 1954, Amador County, Black Wonder, Contini, Peterson, Rainbow, and Red Hill mines: California Jour. Mines and Geology, vol. 50, pp. 172, 177–178, 189, and 192.

Turner, H. W., 1894, Jackson folio: U. S. Geol. Survey Geol. Atlas of the U. S., folio 11, 6 pp.

Photo 54. Contini Mine, Pine Grove District. This Amador County mine is a typical Sierra Nevada East Gold Belt pocket mine. It has been worked on a small scale since the 1940s. The view, in 1965, looks west. *Photo by Jeffrey Schweitzer.*

Photo 55. Red Hill Mine, Pine Grove District. This 1952 photo shows an Amador County pocket mine that was active in the 1940s. Much of the area around this mine has now been subdivided. *Photo by Jeffrey Schweitzer.*

Photo 56. Ground Sluicing, Placerville District. This photo of an operation in the district, in El Dorado County, was taken in about 1849. *From the collection of Mrs. T. J. Lobbard, San Francisco.*

Piute Mountains

Location. The Piute Mountains are in the southern Sierra Nevada, in east-central Kern County near Claraville, about 14 miles southeast of Bodfish. Gold was probably discovered here during the 1850s, but the principal periods of mining were 1870 to 1900 and the 1930s and early 1940s. Some tungsten has been produced in the district.

Geology. Most of the district is underlain by Mesozoic granitic rocks. In the northwest portion a roof pendant of Mesozoic metasedimentary rock crops out. Most of the gold deposits are confined to a two-mile-wide belt that extends northwest through the Claraville area in granitic rock and then north in the metamorphic rocks. The deposits consist of gold-quartz veins in shear zones. Some sulfides and also scheelite, in places, are present. Milling ore averaged about ½ ounce of gold per ton. Silver and antimony also are present.

Mines. Amy, Blue Jay, Bright Star $600,000, Dearborn, Donnie, French, Gold Standard, Gwynne $770,000, Henry Ford, Hilltop, Jeannette, Jeanette-Grant, Jerry, Little Joe, Lone Star, Mary Ellen, Retreat, Shellenberger, Simon, Surprise.

Bibliography

Brown, G. C., 1916, Kern County, Bright Star mine: California Min. Bur. Rept. 14, p. 490.

Troxel, B. W., and Morton, P. K., 1962, Kern County, Piute Mountains district: California Div. Mines and Geology, County Rept. 1, pp. 45–46.

Tucker, W. B., and Sampson, R. J., 1933, Kern County, Gwynne and Jeanette mines: California Div. Mines Rept. 29, pp. 307–309.

Placerville

Location. Placerville is in west-central El Dorado County. The district includes the lode mines of the Mother Lode belt, which extends north through the district, and the placer deposits here and in the adjacent Smith Flat, Diamond Springs, Texas Hill, Coon Hollow, and White Rock areas.

History. Gold was discovered in the Placerville area in July 1848. The town was first known as Dry Diggings but had the nickname of Old Hangtown; three robbers were hanged here on October 17, 1849. From the middle 1850s through the 1870s, the hydraulic and drift mines in the district were extremely rich. One 20-acre claim at Coon Hollow yielded $5 million, and the Spanish Hill area yielded $6 million. Quartz mining began in 1852 at the Pacific mine, but the chief period of lode mining was from the 1880s until about 1915. There was some mining in the district again in the 1930s, but there has been little activity since. Many of the mines in the district came under the control of the Placerville Gold Mining Company. The value of the total output for the district is unknown, but the placer mines are estimated to have yielded at least $25 million.

Geology. A belt of gray to black slate of the Mariposa Formation (Upper Jurassic) one to two miles wide extends north through the central portion of the district. Greenstone and amphibolite are to the west, and schist and slate of the Calaveras Formation (Carboniferous to Permian) and granodiorite lie to the east. The Tertiary South Fork of the American River, which has numerous tributaries, entered the Placerville basin from Newtown. In places the Tertiary gravels are overlain by thick beds of rhyolite tuff and andesite.

Ore Deposits. Of the numerous tributaries of the main Tertiary channel in this district, probably the best known and one of the richest was the Deep Blue Lead. This channel extended south from White Rock to Smith's Flat and then west-southwest through the Texas Hill area (fig. 21). The lode-gold deposits are massive quartz veins as much as 20 feet thick with numerous parallel stringers. The ore bodies are low to moderate in grade ($\frac{1}{7}$ to $\frac{1}{4}$ ounce of gold per ton), but the veins have been mined to depths of 2000 feet. The ore contains finely disseminated free gold and small amounts of pyrite. The veins occur chiefly in slate.

Mines. Lode: Elliott, Epley $100,000+, Griffith, Guildford $200,000+, Harmon $100,000+, Larkin $125,000, Margurite, Oregon $100,000+, Pacific $1,486,000, River Hill, Sherman $136,000, Superior, True Cons. $100,000, Van Hooker $100,000+.

Placer: Coon Hollow $10 million, Diamond Springs, Green Mountain, Negro Hill, Sacramento Hill, Spanish Hill $6 million, Smith's Flat $2 million+, Texas Hill, White Rock $5 million. Drift: Benfield, Cedar Spring, Clark, Kumfa, Landecker, Lyon, Pascoe, Rivera, Texas Hill, Try Again, Union.

Bibliography

Clark, W. B., and Carlson, D. W., 1956, El Dorado County, Placerville Gold Mining Company and Placerville area placer deposits: California Jour. Mines and Geology, vol. 52, pp. 422–423 and 432–434.

DeGroot, Henry, 1890, Smiths Flat mines: California Min. Bur. Rept. 10, pp. 179–180.

Irelan, William, Jr., 1888, Van Hooker, Pacific, and Epley, Cons. mines: California Min. Bur. Rept. 8, pp. 181–187.

Lindgren, Waldemar, and Turner, H. W., 1894, Placerville folio: U. S. Geol. Survey Geol. Atlas of the U. S., folio 3, 3 pp.

Lindgren, Waldemar, 1911, Tertiary gravels of the Sierra Nevada: U. S. Geol. Survey Prof. Paper 73, pp. 171–180.

Logan, C. A., 1935, Harmon group: California Div. Mines Bull. 108, pp. 26–27.

Rowlands, R., 1894, Map of the principal gravel mines in the vicinity of Placerville: California Min. Bur. Rept. 12, p. 100.

Tucker, W. B., 1919, El Dorado County, Pacific mine: California Min. Bur. Rept. 15, pp. 293–295.

Poker Flat

Location. This district is in northern Sierra County about 10 miles north of Downieville. It includes the Howland Flat, Table Rock, Deadwood, Mt. Fillmore, Potosi, and Rattlesnake Peak areas. It is mainly a placer-mining district.

History. The streams were first mined during the gold rush. The locality was extremely rich then; in one month gold valued at $700,000 was produced. Hydraulic mining was done on a major scale from

Photo 57. California Gold Mine. The locale is uncertain, but it probably was in the Placerville district. *Photo courtesy of Bancroft Library.*

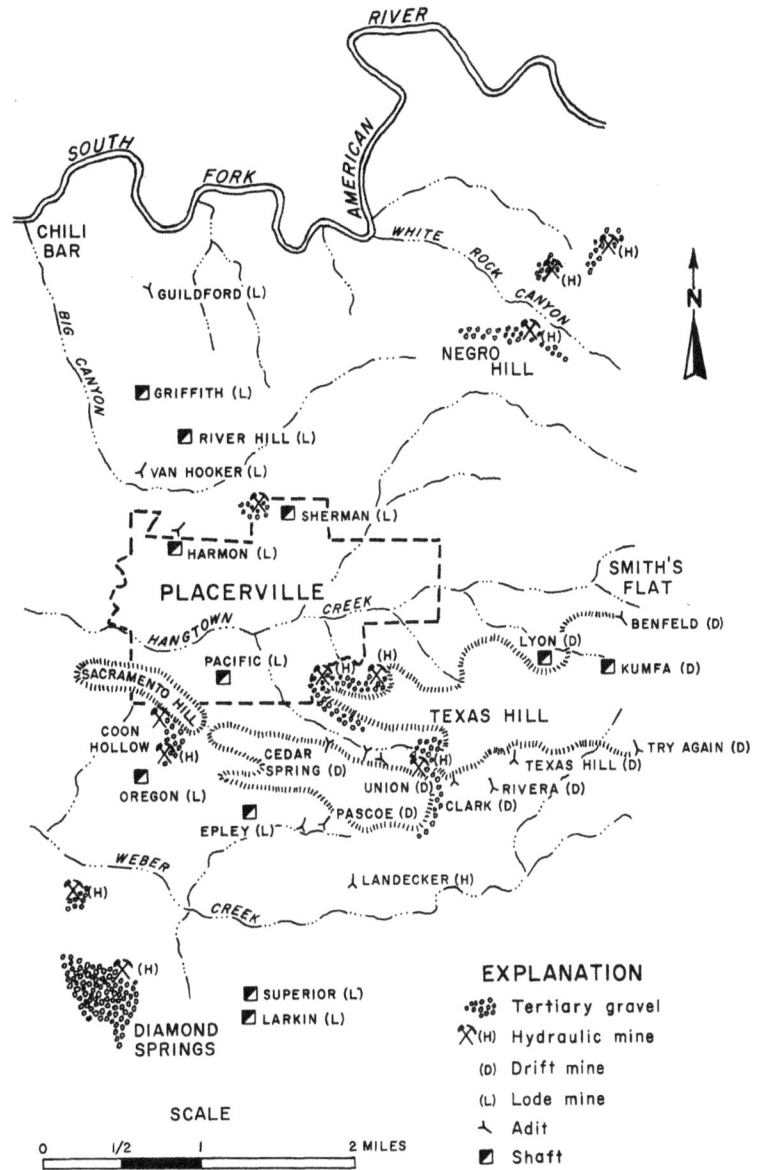

Figure 21. Map of Placerville District, El Dorado County. The map shows mine locations. *After Clark and Carlson, 1956.*

the late 1850s through the 1880s. Some lode mining and drift mining continued through the early 1900s, and the area was prospected in the 1920s and 1930s. The district was made famous by Bret Harte's tale, *The Outcasts of Poker Flat*. This district has been highly productive, the mines at Howland Flat alone being credited with an output valued at $14 million.

Geology. The northern part of the district is underlain by amphibolite with some serpentine. To the south and east there are slates of the Blue Canyon Formation (Carboniferous). Substantial portions of the area are capped by andesite. Extensive deposits of Tertiary auriferous quartz gravels are part of the Port Wine channel, which extends west and northwest through this district and then west and southwest into the Port Wine district. The lower quartz-rich gravels were also gold-rich. Portions of the channel have been

faulted. Some narrow gold-quartz veins occur in amphibolite and slate.

Mines. Placer: Caledonia, California, Clippership, Deadwood, Forest Queen, Gibraltar, Hawkeye, Herkimer and Bunker Hill, Manchester, Miners Home, Pacific, Poker Flat, Potosi, Rattlesnake, Scott, Virginia, Tennessee, Winkeye. Lode: Alhambra, Mammoth, Mt. Fillmore Cons., New York.

Bibliography

Lindgren, Waldemar, 1911, Tertiary gravels of the Sierra Nevada: U. S. Geol. Survey Prof. Paper 73, pp. 108–109.

Logan, C. A., 1924, Gravel mines of Howland Flat Ridge: California Min. Bur. Rept. 20, pp. 362–367.

Logan, C. A., 1929, Sierra County, placer mines: California Div. Mines Rept. 25, pp. 184–211.

MacBoyle, Errol, 1920, Sierra County, Poker Flat district: California Min. Bur. Rept. 16, pp. 15–19.

Turner, H. W., 1897, Downieville folio: U. S. Geol. Survey Geol. Atlas of the U. S., folio 37, 8 pp.

Photo 58. Early View of the Town of St. Louis. In this eastward view of the Sierra County town, the placer diggings are in the left and right background. Little Table Rock Peak rises at left. Photo courtesy of Calif. State Library.

Polk Springs

This is a small placer-mining district in eastern Tehama County just south of Polk Springs. It is about 30 miles east-southeast of Red Bluff. Some gold was recovered by hydraulicking years ago, the last work apparently having been done in the 1930s. The deposit consists of gravels up to 50 feet in thickness that contain pebbles of quartz, granitic rocks, and volcanic rocks. Bedrock is schist, which occurs with granitic rocks in a window completely surrounded by volcanic rocks.

Port Wine

Location. The Port Wine district is in northwestern Sierra County about 10 miles northwest of Downieville. The site was named by a party of prospectors who found a keg of port concealed in the bushes. The La Porte district adjoins on the northwest, the Poverty Hill district on the southwest, and the Poker Flat-Howland Flat district on the northeast. This district includes the "diggings" at Grass Flat, Queen City, and St. Louis. The area was extensively mined by hydraulicking and drifting during and after the gold rush and has been intermittently prospected since.

Geology. The Port Wine channel, a major branch of the La Porte channel, extends southwest through this area. It is roughly parallel to the La Porte channel for some miles and joins it at Scales. The channel is well-defined and several hundred feet wide. The gravels are quartz-rich and in places are covered by "pipe" clay and andesite. Bedrock is slate, quartzite, amphibolite, and greenstone.

Bibliography

Lindgren, Waldemar, 1911, Tertiary gravels of the Sierra Nevada: U. S. Geol. Survey Prof. Paper 73, pp. 108–110.

MacBoyle, Errol, 1920, Sierra County, Port Wine mining district: California Min. Bur. Rept. 16, pp. 19–23.

Turner, H. W., 1897, Downieville folio, California: U. S. Geol. Survey Geol. Atlas of the U. S., folio 17, 8 pp.

Poverty Hill

Location and History. This district is in western Sierra County, 10 miles northwest of Downieville and five miles south of La Porte. The Port Wine district is to the northeast, and the Brandy City district is to the south-southwest. Poverty Hill district includes the Scales and Mount Pleasant areas. The area was mined by hydraulicking and drifting during and after the gold rush. Mining here again during the 1930s and early 1940s included an attempt to work the Poverty Hill pit with a bucket-line dredge.

Geology. The main channel of the Tertiary North Fork of the Yuba River, or La Porte channel, enters this district from the north and continues south and southwest to the Brandy City and Indian Hill districts. The Port Wine channel, a branch of the La Porte channel, enters the area from the northeast at Scales. This smaller branch channel parallels the main channel for some miles. At the Poverty Hill pit, the channel is up to 1500 feet wide and 150 feet deep. The lower "blue" gravels are quartz-rich and cemented in places. There are a number of large boulders. At Scales the

channel is similar. Bedrock consists of slate, quartzite, amphibolite and serpentine. Lindgren estimated in 1911 that, at Poverty Hill, 2.25 million yards had been removed and 5 million yards were ultimately available, while at Scales and Mt. Pleasant, 4.05 million yards had been excavated and 60 million yards were ultimately available.

Bibliography

Averill, C. V., 1942, Sierra County, Poverty Hill properties: California Div. Mines Rept. 38, pp. 35–37.

Lindgren, Waldemar, 1911, Tertiary gravels of the Sierra Nevada: U. S. Geol. Survey Prof. Paper 73, pp. 104–108.

MacBoyle, Errol, 1920, Sierra County, Poverty Hill mining district: California Min. Bur. Rept. 16, pp. 23–26.

Turner, H. W., 1897, Downieville folio, California: U. S. Geol. Survey Geol. Atlas of the U. S., folio 37, 8 pp.

Quincy

Location. This district is in central Plumas County in the general vicinity of Quincy, the county seat. It includes the Elizabethtown and Butterfly Valley areas. The Meadow Valley district lies just to the west and the Sawpit Flat district to the south. The district was first mined during the gold rush, and there has been intermittent prospecting and development work ever since.

Geology. The principal rocks that underlie the district are slate, mica schist, and quartzite. Greenstone lies to the northeast and serpentine to the southwest. American and Thompson Valleys are underlain by Recent and Pleistocene alluvium. A few isolated peaks in the area are capped by basalt.

Ore Deposits. Gold-quartz veins occur principally in slate and mica schist; some are as thick as 15 feet. The veins may be massive or consist of numerous parallel stringers. The ore contains free gold and varying amounts of sulfides, chiefly pyrite. Although some of the veins have been developed for horizontal distances of several thousand feet, none has been worked to depths of greater than a few hundred feet. There are a few small Tertiary channel gravel deposits to the south. The Recent and Pleistocene valley alluvium is gold-bearing in places.

Mines. Placer: Bushman, Carr, Cascade, Elizabethtown Flat, Imperial, Manhattan, Mill Creek, Newton Cons., Newton Flat, Riverdale. Lode: Bell, Butterfly, Fairplay, Gold Leaf Cons., Homestake, King Solomon, St. Nicolas, Tefft, White Oak.

Bibliography

Averill, C. V., 1937, Plumas County, gold: California Div. Mines Rept. 33, pp. 103–124.

Lindgren, Waldemar, 1911, Tertiary gravels of the Sierra Nevada: U. S. Geol. Survey Prof. Paper 73, pp. 111–113.

MacBoyle, Errol, 1920, Plumas County, Quincy mining district: California Min. Bur. Rept. 16, pp. 36–41.

Turner, H. W., 1897, Downieville folio, California: U. S. Geol. Survey Geol. Atlas of the U. S., folio 37, 8 pp.

Railroad Flat

Location. This district is in central Calaveras County in the vicinity of the town of Railroad Flat, seven miles south of West Point and 13 miles east of Mokelumne Hill. The district was named by an early-day placer miner who had laid a few hundred feet of wooden railroad track on his claim.

Geology. The area is underlain chiefly by graphitic schist, slate, quartzite, and metamorphosed chert (see fig. 25, p. 129). A number of narrow north-striking quartz veins contain free gold and often abundant sulfides, especially galena. The ore shoots are small and usually do not extend to depths of greater than 200 feet, but they often are rich.

Several deposits of quartz-rich gravels have been mined by drifting or hydraulicking. These gravels were deposited by the south-trending Tertiary Fort Mountain channel. In places the gravels are overlain by rhyolite.

Mines. A.V.G., Bald Eagle, Banner Blue (placer), Clary, Fine Gold $200,000, Jeff Davis, Kaiser Wilhelm, Lampson (placer), Mohawk, Old Gray, Petticoat, Poe, Prussian Hill, Sanderson $100,000+, Summit, Swiss.

Bibliography

Clark, W. B., and Lydon, P. A., 1962, Mines and mineral resources of Calaveras County: California Div. Mines and Geology, gold, pp. 32–93.

Lindgren, Waldemar, 1911, Tertiary gravels of the Sierra Nevada: U. S. Geol. Survey Prof. Paper 73, pp. 210–212.

Storms, W. H., 1894, Ancient channel system of Calaveras County: California Min. Bur. Rept. 12, pp. 482–492.

Turner, H. W., 1894, Jackson folio: U. S. Geol. Survey Geol. Atlas of the U. S., folio 11, 6 pp.

Turner, H. W., and Ransome, F. L., 1898, Big Trees folio: U. S. Geol. Atlas of the U. S., folio 51, 8 pp.

Ralston Divide

Location. This is a placer-gold district in southeastern Placer County, 25 miles east of Forest Hill. It is an extensive region that includes the Ralston Ridge, Long Canyon, and Nevada Point Ridge areas. It is just south of the Duncan Peak district.

Geology. The deposits are along a west- and southwest-trending Tertiary gravel channel known as the Long Canyon channel, the eastward continuation of the main Michigan Bluff-Forest Hill channel. The gravels are interbedded with rhyolite tuffs and contain granitic boulders. The main Long Canyon channel is fairly broad and flat and covers large areas, but it is generally of low grade. The gold usually is fine. Bedrock is slate and quartz-bearing schist of the Blue Canyon Formation (Carboniferous), with granodiorite to the east. The gravels are capped by andesite and rhyolite. Mining was done by hydraulicking and drifting.

Mines. Blacksmith Flat, Clydesdale, Goggins, Granite, Ralston, Russian Ravine, Zuver.

Bibliography

Lindgren, Waldemar, 1911. Tertiary gravels of the Sierra Nevada: U. S. Geol. Survey Prof. Paper 73, pp. 152–153.

Lindgren, Waldemar, 1900, Colfax folio: U. S. Geol. Survey Geol. Atlas of the U. S., folio 66, 10 pp.

Lindgren, Waldemar, and Turner, H. W., 1894, Placerville folio: U. S. Geol. Survey Geol. Atlas of the U. S., folio 3, 3 pp.

Logan, C. A., 1936, Gold mines of Placer County, placer mines: California Div. Mines Rept. 32, pp. 49–96.

Rattlesnake Bar

Location and History. Rattlesnake Bar is in northwestern El Dorado County and southern Placer County. The placer mines here along the American River were highly productive during the gold rush. The town was established in 1849 and became good-sized until 1864, when it was destroyed by fire. The Zantgraff mine, the principal lode mine in the district with a reported production of $1 million, was active from 1880 to 1901 and again in the 1930s. Dragline dredging was done in the region during the 1930s. Part of the district is covered by the Folsom Reservoir.

Geology. The district is on the eastern flank of a major granodiorite stock that is intrusive into greenstones and amphibolite. A major body of serpentine and a limestone lens crop out in the area. Several extensive deposits of Pleistocene shore gravels along the American River were hydraulicked. The Zantgraff vein contains abundant sulfides, including galena and chalcopyrite, and was mined to a depth of 1100 feet. This district also has yielded substantial amounts of chromite and limestone and some copper.

Bibliography

Clark, W. B., and Carlson, D. W., 1956, El Dorado County, Zantgraff mine: California Jour. Mines and Geology, vol. 52, p. 429.

Lindgren, Waldemar, 1894, Sacramento folio: U. S. Geol. Survey Geol. Atlas of the U. S., folio 5, 3 p.

Rich Bar

Rich Bar is in western Plumas County near the junction of the North Fork and the East Branch of the North Fork of the Feather River. During the gold rush this was an extremely rich placer-mining district; Rich Bar alone is credited with an output of $9 million. Later the river was mined by Chinese. The river here goes around a series of sharp bends, a course that has resulted in the formation of wide gravel bars. There are several narrow gold-quartz veins in the area.

Rich Gulch

Location. Rich Gulch is in north-central Calaveras County adjacent to the Mokelumne Hill district on the east and west of the West Point and Railroad Flat districts. It includes the Jesus Maria area.

Geology and Ore Deposits. The area is underlain by graphite schist, quartzite, slate and numerous limestone lenses of the Calaveras Formation (Carboniferous to Permian) that have been intruded by several small granodiorite stocks. The quartz veins usually are narrow and contain varying amounts of gold and sulfides. Much high-grade ore was recovered from shallow workings during the early days, but at depth the ore usually contains only a few dollars per ton in gold.

Mines. Blue Jay, Ilex, Quartz Glen $300,000+, Rindge No. 1, No. 2, No. 3, Salvador.

Bibliography

Clark, W. B., and Lydon, P. A., 1962, Calaveras County, gold: California Div. Mines and Geology County Rept. 2, pp. 32–93.

Irelan, William, Jr., 1888, Ilex Gold Mining Company: California Min. Bur. Rept. 8, pp. 135–138.

Photo 59. Lee Drift Mine, Rocklin District. This 1956 view, looking east, shows the headframe and washing plant at the mine, in Placer County.

Logan, C. A., and Franke,.H., 1936, Calaveras County, Rich Gulch district: California Div. Mines Rept. 32, p. 238.

Turner, H. W., 1894, Jackson folio: U. S. Geol. Survey Geol. Atlas of the U. S., folio 11, 6 pp.

Rocklin

Location. This is a placer-mining district in southwestern Placer County, two miles east of Rocklin and two miles south of Loomis. A gravel channel of the Tertiary American River trends southwestward through the area. There are actually two channels: an upper loosely consolidated intervolcanic channel that contains some gold and a lower well-cemented quartz-rich channel that in places was rich. The lower gravels yielded $1 or more per yard in gold. The gold is fine, flat, and sometimes rusty. The lower channel is as much as 1500 feet wide. Bedrock is granodiorite, and in places the gravels are capped by andesite. The Lee drift mine, one of the principal sources of gold in the district, has been prospected in recent years.

Bibliography

Lindgren, Waldemar, 1894, Sacramento folio: U. S. Geol. Survey Geol. Atlas of the U. S., folio 5, 3 pp.

Lindgren, Waldemar, 1911, Tertiary channels of the Sierra Nevada: U. S. Geol. Survey Prof. Paper 73, pp. 163–164.

Waring, C. A., 1919, Placer County, The Rocklin district: California Min. Bur. Rept. 15, p. 319.

Rough-and-Ready

Location and History. Rough-and-Ready is in western Nevada County about five miles west of Grass Valley. Placer mining began here during the gold rush. The town was founded in 1849 by the Rough and Ready military company led by Captain A. A. Townsend. He had once served under General Zachary Taylor, who was known as "Old Rough and Ready". Drift and hydraulic mining began in the late 1850s, and lode mining became important in the 1860s. There was considerable activity that lasted until about 1900. Some work was done in the district in the 1930s, and there has been minor prospecting since.

Geology and Ore Deposits. A north-trending belt of amphibolite one to two miles wide traverses the central portion of the district. Gabbro and diorite lie to the east and granodiorite to the north. Some orbicular gabbro is present. A west-trending Tertiary gravel channel of the Yuba River crosses the north portion of the district. A number of north-striking quartz veins occur chiefly in amphibolite. The veins are one to five feet thick and contain free gold with pyrite and other sulfides. Considerable high-grade ore was taken out. None of the veins has been developed to depths of more than 500 feet.

Mines. Lode: Alcade (Kenosha) $500,000, Black Bear, California, Ironclad, Mystery, Niagara, Normandie-Dulmaine $100,000, Osceola, Seven-Thirty, Vulcan-Grey Eagle. Placer: Alta-California, Jenny Lind.

Bibliography

Lindgren, Waldemar, 1895, Smartsville folio, California: U. S. Geol. Survey Geol. Atlas of the U. S., folio 18, 6 pp.

Lindgren, Waldemar, 1911, Tertiary gravels of the Sierra Nevada: U. S. Geol. Survey Prof. Paper 73, pp. 120–124.

MacBoyle, Errol, 1919, Nevada County, Rough-and-Ready mining district: California Min. Bur. Rept. 16, pp. 54–57.

Sampson Flat

Location. Sampson Flat is in eastern Fresno County about eight miles north of Dunlap and 45 miles east of Fresno. The district includes the area known as Davis Flat. Sometimes it has been included in the Mill Creek District. It was placer-mined in the early days, and the lode mines were active from the 1880s until about 1915.

Geology. The district is underlain by granodiorite, gabbro and small amounts of pyroxenite. Several north-trending quartz veins two to three feet thick contain free gold and varying amounts of sulfides. There is some tungsten mineralization in bodies of tactite.

Mines. Davis Flat, Delilah (Black Jack & Hercules), Little Monitor, Oro Fino, Sampson.

Bibliography

Bradley, Walter W., 1916, Fresno County, Delilah Mining Co.: California Min. Bur. Rept. 14, p. 443.

San Andreas

Location. This district is in west-central Calaveras County. It consists of that part of the Mother Lode gold belt that extends from the North Branch-Cottage Springs area southeast through San Andreas and Kentucky House to the vicinity of Fourth Crossing. It is both a lode- and placer-mining district.

History. The streams were mined during the early part of the gold rush. San Andreas Gulch was first settled by Mexicans in 1848 or 1849. The Tertiary channel deposits were mined by hydraulicking and drifting from the 1850s through the 1880s, and some drift mining has continued until the present time. Lode mining also began in the 1850s. Numerous lode mines were active from the 1870s until about 1900 and again during the 1930s. The Union gold mine was prospected for uranium in the middle 1950s.

Geology. The area is underlain by northwest-trending belts of greenstone, amphibolite schist, slate, and lenticular serpentine bodies. At Kentucky House, a large body of dolomitic limestone contains lenses of high-calcium limestone mined by the Calaveras Cement Company. To the east of San Andreas is a granodiorite stock. Overlying portions of the district are numerous deposits of auriferous Tertiary channel gravels.

Ore Deposits. The lode deposits consist of northwest-trending quartz veins with stringers that contain finely divided free gold and pyrite. The veins occur principally in amphibolite and greenstone. Most of the milling ore averages ⅓ ounce of gold per ton or less, but several of the ore shoots have been mined to depths of 700 feet. Considerable high-grade ore was taken from shallow workings during the early days. Small amounts of tellurides and uranium minerals have been found in this district.

Mines. Lode: Commodore, Etna, Everlasting, Fellowcraft, Ford, Gottschalk, Golden Hill, Helen, Holland, Kate Hageman, Lookout Mtn., Mester, Pioneer Chief, Rathgeb, Thorpe, Union $200,000+. Placer: Benson, Central Hill, Lloyd, North Branch, Rising Star, Wheats.

Bibliography

Clark, W. B., and Lydon, P. A., 1962, Calaveras County, gold: California Div. Mines and Geology County Rept. 2, pp. 32–93.

Knopf, Adolph, 1929, The Mother Lode system of California: U. S. Geol. Survey Prof. Paper 157, pp. 70–71.

Lindgren, Waldemar, 1911, Tertiary gravels of the Sierra Nevada: U. S. Geol. Survey Prof. Paper 73, pp. 209–210.

Logan, C. A., and Franke, H., 1936, Calaveras County, gold: California Div. Mines Rept. 32, pp. 235–364.

Ransome, F. L., 1900, Mother Lode district folio: U. S. Geol. Survey Geol. Atlas of the U. S., folio 63, 11 pp.

Storms, W. H., 1894, Ancient channel system of Calaveras County: California Min. Bur. Rept. 12, pp. 482–492.

Turner, H. W., 1894, Jackson folio: U. S. Geol. Survey Geol. Atlas of the U. S., folio 11, 6 pp.

Sawpit Flat

Location. This is an extensive gold-bearing region in southern Plumas County. It is contiguous with the Quincy district to the north and the Gibsonville district in Sierra County to the south. It includes the Last Chance, Sawmill Flat, Monitor Flat, Onion Valley, Harrison Flat, Blue Nose Mountain, and Nelson Point areas. The district was named in 1850 for a pit that was dug for the use of a whipsaw. The region was extensively mined during the early days and has been intermittently prospected since. In recent years there has been some placer mining at Monitor Flat. Skin divers have prospected in the Middle Fork of the Feather River.

Geology. The east portion of the district is underlain by slate, schist, and quartzite of the Calaveras Formation (Carboniferous to Permian). Also there are several limestone lenses. To the west the district is underlain by serpentine and amphibolite. The ridges are capped by Tertiary andesite and basalt.

Ore Deposits. The Tertiary gravels are largely quartz-rich, and in places, rich in gold. At Sawpit Flat the gravels are part of the Richmond Hill-Onion Hill channel, while those to the east at Bunker Hill and Blue Nose Mountain are in the northeast end of the famous La Porte channel. A number of gold-quartz veins are in the district, some of which are part of a vein system that extends along the contact zone between serpentine and schist and slate.

Mines. Placer: Boulder West, Bunker Hill, Fordham, Golden Gate, H & G, Kelly, King Solomon, Mayflower, Morning Star Cons., Nelson Creek, Red Slide, Richmond Hill, Rio Vista, Smith, Turkeytown, Union Hill, Zumwalt. Lode: Bainbridge, Belfrin, Dean, Five Bear, Gold Point, Gold Run, Independence, Oro Fino, Oversight, Pilot, Pilot Peak, Plumas Bonanza, Rose Quartz, Sugar Pine, Wilson-Gomez.

Bibliography

Averill, C. V., 1937, Plumas County, gold: California Div. Mines Rept. 33, pp. 103–124.

Lindgren, Waldemar, 1911, Tertiary gravels of the Sierra Nevada: U. S. Geol. Survey Prof. Paper 73, p. 110.

MacBoyle, Errol, 1920, Plumas County, Sawpit Flat mining district: California Min. Bur. Rept. 16, pp. 42–46.

Turner, H. W., 1897, Downieville folio, California: U. S. Geol. Survey Geol. Atlas of the U. S., folio 37, 8 pp.

Scotts Flat

Location and History. This district is in west-central Nevada County about seven miles due east of Nevada City. It includes the Tertiary placer "diggings" at Scotts Flat, Quaker Hill, Hunts Hill, Buckeye Hill, and Burrington Hill. The You Bet-Red Dog district lies immediately to the south and the Nevada City district to the west. The various mines were extensively hydraulicked from the 1850s through the

1880s, and later the tailings were reworked by Chinese miners. Also there was drift mining in the district. The area was prospected during the 1930s.

Geology. These deposits are in the north-northwest trending Tertiary gravel channel that extends from You Bet-Red Dog to North Columbia. A southwest-trending tributary comes into the area from Burrington Hill and joins this channel at Hunts Hill. At Hunts Hill and Quaker Hill the main channel is nearly 600 feet deep with bench gravels up to 300 feet in depth. The deep gravels are well-cemented and quartz-rich and, in places, were very rich in gold. The upper gravels usually are fine and contain abundant sand. The deep channel is believed to be continuous all the way from Hunts Hill to the Blue Tent district, a distance of seven miles. Bedrock in the east portion is slate and in the west, phyllite and greenstone. On the major ridges the gravels are capped by Tertiary andesite and rhyolite.

In 1911 Lindgren estimated that 12 million cubic yards of gravel had been removed from Scotts Flat and 35 million from Quaker Hill. He also estimated that a vast amount (140 million cubic yards) remained at Quaker Hill. The U. S. Army Engineers (Jarman, 1927) estimated 50 million to 90 million cubic yards remained at Quaker Hill. They also estimated that 6.75 million cubic yards had been removed, and 4 million to 5 million remained at Hunts Hill.

Bibliography

Jarman, Arthur, 1927, Hunts Hill, Quaker Hill, and Buckeye Hill: California Min. Bur. Rept. 23, pp. 100–101.

Lindgren, Waldemar, 1911, Tertiary gravels of the Sierra Nevada: U. S. Geol. Survey Prof. Paper 73, pp. 143–144.

Lindgren, Waldemar, 1900, Colfax folio: U. S. Geol. Survey Geol. Atlas of the U. S., folio 66, 10 pp.

Sheep Ranch

Location. The Sheep Ranch district is in south-central Calaveras County about 16 miles east of San Andreas and eight miles north of Murphys. It includes the old Washington district just to the south and the El Dorado area to the west.

History. The streams in the area were first mined during the gold rush. Drift mining of the Tertiary channel gravels began in the late 1850s and continued intermittently through the early 1900s. The Sheep Ranch mine, the largest source of gold in the district and the largest mine of the Sierra Nevada east gold belt, was first worked in 1868. Senator George Hearst had an interest in this mine from 1877 to 1895. It was operated on a major scale until around 1907 and again from 1936 to 1942. The Right Bower mine has been intermittently worked since 1946.

Geology. The principal rocks underlying the area are slate, impure quartzite and graphite schist of the Calaveras Formation (Carboniferous to Permian), which have been intruded by gabbroic stocks. The gravel deposits are quartz-rich and occur as patches. They are part of the Tertiary Fort Mountain channel, which extends south from the Railroad Flat district. In places the gravels are overlain by rhyolite.

Photo 60. Sheep Ranch Mine, Sheep Ranch District. This westward view of the Calaveras County mine was taken in about 1905. With a total output of more than $7 million, the Sheep Ranch mine was the most productive operation in the Sierra Nevada East Gold Belt. *Photo courtesy of Hillcrest Studio, Angels Camp.*

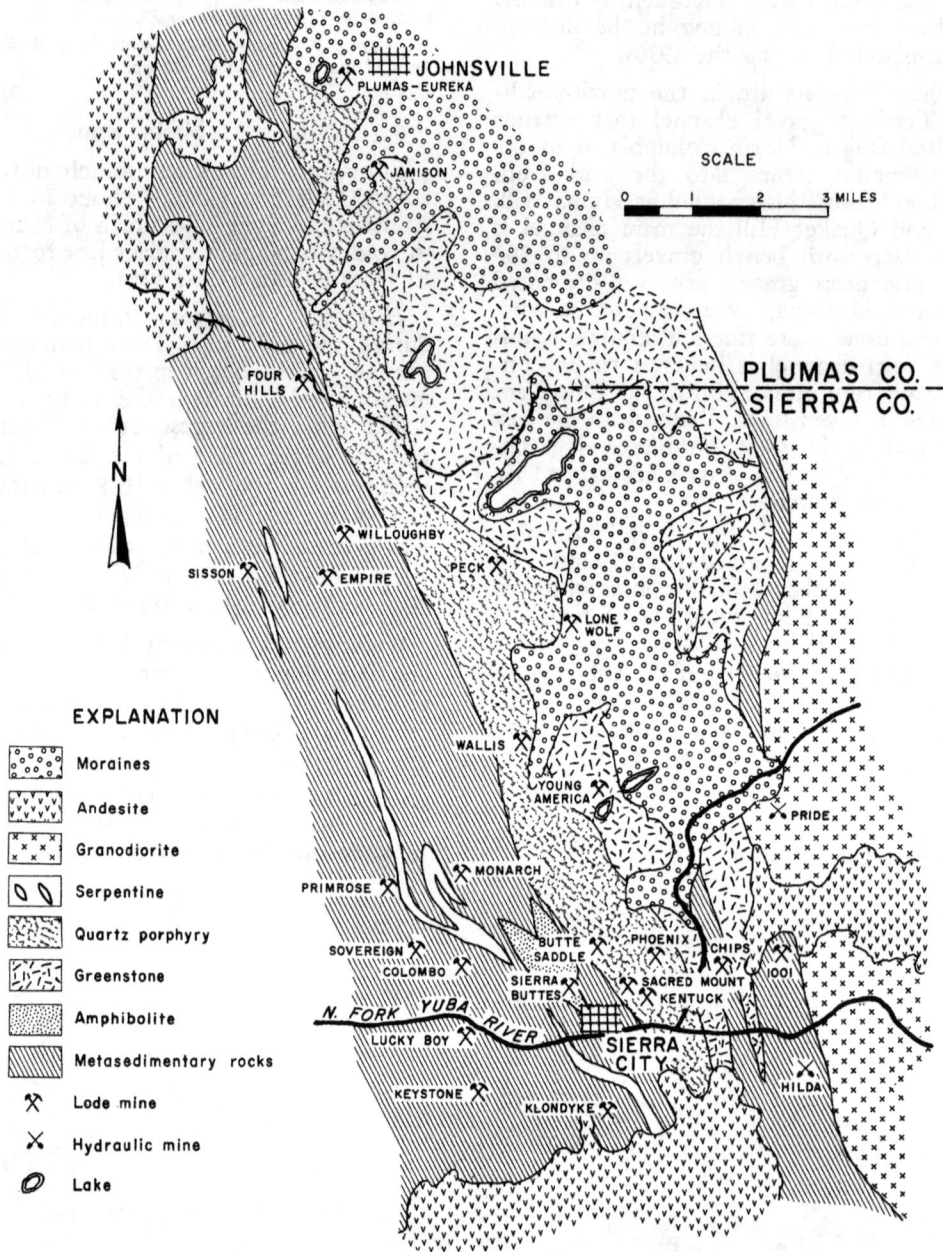

Figure 22. Geologic Map of Sierra City and Johnsville Districts, Sierra and Plumas Counties.
After Turner, Downieville folio, 1897.

Ore Deposits. The quartz veins usually are a few feet thick and range from white to dark smoky gray in color. Smoky vein quartz is characteristic of some of the mines in this district. The ore contains free gold, which often is coarse, and varying amounts of sulfides. Appreciable quantities of high-grade ore have been recovered. The Sheep Ranch vein has been mined to a depth of about 3000 feet. The gravels usually are fairly well-cemented and were rich in places.

Mines. Lode: Bon Ton, Fenian, Mar John $360,-000, Right Bower, Sheep Ranch $7 million, Sonoma, Washington $600,000. Placer: Brassila, Lava Bed.

Bibliography

Clark, W. B., and Lydon, P. A., 1962, Calaveras County, gold: California Div. Mines and Geology County Rept. 2, pp. 32–93.

Irelan, William, Jr., 1888, Sheep Ranch mine: California Min. Bur. Rept. 8, pp. 131–133.

Lindgren, Waldemar, 1911, Tertiary gravels of the Sierra Nevada: U. S. Geol. Survey Prof. Paper 73, pp. 210–212.

Logan, C. A., and Franke, H., 1936, Calaveras County, Sheep Ranch mine: California Div. Mines Rept. 32, p. 288.

Storms, W. H., 1900, Sheep Ranch mine: California Min. Bur. Bull. 18, pp. 104–105.

Tucker, W. B., 1916, Calaveras County, Sheep Ranch mine: California Min. Bur. Rept. 14, pp. 104–105.

Turner, H. W., and Ransome, F. L., 1898, Big Trees folio: U. S. Geol. Survey Geol. Atlas of the U. S., folio 51, 8 pp.

Shingle Springs

Location and History. This district is in the Sierra Nevada west gold belt in western El Dorado County. A belt of lode-gold mines extends from the Pyramid mine south through Shingle Springs to the vicinity of Brandon Corner, a distance of 10 miles. The district was first worked during the gold rush. The town was settled in 1850; the name derived from "a shingle machine used at a cluster of springs". There was extensive mining activity here during the 1930s.

Geology and Ore Deposits. A north-trending belt of greenstone, green schist, and slate four to six miles wide, which has been intruded by numerous serpentine bodies, both large and small, extends through the central part of the district. A granodiorite-gabbro intrusion lies to the west. The ore deposits consist chiefly of large but low-grade bodies of mineralized talcose, amphibolite-chlorite schist or greenstone with numerous quartz veinlets and stringers. The values occur in disseminated auriferous pyrite found in both the wall rock and the quartz veins and stringers. The ore bodies commonly are found in both the wall rock and the quartz. Some quartz veins with high-grade pockets and abundant sulfides exist. Some of the deposits were mined by open-pit methods.

Mines. Big Canyon $3 million+, Brandon, Bugtown, Crystal $100,000+, French Creek, Greenstone, Marcelias, Pyramid $1 million, Sugar Loaf, Vandalia $100,000+.

Bibliography

Clark, W. B., and Carlson, D. W., 1956, El Dorado County, lode-gold deposits: California Jour. Mines and Geol., vol. 52, pp. 401–429.

Irelan, William, Jr., 1888, Vandalia and Big Canyon mines: California Min. Bur. Rept. 8, pp. 172–175.

Lindgren, Waldemar, and Turner, H. W., 1894, Placerville folio: U. S. Geol. Survey Geol. Atlas of the U. S., folio 3, 3 pp.

Logan, C. A., 1938, El Dorado County, Big Canyon, Pyramid, and Vandalia mines: California Div. Mines Rept. 34, pp. 219–223, 244–246, and 254.

Sierra City

Location. The Sierra City district covers an extensive area in central Sierra County. It includes the Furnier, Loganville, Church Meadows, Gold Valley, and Sierra City-Sierra Buttes areas. It is located at the south end of the major belt of gold mineralization that extends north-northwest to the Johnsville district in Plumas County (fig. 22).

History. This district was placer-mined soon after the beginning of the gold rush. Many coarse nuggets were recovered at that time. Sierra City was founded in 1850; destroyed by an avalanche in 1852, it was soon rebuilt. Many Indians lived in this district in those days. The famous roistering society E. Clampus Vitus originated in Sierra City. It was reorganized several years ago by the California Historical Society.

The Sierra Buttes mine was opened in 1850, and most of the other important lode mines soon afterward. A number of very rich high-grade surface pockets were discovered, including one at the Four Hills mine that yielded between $250,000 and $500,000. The district was highly productive from about 1870

until 1914. There was some mining activity again during the 1920s and 1930s, and intermittent prospecting and development work has continued until the present time. The value of the total output of the district is unknown, but it is estimated to be at least $30 million.

Geology. A northwest-trending belt of Calaveras (Carboniferous to Permian) slate, schist, and quartzite with limestone lenses runs through the west portion; a quartz porphyry (metarhyolite) belt lies in the central portion, and greenstone and amphibolite schist, to the east. A belt of meta-tuff of the Milton Formation (Jurassic) extends along the east margin. A few serpentine lenses are present. The northeastern area is overlain by glacial moraines. Tertiary andesite caps some of the ridges and rhyolite caps some Eocene auriferous channel gravels.

Ore Deposits. An extensive series of north- to northwest- and occasionally northeast-trending quartz veins range from a few feet to as much as 40 feet in thickness. The veins occur in all of the metamorphic rocks. The ore bodies contain free gold, pyrite, and minor galena and chalcopyrite. The ore shoots often were large (up to 300 feet long and several thousand feet deep) and usually averaged ⅓ to ½ ounce of gold per ton. Much high-grade ore was taken from the district. Several fairly extensive magnetite deposits are found in the north end of the district.

Mines. Lode: Buffalo, Butcher Ranch, Butte Saddle $100,000+, Chipps, Cleveland, Colombo $400,000?, Empire, Four Hills $2 million, Great Northern, Kentuck $100,000+, Keystone, Klondyke, Loeffler, Lucky Boy, Monarch $100,000+, Peck, Primrose, Phoenix $160,000+, 1001, Roman, Sacred Mount $100,000+, Sebastopol, Sierra Buttes $17 million to $20 million, Sisson, Sovereign, Wallis, William Tell, Willoughby, Young America $1.5 million. Placer: Hilda, Ladies Canyon $500,000, Pride.

Bibliography

Averill, C. V., 1942, Mines and mineral resources of Sierra County: California Div. Mines Rept. 38, pp. 7–67.

Goldstone, L. P., 1890, Sierra County: California Min. Bur. Rept. 10, pp. 642–654.

Irelan, William, 1888, Sierra Buttes mine: California Min. Bur. Rept. 8, pp. 573–577.

Lindgren, Waldemar, 1911, Tertiary gravels of the Sierra Nevada: U. S. Geol. Survey Prof. Paper 73, pp. 112–113.

Logan, C. A., 1929, Sierra County, Sierra City district: California Div. Mines Rept. 25, pp. 155–156.

MacBoyle, Errol, 1919, Sierra County, Sierra City district: California Min. Bur. Rept. 16, pp. 26–28, 119–121.

Turner, H. W., 1897, Downieville folio: U. S. Geol. Survey, folio no. 37, 10 pp., 4 maps, 1 pl.

Sierra Nevada Copper Belts

Extending along the foothills of the west slope of the Sierra Nevada from Butte County on the north to Fresno County on the south is a discontinuous belt of copper and zinc mineralization. This belt also has been the source of substantial amounts of gold. Gold-bearing gossans in the oxidized zones overlying the copper-zinc deposits were mined during the gold rush. Later, during the copper "booms" of the Civil War

Photo 61. Young America Mine, Sierra City District. This 1885 view of the mine, in Sierra County, looks southwest. The Sierra Buttes are in the background and Lower Sardine Lake is to the left. Photo courtesy of Calif. State Library.

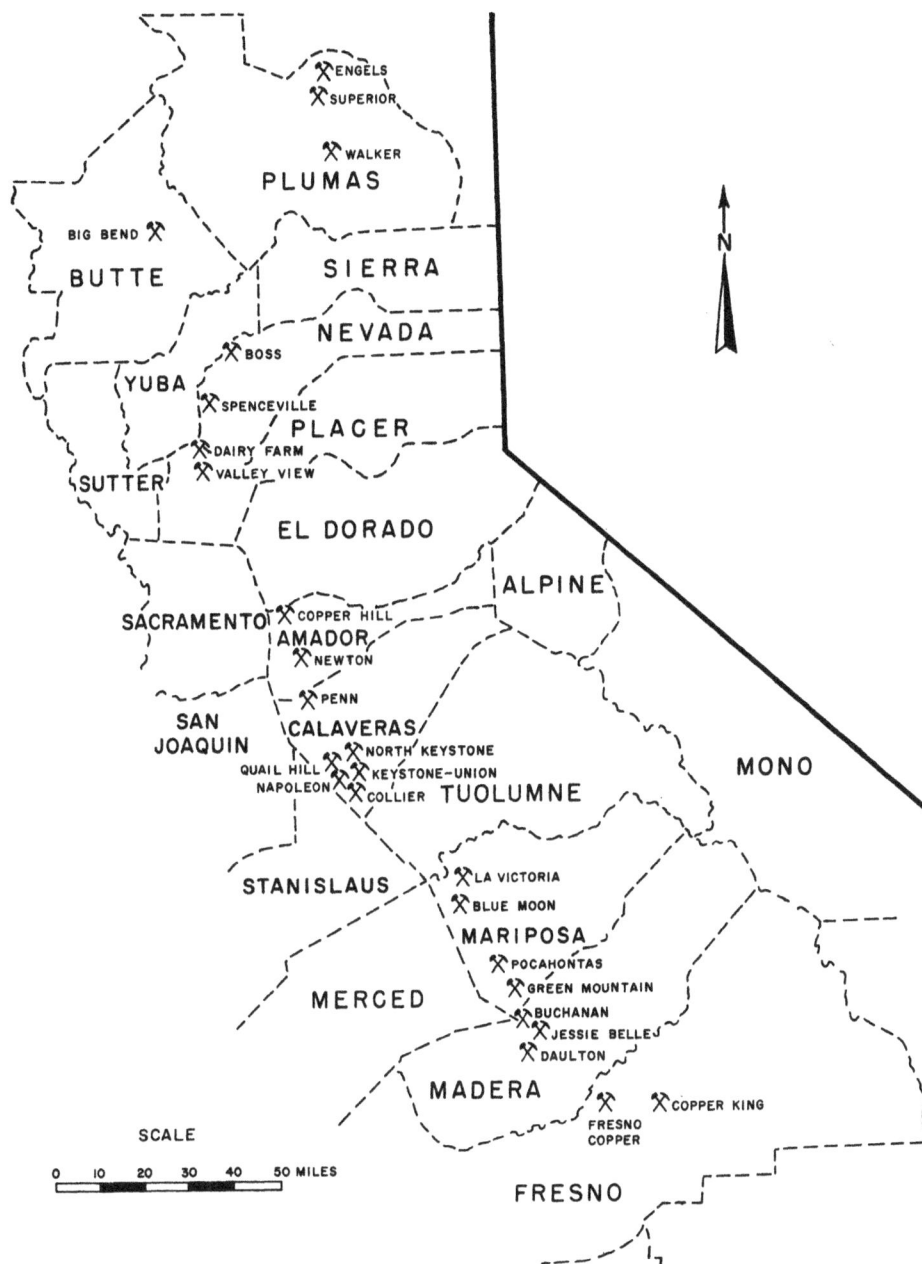

Figure 23. Map of Copper and Zinc Belts, Sierra Nevada. The foothill copper-zinc belt and the Plumas County copper belt are shown. The principal mines are marked.

and World Wars I and II, considerable amounts of gold were recovered as a by-product. During the 1930s a few gossan deposits in this belt were again mined for gold.

The primary copper and zinc deposits consist of lenticular sulfide bodies in zones of alteration in greenstones and various types of schists. The ore bodies contain abundant pyrite with associated chalcopyrite, sphalerite and some gold and silver. Most of the ore contains only a small fraction of an ounce of gold per ton, but a few ore bodies have yielded as much as one ounce of gold per ton. Also present are galena, bornite, tetrahedrite, covellite, and chalcocite.

The most important mines in the foothill belt have been the Big Bend mine, Butte County; Spenceville and Boss mines, Nevada County; Dairy Farm and Valley View mines, Placer County; Copper Hill and Newton mines, Amador County; Penn, Quail Hill, Napoleon, Collier, Keystone-Union, and North Keystone mines, Calaveras County; Blue Moon, Pocahontas, Green Mountain and La Victoria mines, Mariposa County; Buchanan, Jessie Belle, and Daulton mines, Madera County; and Fresno Copper and Copper King mines, Fresno County.

Considerable by-product gold has been recovered from copper mines in northeastern Plumas County, the principal sources having been the Walker, Engels, and

Superior mines. However, few production figures are available, so the total gold output of these mines is unknown. In 1931, the Walker mine was the source of 432,000 tons of copper ore that had an average gold content of .05 ounces per ton. At the Walker mine, the ore bodies consist of wide chalcopyrite-bearing quartz veins in schist and hornfels near granitic rocks. At the Engels and Superior mines, the ore bodies are bands of chalcopyrite and bornite in sheared granitic rocks.

Silver King

Silver King is in southeastern Alpine County near the headwaters of Silver King Valley about 17 miles southeast of Markleeville. The area was prospected during the 1860s with apparently some production, but little or no mining has been done since. The country rock consists of slate, schist, and granite which in places are capped by andesite. A few bodies of schist contain disseminated pyrite, which in a few places is gold-bearing. Also there are a few narrow quartz veins.

Bibliography

Eakle, A. S., 1919, Alpine County, Silver King district: California Min. Bur. Rept. 15, pp. 26–27.
Irelan, William, Jr., 1888, Silver King district: California Min. Bur. Rept. 8, p. 39.

Silver Mountain

Location and History. Silver Mountain is in south-central Alpine County about 10 miles south of Markleeville. Gold and silver were discovered here around 1860, and a rush began soon afterward. Kongsburg, later known as Silver Mountain City, was established by Scandinavian miners in 1862. The town grew rapidly and had a population of nearly 3000 in the following year. It was the first seat of government in Alpine County.

However, the district was not too productive. Hundreds of claims were located, and vast sums of money were spent on long tunnels and unsuccessful reduction works. Many of the mines were controlled by the Isabel Mining Company of England. Activity declined in the 1870s, and by 1886 the town had been abandoned. Some of the buildings were moved to Markleeville. The area has been prospected since, but there has been little recorded production. The total output of the district is estimated to be about $200,000 worth of gold and silver.

Geology and Ore Deposits. The gold- and silver-bearing deposits occur chiefly in altered volcanic rocks. The deposits consist of veins of silicified breccia containing pyrite, chalcopyrite, and other sulfides. Some deposits contain a series of parallel veins. The values usually are erratically distributed, although a few high-grade pockets were found in the early days. A few quartz veins are found in granitic rocks in the western part of the district.

Mines. Exchequer, Garfield, Isabella, I.X.L., Lady Franklin, Pennsylvania, Raymond Meadows.

Bibliography

Eakle, A. S., 1919, Alpine County, Silver Mountain district: California Min. Bur. Rept. 15, pp. 22–23.

Irelan, William, Jr., 1888, Silver Mountain district: California Min. Bur. Rept. 8, pp. 38–39.

Slate Mountain

A few small lode mines and prospects exist in the general vicinity of Slate Mountain, which is in north-central El Dorado County about 10 miles northeast of Placerville and 10 miles southeast of Georgetown. The veins are narrow and occur in slate and schist. Years ago several small but rich surface pockets were recovered. There are several tungsten prospects in the area.

Bibliography

Lindgren, Waldemar, and Turner, H. W., 1894, Placerville folio: U. S., folio 3, 3 pp.

Smartsville

Location. This district is located in western Nevada and eastern Yuba Counties 20 miles east of Marysville and 15 miles west of Grass Valley. It includes the Mooney Flat, Sicard Flat and Timbuctoo areas. It is mainly a placer-mining district.

History. The streams were placer-mined during the gold rush. The town was named for James Smart, who built a hotel there in 1856. The area was extensively hydraulicked from around 1855 to 1877. Some drift mining also was done during these years and continued through the early 1900s, but little work has been done here since. The value of the total output of the district is unknown. In 1877, it was reported to have been $13 million.

Geology. The main channel of the Tertiary Yuba River enters the area from the north and goes through Mooney Flat. It then curves west and northwest through Smartsville and Timbuctoo and then west to Sicard Flat. The gravel deposits are extensive and up to 200 feet thick. The lowest "blue" gravel on bedrock was the richest. The average yield during the major early operations was 37 cents in gold per yard, but some drifting yielded up to $3 per yard at the old price of gold. It has been estimated that 46.5 million yards were removed prior to 1891. Bedrock is greenstone, and in places the gravel is overlain by andesite.

Bibliography

Hobson, J. B., and Wiltsee, E. A., 1893, Smartsville mining district: California Min. Bur. Rept. 11, pp. 314–316.
Lindgren, Waldemar, 1895, Smartsville folio: U. S. Geol. Survey Geol. Atlas of the U. S., folio 18, 6 pp.
Lindgren, Waldemar, 1911, Tertiary gravels of the Sierra Nevada: U. S. Geol. Survey Prof. Paper 73, pp. 121–130.

Snelling

Location and History. The Snelling district is in eastern Merced County along the Merced River between the towns of Snelling and Merced Falls. It is principally a dredging field. Some placer mining and hydraulic mining of the terrace deposits along the river were practiced during the gold rush. The town, named in 1851 for Charles Snelling, who operated a hotel and ranch here, was the governmental seat of Merced County from 1857 until 1872. Gold dredging began in 1907 and continued until 1919. There was dredging again from 1929 until 1942 and 1946 to 1952. The value of the total output of the district is unknown, but the dredges are estimated to have produced about $17 million.

Geology. The values were recovered from stream gravels and flood plain and terrace deposits in and adjacent to the Merced River. The gravels are loose with very little clay and range from 20 to 35 feet in depth. The dredged area is about nine miles long and ½ to 1½ miles wide. Bedrock is slate in the east and volcanic ash in the west. The gold is fairly fine and about 890 in fineness. A small amount of platinum is present. Dredge recoveries ranged from 10 to 30 cents in gold per yard, with the average close to 10 cents.

Dredging Concerns. Merced Dredging Co., 1934–42 and 1945–49, one dredge; San Joaquin Mining Co., 1936–42, one dredge; Snelling Gold Dredging Co., 1932–42 and 1946–52, two dredges; Yosemite Mining & Dredging Co., 1907–19, one dredge; Yuba Cons. Goldfields, 1930–41, two dredges.

Bibliography

Davis, F. F., and Carlson, D. W., 1952, Merced County, gold: California Jour. Mines and Geology, vol. 48, pp. 220–227.

Lowell, F. L., 1916, Merced County, gold: California Min. Bur. Rept. 14, p. 606.

Winston, W. B., 1910, Merced County, dredging: California Min. Bur. Bull. 57, pp. 211–213.

Sonora

History. This is one of the famous pocket-mining districts of the Sierra Nevada east gold belt. Sonora, the seat of Tuolumne County, was founded in 1848 soon after the discoverey of rich placer deposits here and at Shaws Flat to the north. It was named for the state of Sonora in Mexico. The placers were extremely rich; the Sonora placers were credited with an output of $11 million and those of Shaws Flat yielded $6 million. The famous Holden Chispa nugget, which weighed over 28 pounds, was taken from within the city limits of Sonora in Holden's Gardens. In 1879 the Bonanza mine, also in town, yielded a pocket that contained $300,000. Later, large amounts of beautifully crystallized gold with tellurides were recovered from the Sugarman and Negro mine. Pocket mining continued almost steadily until World War II, and there has been some prospecting and development work since.

Geology. The central part of the Sonora district is underlain by a belt of crystalline limestone, which extends south from Columbia. The limestone is associated with slate, schist, and quartzite; all are part of the Calaveras Formation (Carboniferous to Permian). To the west is amphibolite and to the east is granodiorite.

Ore Deposits. The gold-quartz veins are largely confined to the slate, schist, and amphibolite. Much of the output from the lode mines has been from small but extremely rich pockets. The high-grade ore commonly contains crystallized gold, and in a few places

the telluride minerals petzite and sylvanite are present. The veins, usually only a few feet wide, are often associated with diorite and aplite dikes. The rich early-day placer deposits were in deep crevices and potholes in the limestone.

Mines. Aetna, Bonanza $1.5 million, Eureka, Fairview, Gerrymander, Golden Gate $100,000+, Hope $200,000+, Josephine, Lazar $100,000+, Lewis, Manzanita, O'Hara $100,000+, Rainbow, San Guiseppe, Sell, Stockton, Stuart and Morris, Sugarman $1 million, Tanzy, Vandelier.

Bibliography

Eric, J. H., Stromquist, A. A., and Swinney, C. M., 1955, Geology and mineral deposits of the Angels Camp and Sonora quadrangles: California Div. Mines Spec. Rept. 41, 55 pp.

Logan, C. A., 1949, Tuolumne County, gold: California Jour. Mines and Geology, vol. 45, pp. 54–75.

Tucker, W. B., 1916, Tuolumne County, gold: California Min. Bur. Rept. 14, pp. 135–168.

Turner, F. L., and Ransome, F. L., 1897, Sonora folio: U. S. Geol. Survey Geol. Atlas of the U. S., folio 41, 7 pp.

Soulsbyville

Location. The Soulsbyville district is in west-central Tuolumne County in the general vicinity of the towns of Soulsbyville and Toulumne. It includes the Arrastraville and Buchanan areas.

History. This district was placer-mined during the gold rush. Lode mining began in the early 1850s, and there was a rush to the district that began in 1858 after Ben Soulsby discovered rich ores. The mines were worked on a major scale until about 1915. There was some activity during the 1920s and 1930s, and there has been minor prospecting and development in a few of the mines since. This has been the most productive district in the Sierra Nevada east gold belt, with a total output value to be at least $20 million.

Geology. Granitic rocks, of which granodiorite predominates, underlie the west-central portion of the district (fig. 24). These granitic rocks are intrusive into slate, schist, phyllite, and quartzite of the Calaveras Formation (Carboniferous to Permian). Limestone is to the south and west, and the interstream ridges to the north are capped by andesite. Numerous dioritic and aplitic dikes are present, often associated with gold-quartz veins.

Ore Deposits. A large number of unoriented gold-quartz veins occur in both the granitic and metamorphic rocks, usually ranging from one to five feet in thickness. The ore bodies are often lenticular in shape and contain native gold and often abundant sulfides, especially galena, which is nearly always associated with the gold. Milling-grade ore usually ranged from ½ to one ounce of gold per ton, and considerable high-grade ore has been mined in the district. The maximum depth of development is 1500 feet.

Mines. Agua, Caliente, Black Oak $3.5 million, Black Hawk, Blue Lead, Buchanan $600,000, Carlotta, Chickenfeed, Columbus $100,000+, Consolidated Eureka, Dead Horse, Draper $1 million, Drei-

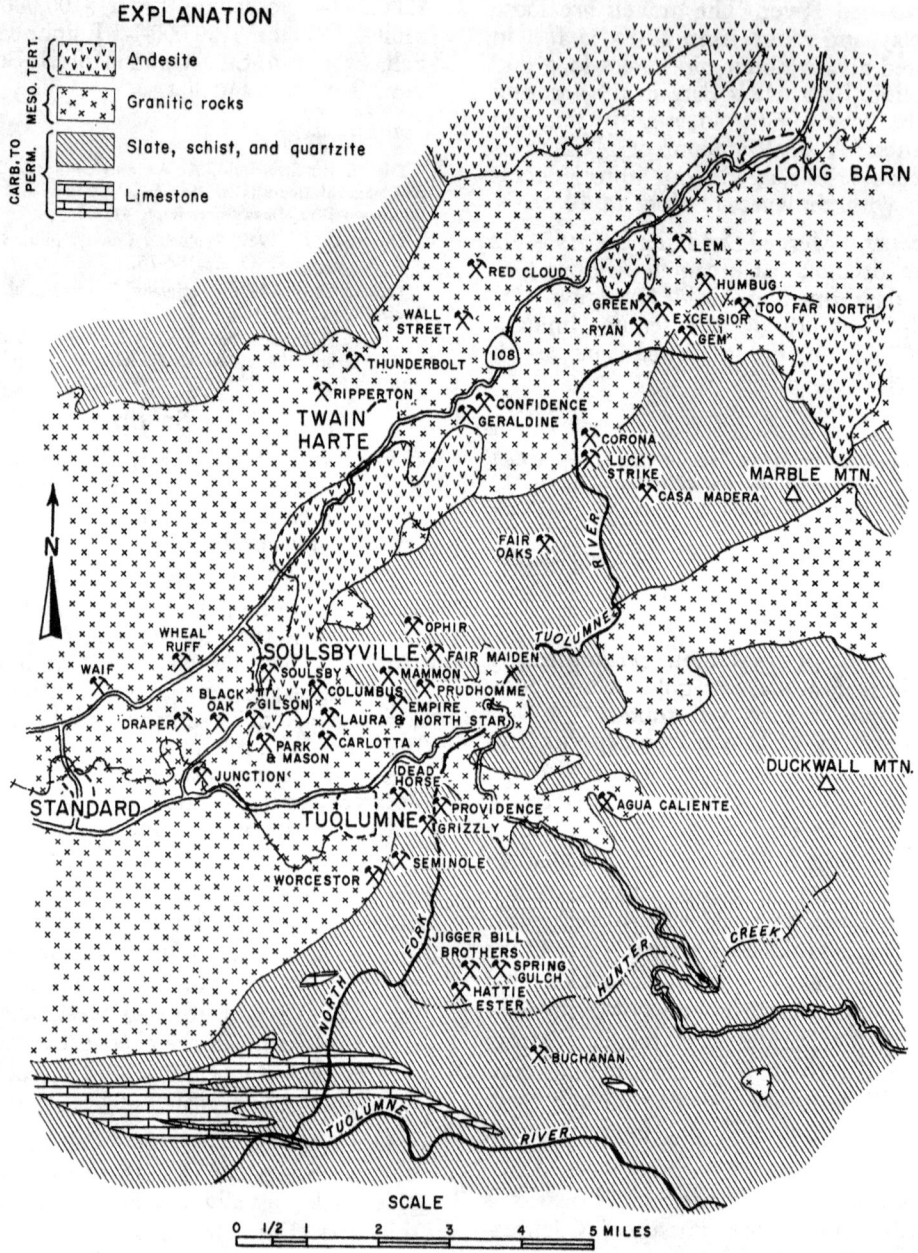

Figure 24. Geologic Map of Soulsbyville and Confidence Districts, Tuolumne County. The locations of the mines are shown. *After Turner and Ransome, 1897 and 1898.*

Photo 62. Soulsby Mine, Soulsbyville District. This is an early view of the highly productive mine, in Tuolumne County. *Photo courtesy of Tuolumne County Museum.*

sen, Empire, Fair Maiden, Fair Oaks, Garfield, Gilson (Platt & Gilson) $1.25 million, Grizzly $1.5 million, Hattie Ester, Hunter $300,000, Jigger Bill Brothers, Junction, Lady Washington, Laura & North Star, Louisiana, Mammon, New Albany, Ophir, Park and Mason, Phoenix, Providence $700,000, Prudhomme, Seminole, Spring Gulch $250,000, Soulsby $5.5 million, Starr King $100,000, South United $1.7 million, Waif, Wheal Ruff, Worcestor.

Bibliography

Goldstone, L. D., 1890, Soulsbyville mining district: California Min. Bur. Rept. 10, pp. 742–755.

Irelan, William, Jr., 1888, Black Oak and Buchanan mines: California Min. Bur. Rept. 8, pp. 665–669.

Logan, C. A., 1928, Tuolumne County, gold-quartz mines: California Div. Mines and Mining Rept. 24, pp. 8–41.

Storms, W. H., 1900, Black Oak mine: California Min. Bur. Bull. 18, pp. 137–138.

Tucker, W. B., 1916, Tuolumne County, lode gold: California Min. Bur. Rept. 14, pp. 136–166.

Turner, H. W., and Ransome, F. W., 1897, Sonora folio: U. S. Geol. Survey Geol. Atlas of the U. S., folio 41, 7 pp.

Turner, H. W., and Ransome, F. L., 1898, Big Trees folio: U. S. Geol. Survey Geol. Atlas of the U. S., folio 51, 8 pp.

Spanish Flat

Location and History. This well-known high-grade district is in west-central El Dorado County in the vicinity of the old mining town of Spanish Flat. It is about 10 miles north of Placerville and 3 miles northeast of Kelsey. The district was placer-mined soon after the beginning of the gold rush, and there has been intermittent mining ever since. Many of the early-day miners in this area, from South America, Mexico, and Portugal, were that time collectively known as "Spanish". The Alhambra mine, the largest source of gold in the district with a total output of $1.25 million, was active in the 1930s and 1940s and yielded much high-grade ore. Other mines in this district include the Brust, Shumway, and Timm mines.

Geology. The gold-bearing veins in this district are in a northwest-trending belt or zone about two miles east of the Mother Lode belt (see fig. 5). The deposits consist of northwest-striking quatrz veins with numerous parallel stringers that occur in shear zones

Photo 63. Alhambra Mine, Spanish Flat District. This 1940 view of the mine, in El Dorado County, looks west. At about the time the photo was taken, miners discovered an ore pocket that held $550,000 in gold. *Photo by Olaf P. Jenkins.*

with gouge. Country rock is amphibolite, chlorite, and graphite-quartz schist and slate. The ore shoots are not usually too extensive, but some have been extremely rich. A high-grade pocket discovered in the Alhambra mine in 1939 yielded $550,000. This pocket was a mass of native gold in quartz nearly 5 feet wide. The greatest depth of development is about 500 feet.

Bibliography

Clark, W. B., and Carlson, D. W., 1956, El Dorado County, lode gold: California Jour. Mines and Geology, vol. 52, pp. 401–429.

Lindgren, Waldemar, and Turner, H. W., 1894, Placerville folio, California: U. S. Geol. Survey Geol. Atlas of the U. S., folio 3, 3 pp.

Logan, C. A., 1938, El Dorado County, gold: California Div. Mines Rept. 34, pp. 215–272.

Spring Garden

A number of small lode mines and prospects occur in the general area of Spring Garden and Argentine Rock in south-central Plumas County. A patch of Tertiary auriferous gravel was mined by hydraulicking years ago and the area has been intermittently prospected in recent years. The country rock is greenstone, slate, and quartzite that is overlain to the south by andesite.

Bibilography

Turner, H. W., 1897, Downieville folio: U. S. Geol. Survey Geol. Atlas of the U. S., folio 37, 8 pp.

Sweet Oil

The Sweet Oil "diggings" in southwestern Plumas County about eight miles north of La Porte were mined by hydraulicking years ago. The gravel deposits are believed to be located on a branch of the Tertiary La Porte channel. Bedrock is slate, and to the south the gravels are capped by andesite and basalt.

Bibliography

Turner, H. W., 1898, Bidwell Bar folio: U. S. Geol. Survey Geol. Atlas of the U. S., folio 43, 6 pp.

Sycamore Flat

Location. Sycamore Flat is in east-central Fresno County just north of Piedra and about 25 miles due east of Fresno. It also is known as the Hughes Creek district. Superficial placer mining was done here during the gold rush, and the lode mines were active from the 1880s until about 1915. There has been minor prospecting since then.

Geology. The area is underlain by schist on the west, gabbro in the central portion, and granite in the east. There are a number of aplite dikes. Several narrow north-trending quartz veins with gentle to steep dips contain free gold and varying amounts of pyrite, chalcopyrite, and galena. A few high-grade pockets have been found here. One of the veins was mined to a depth of 300 feet.

Mines. Eliza Jane $100,000+, Independence, Sunnyside.

Bibilography

Bradley, W. W., 1916, Fresno County, Eliza Jane and Sunnyside gold mines: California Min. Bur. Rept. 14, pp. 444–445, 449–451.

Irelan, Wm., Jr., 1888, Sycamore mining district: California Min. Bur. Rept. 8, pp. 206–207.

Tahoe

Location. This district is in eastern Placer County west of and north of Lake Tahoe. It includes the areas known as the Squaw Valley and Red, White, and Blue or Elizabethtown districts north of the lake and a few scattered lode-gold mines and prospects west of the lake.

History. Gold and silver were discovered north of Lake Tahoe in 1861 and soon brought many miners to the area. Settlements known as Elizabethtown and Neptune City were established a few miles northwest of what is now Kings Beach, and Claraville and Knoxville were founded near the mouth of Squaw Creek. All of the prospects and these settlements were abandoned after 1864. In 1932 gold was discovered at the Tahoe Treasure mine a few miles west of Chambers Lodge. This mine has been worked intermittently since.

Geology and Ore Deposits. North of the lake lie massive andesite flows with andesitic tuffs and breccias. In places, zones of bleaching and silicification with impregnation of disseminated pyrite contain traces of gold and silver. West of the lake a few narrow gold-quartz veins occur in granodiorite and pyritic bodies in hornfels and schist.

Bibliography
Lindgren, Waldemar, 1897, Truckee folio: U. S. Geol. Survey Geol. Atlas of the U. S., folio 39, 8 pp.
Logan, C. A., 1936, Gold mines of Placer County, Tahoe Treasure mine: California Div. Mines Rept. 32, pp. 37–38.

Taylorsville

Location. This district is part of the Crescent Mills-Taylorsville-Genesee gold belt of east-central Plumas County. It has not been as productive as the other two districts in this belt. The general region was first mined during the gold rush, and there has been intermittent prospecting and development work ever since. It was named for J. T. Taylor, who built a mill and hotel there in 1852.

Geology. The Taylorsville area is underlain by a series of northwest-trending belts of Paleozoic and Mesozoic metamorphic rocks, serpentine, and granodiorite. The gold-bearing quartz veins are narrow and strike in a northwest direction. The veins usually occur in and near the granodiorite. The ore contains free gold and varying amounts of pyrite and chalcopyrite.

Mines. Buster, California, Deadman, Iron Dike, King Solomon (placer), Pettinger, Premium $180,000.

Bibliography
Averill, C. V., 1937, Plumas County, gold: California Div. Mines Rept. 33, pp. 103–124.
Diller, J. S., 1908, Geology of the Taylorsville region, California: U. S. Geol. Survey Bull. 353, 128 pp.
Diller, J. S., 1909, Mineral resources of the Indian Valley region: U. S. Geol. Survey Bull. 260, pp. 45–49.
MacBoyle, Errol, 1920, Plumas County, Taylorsville mining district: California Min. Bur. Rept. 16, pp. 49–52.

Tehachapi

Gold has been recovered from the Tehachapi Mountains a few miles south of the town of that name in south-central Kern County. Most of it came from the Pine Tree mine, which was active from 1876 to 1907 and had a reported total production of $250,000. The gold occurs in faulted and sheared quartz veins in granitic rocks. Scheelite also occurs locally in the quartz veins.

Bibliography
Troxel, B. W., and Morton, P. K., 1962, Kern County, Tehachapi district: California Div. Mines and Geology County Rept. 1, p. 52.

Temperance Flat

Location. Temperance Flat is in northeastern Fresno County on the south side of Millerton Lake. It is 10 miles northeast of Friant and about 25 miles northeast of Fresno. The area was placer-mined in the early days. Lode mining began at the Sullivan mine in 1853 and continued intermittently until about 1915. The area was prospected again during the 1930s.

Geology and Ore Deposits. The chief rock types are coarse-grained granite and granodiorite with diorite inclusions. Portions of the area are capped by thick flat beds of basalt of Table Mountain. A number of north-trending quartz veins, in shear zones in granitic rock, contain free gold and often abundant pyrite. Small amounts of other sulfides are present. A few small high-grade pockets containing leaf gold have been found here.

Mines. Henrietta, Keno, Quien Sabe, Providence, Rattlesnake, San Joaquin, Sullivan $100,000, Temperance, White Mule.

Bibliography
Bradley, W. W., 1916, Fresno County, John L. mine: California Min. Bur. Rept. 14, p. 446.
Crawford, J. J., 1896, Inyo, Keno, and Temperance mines: California Min. Bur. Rept. 13, pp. 167–170.
Irelan, Wm., Jr., 1888, Temperance Flat mining district: California Min. Bur. Rept. 8, pp. 214–215.

Tioga

Location. This district is at the crest of the Sierra Nevada in the vicinity of the Tioga Pass in eastern Tuolumne and western Mono Counties.

History. Gold-bearing outcrops were discovered here as early as 1860, and the area was intermittently prospected during the next 20 years. A boom was on from 1880 to 1884 when the Great Sierra Consolidated Silver Company was driving the Great Sierra tunnel. During that time the towns of Dana City and Bennettville existed, and the Tioga Road (now State Highway 120) was built, extending nearly 100 miles west to Groveland. The company failed in 1884. The tunnel was extended in 1933–34 to the projected extension of the ore body, but no values were encountered. Historically this is an interesting area, but it is doubtful if the district has yielded more than a few thousand dollars. The only property that has had any development is the Great Sierra mine, where more than $300,000 was expended. Nearly 350 claims were located in the district.

Geology. A number of narrow to thick northwest-striking quartz veins and mineralized metamorphic rocks contain pyrite, which is abundant in places. Traces of gold and silver are present. If there was any production, it probably came from oxidized surface material.

Bibliography
Bowen, O. E., Jr., 1962, Mines near Yosemite: California Div. Mines and Geology Mineral Information Service, vol. 15, no. 3, pp. 1–4.

DeGroot, Henry, 1890, Tioga district: California Min. Bur. Rept. 10, pp. 342–343.

Hubbard, Douglass, 1958, Ghost mines of Yosemite, Awani Press, Fresno, California, 40 pp.

Sampson, R. J., and Tucker, W. B., 1940, Mono County, Tioga mine: California Div. Mines Rept. 36, p. 139.

Whiting, H. A., 1888, Tioga district: Calif. Min. Bur. Rept. 8, pp. 371–373.

Tuttletown

Location. This district is in the Mother Lode belt in northwestern Tuolumne County. The Carson Hill district lies to the north and the Jamestown-Rawhide district to the south. It includes the Jackass Hill and French Gulch areas.

History. The streams and rich surface placers were worked during the gold rush. The area was known as Mormon Gulch in 1848 but was renamed for Judge Anson H. Tuttle, who built the first log cabin there. The Gillis brothers came here from Virginia City and it was in their cabin that Mark Twain stayed for five months in 1864–65 and Bret Harte for a night. A replica of the Gillis cabin is now a tourist attraction. This area became a well-known pocket-mining district during the 1860s and was mined almost continuously until World War II. Some work has been done at the Gross and Street mine since then.

Geology and Ore Deposits. A northwest-trending belt of amphibolite traverses the central portion of the district. Phyllite and slate are to both the northeast and southwest and serpentine lies to the south. Although this district lies between the Carson Hill and Jamestown districts, where large ore bodies were mined, most of the output here has been from small, rich pockets. These pockets occur in quartz veins and stringers and contain native gold, abundant pyrite, galena, and tellurides. The country rock adjacent to the veins commonly contains disseminated pyrite and ankerite.

Mines. Albion Cons., Alta, Anti-Chinese, Arbona, Ball, Bown, Cardinelle, Chileno, Gagnere, Gross and Street, Marryatt, Norwegian $200,000+, Patterson, Tarantula, Toledo.

Bibliography

Eric, J. H., Stromquist, A. A., and Swinney, C. M., 1955, Geology and mineral deposits of the Angels Camp and Sonora quadrangles: California Div. Mines Spec. Rept. 41, 55 pp.

Logan, C. A., 1949, Tuolumne County, Gross and street mines: California Jour. Mines and Geology, vol. 45, pp. 66–67.

Nolan, T. B., 1929, Norwegian and Chileno mines: U. S. Geol. Survey Prof. Paper 157, pp. 77–78.

Ransome, F. L., 1900, Mother Lode district folio: U. S. Geol. Survey Geol. Atlas of the U. S., folio 63, 11 pp.

Turner, H. W., and Ransome, F. L., 1897, Sonora folio: U. S. Geol. Survey Geol. Atlas of the U. S., folio 41, 7 pp.

Vallecito

Location. Vallecito is in south-central Calaveras County about five miles east of Angels Camp and five miles south of Murphys. It a placer-mining district and includes the Douglas Flat and Dead Horse Flat areas.

History. Daniel and John Murphy found gold here in 1848, and the district was first known as Murphys' Old Diggins. The name was changed to Vallecito in 1854. The hydraulic mines were active from the late 1850s through the 1880s. Drift mining was done at this time and again during the 1930s, when the Vallecito Western drift mine was operated on a fairly large scale. Dragline dredging was active in the district during the 1930s.

Geology. The Tertiary Central Hill channel enters the district from the north from Murphys and then extends west toward Altaville and Angels Camp. Here the channel is joined by two small tributaries from the east and south, the one from the east known as the Murphy's Gulch channel. Farther east is the south-trending Cataract channel. The gravels, consisting of granitic material and quartz, were richest near bedrock. The gold was fairly coarse. The gravels are overlain in places by rhyolite and andesite and also by some Pleistocene gravels. Bedrock is limestone, schist, slate, and amphibolite, and to the west there is granodiorite.

Bibliography

Clark, W. B., and Lydon, P. A., 1962, Calaveras County, gold: California Div. Mines and Geology County Rept. 2, pp. 32–93.

Lindgren, Waldemar, 1911, Tertiary gravels of the Sierra Nevada: U. S. Geol. Survey Prof. Paper 73, pp. 199–201.

Logan, C. A., and Franke, H., 1936, Calaveras County, Vallecito-Western mine: California Div. Mines Rept. 32, pp. 353–355.

Steffa, Donald, 1932, Gold mining and milling methods and costs at the Vallecito Western drift mine: U. S. Bur. Mines Inf. Circ. 6612, 13 pp.

Storms, W. H., 1894, Ancient channel system of Calaveras County: California Min. Bur. Rept. 12, pp. 482–492.

Turner, H. W., and Ransome, F. L., 1898, Big Trees folio: U. S. Geol. Survey Geol. Atlas of the U. S., folio 51, 8 pp.

Volcano

Location. This district, in north-central Amador County, is in the vicinity of the old mining town of Volcano, 15 miles northeast of Jackson.

History. The creeks were first mined during the gold rush, and the town was settled by soldiers from Stevenson's 1st New York Volunteer Regiment. The town received its name in 1850, as it was believed then that the limestone caves were related to a volcano. During the 1850s this was one of the richer placer-mining districts in the state. Since the 1930s there has been a small output from dragline dredging in some of the creeks, and some of the channel gravels have been prospected in recent years. Many of the buildings in the old town are well-preserved, and the town is now a well-known tourist attraction.

Geology. The central portion of the district is underlain by crystalline limestone of the Calaveras Formation (Carboniferous to Permian), which has many potholes and crevices that contained much rich gravel. The rest of the area is underlain by graphitic slate and schist. To the north and east several deposits of early Tertiary quartz-rich gravels were mined by hydraulicking and drifting. Those to the north are parts of the deposits of the ancestral Cosumnes River that extended west through the area and west-north-west towards the Fiddletown district. In places the gravels are capped by andesite. A few narrow gold-quartz veins are found in the district.

Photo 64. Early Placer Mining, Volcano District. This scene, in Volcano, Amador County, shows sluice boxes, a pump and a deep crevice in lime-stone bedrock.

Bibliography

Carlson, D. W., and Clark, W. B., 1954, Amador County, placer gold: California Jour. Mines and Geology, vol. 50, pp. 197–200.

Haley, C. S., 1923, Gold placers of California: California Min. Bur. Bull. 92, pp. 146–147.

Lindgren, Waldemar, 1911, Tertiary gravels of the Sierra Nevada: U. S. Geol. Survey Prof. Paper 73, p. 199.

Volcanoville

Location. The Volcanoville district is in north-central El Dorado County and south-central Placer County, about eight miles northeast of Georgetown and 30 miles east of Auburn. It includes the Kentucky Flat area. It is both a lode- and placer-mining district.

Geology and Ore Deposits. A number of patches of andesite-capped Tertiary gravel were deposited by a channel that extended north through Kentucky Flat and then west through Volcanoville. An older well-cemented "white" channel contains much quartz gravel; there is a younger channel. The gold is coarse. Several gold-quartz veins in schist and slate, near or adjacent to a belt of serpentine, crop out in the west portion of the district. Some of the quartz veins have yielded high-grade pockets and well-developed quartz crystals.

Mines. Placer: Bedrock, Buckeye Point, Kenna, Kenny, Kentucky Flat, Morris, Tiedeman. Lode: Boedner, Bootjack, Green, Josephine, Paymaster.

Bibliography

Clark, W. B., and Carlson, D. W., 1956, El Dorado County, placer gold deposits: California Jour. Mines and Geology, vol. 52, pp. 429–435.

Irelan, William, Jr., 1888, The Josephine mine: California Min. Bur. Rept. 8, pp. 165–166.

Lindgren, Waldemar and Turner, H. W., 1894, Placerville folio: U. S. Geol. Survey Geol. Atlas of the U. S., folio 3, 3 pp.

Lindgren, Waldemar, 1911, Tertiary gravels of the Sierra Nevada: U. S. Geol. Survey Prof. Paper 73, pp. 168–169.

Tucker, W. B., 1919, El Dorado County, placer mines: California Min. Bur. Rept. 15, pp. 300–303.

Washington

Location. This district is in east-central Nevada County in the vicinity of the old mining town of Washington, 18 miles northeast of Nevada City. It is in the south end of the Goodyear's Bar-Alleghany belt and includes the "diggings" at Alpha and Omega.

History. The Washington district was first mined during the gold rush, and the placers of the Middle Yuba River were highly productive. The Omega and Alpha hydraulic mines were opened in the middle 1850s and worked on a major scale through the 1880s. Later, Chinese miners reworked the tailings. Lode mining also began in the 1850s and continued steadily until about 1915. There was activity in the district again during the 1930s, and the Red Ledge mine has been worked in recent years. Barite, chromite, and asbestos also have been mined here.

Geology. The district is underlain chiefly by slate, schist and quartzite of the Blue Canyon Formation (Carboniferous). A serpentine body one to two miles wide crops out in the central portion. The Relief Quartzite (Carboniferous) and amphibolite lie to the west and granodiorite to the east. The serpentine is a south extension of a belt that passes north-northwest through Alleghany and Goodyear's Bar in Sierra County. Tertiary andesite overlies the main ridges to the north and south.

Ore Deposits. The auriferous Tertiary channel gravels at Alpha and Omega are part of the main channel that extends west and north to Relief and North Bloomfield. Jarmin (1927) estimated that, at Omega, 13 million yards were mined and yielded 13½ cents in gold per yard. Lindgren (1911) estimated that 40 million yards remained. The quartz veins contain small but rich ore bodies, similar to those of the Alleghany district to the north, but are not as plentiful. Except for arsenopyrite, sulfides are not usually abundant. A number of beautiful specimens of crystallized gold have been found in the Red Ledge mine. The Spanish mine also has yielded large amounts of barite.

Mines. Lode: Giant King, Mexican Cons., Mount Hope, Red Ledge, Red Paint, St. Patrick, Spanish, Treasure Box. Placer: Alpha $2 million+, Centennial, Omega, Phelps, Yuba River.

Bibliography

Averill, C. V., 1946, Placer mining for gold in California, Omega mine: California Div. Mines Bull. 136, pp. 265–266.

Hobson, J. B., 1890, Washington mining district: California Min. Bur. Rept. 10, pp. 389–392.

Irelan, William, Jr., 1888, Washington mining district: California Min. Bur. Rept. 8, pp. 435–444.

Jarman, Arthur, 1927, Washington and Omega hydraulic mines: California Min. Bur. Rept. 23, pp. 112–115.

Lindgren, Waldemar, 1900, Colfax folio, California: U. S. Geol. Survey Geol. Atlas of the U. S., folio 66, 10 pp.

Lindgren, Waldemar, 1911, Tertiary gravels of the Sierra Nevada: U. S. Geol. Survey Prof. Paper 73, pp. 139–141.

MacBoyle, Errol, 1919, Nevada County, Washington mining district: California Min. Bur. Rept. 16, pp. 59–63.

Photo 65. Belden Mine, West Point District. This 1952 view of the mine, in Amador County, looks northeast. The mine was active in the 1930s.

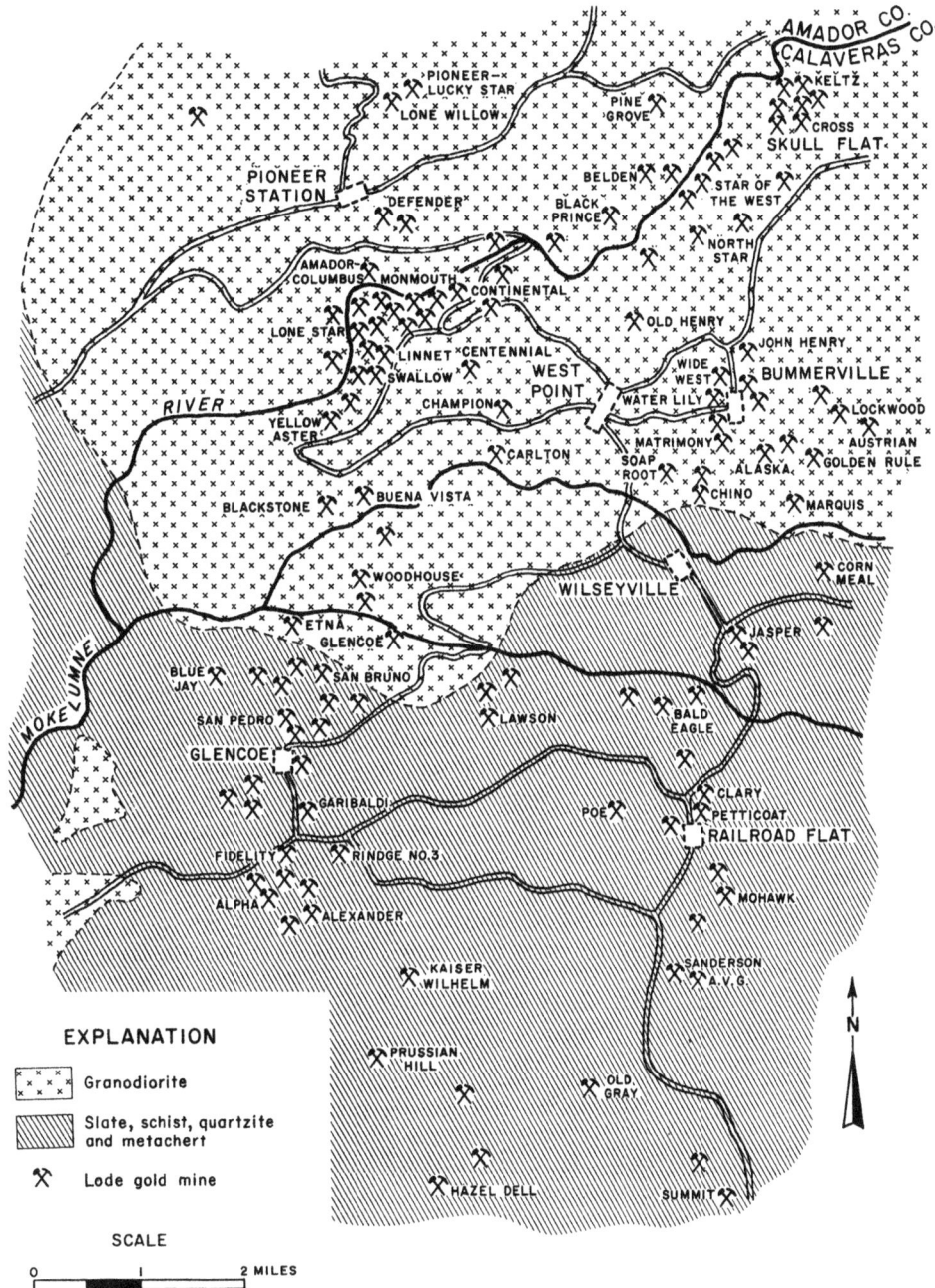

Figure 25. Geologic Map of West Point and Railroad Flat Districts, Calaveras County. The lode-gold mines are marked. *After Carlson and Clark, 1954, and Clark and Lydon, 1962.*

EXPLANATION

Granodiorite

Slate, schist, quartzite and metachert

Lode gold mine

SCALE

0   1   2 MILES

West Point

Location. This extensive Sierra Nevada east gold belt district is in eastern Amador and Calaveras Counties in the general area of the town of West Point. It includes the Skull Flat, Glencoe, Bummerville, Pioneer Station, and Buckhorn areas.

History. The town was first known as Indian Gulch but was renamed West Point after a geographic feature discovered by Kit Carson while he was en-route to Sutter's Fort in 1844. The streams and surface ores were mined extensively during the 1850s, when large amounts of gold were recovered. During the 1860s and 1870s many lode mines and 10 or more custom mills were active, but there was much difficulty with sulfides. Some activity was noted from the 1880s until 1914 and again during the 1920s and 1930s. Several mines have been intermittently worked since World War II, the chief operations having been at the Belden, Blackstone, and Centennial mines. This is one of the more productive districts of the east gold belt, and an extremely large number of mines exist.

Geology. The gold deposits are associated with a west-elongated body of granodiorite five miles wide and 15 miles long (see figs. 4 and 25) that has intruded graphitic slates, quartzites, and schists of the Calaveras Formation (Carboniferous to Permian).

Ore Deposits. Numerous north-trending and west-dipping (a few dip east) quartz veins are found in the granodiorite or in the adjacent metamorphic rocks. The veins usually are one to five feet thick, have persistent strikes, and belong to one of three main vein systems that have not been mapped. Narrow diorite, quartz-diorite, and aplite dikes commonly are associated with the veins. The ore bodies contain free gold and abundant sulfides, especially galena, which is nearly always associated with high-grade ore. The ore shoots usually have horizontal stoping lengths of 150 feet or less, but several were 300 to 400 feet long. Milling-grade ore commonly averages one ounce or more in gold per ton, and much high-grade ore has been recovered. Few of the mines have been developed to depths of more than a few hundred feet. It has been estimated that there are more than 500 mine shafts in the district.

Mines. Amador County: Amador-Columbus, Belden $400,000+, Black Prince $100,000+, Defender $100,000+, Elkhorn, Hageman, Jumbo, Lone Willow $100,000+, Newman $160,000+, Pine Grove, Pioneer-Lucky Strike $300,000+, T.N.T. Calaveras County: Austrian, Billy Williams, Backstone $200,000, Blazing Star, Buena Vista, Carlton, Centennial, Champion $500,000, Chino, Continental $100,000+, Corn Meal, Cross, Etna, Ever Ready, Fidelity, Garibaldi, Gilded Age, Glencoe, Golden Rule, Gold Star, Good Hope, Keltz $300,000+, Lockwood $400,000+, Lone Star, Marquis, Mina Rica, Monte Cristo, North Star, Old Henry, Rindge No. 1, 2, and 3, Riverside, San Bruno, San Pedro, Scorpian, Soap Root, Star of the West, Swallow, Water Lily, Wide West, Woodhouse $100,000+, Yellow Aster $100,000+.

Bibliography

Browne, J. Ross, 1868, Reports upon the mineral resources of the United States: Government Printing Office, Washington (D.C.) (West Point mines), pp. 65–67.

Carlson, D. W., and Clark, W. B., 1954, Amador County, Belden and Black Prince mines: California Jour. Mines and Geology, vol. 50, pp. 170–172.

Clark, W. B., and Lydon, P. A., 1962, Calaveras County, gold: California Div. Mines and Geology County Rept. 2, pp. 32–93.

Logan, C. A., 1923, Notes on the West Point district: California Min. Bur. Rept. 18, pp. 15–21.

Raymond, Rossiter W., 1875, Statistics on mines and mining in the states and territories west of the Rocky Mountains: Government Printing Office, Washington (D.C.) (West Point mines), pp. 63–66.

Tucker, W. B., 1916, Calaveras County, Lockwood, Lone Star, and Star of the West mines: California Min. Bur. Rept. 14, pp. 90–92 and 107.

Turner, H. W., 1894, Jackson folio, California: U. S. Geol. Survey Geol. Atlas of the U. S., folio 11, 6 pp.

Turner, H. W., and Ransome, F. L., 1898, Big Trees folio, California: U. S. Geol. Survey Geol. Atlas of the U. S., folio 51, 8 pp.

Westville

Location. This placer-mining district is in eastern Placer County about 17 miles northeast of Forest Hill and 10 miles due south of Emigrant Gap. It adjoins the Damascus district on the west and the Canada Hill district on the east. The district includes the Macedon Ridge, Whiskey Hill, and Secret Canyon areas.

Geology. The placer deposits lie along a Tertiary intervolcanic channel of the American River, known as the Red Point channel, which extends southwest into the Damascus district. A tributary known as the Whiskey Hill or Black Canyon channel joins the Red Point channel at Westville. This tributary is narrow, steep, and contains coarse gold. The gravels are capped by andesite, and bedrock is quartz-bearing schist and slate. Much of the development in this district has been drift mining.

Mines. Golden Fleece, Greek, Green, Hogsback, Herman, Hungry Hollow, Macedon, Osborne, Union.

Bibliography

Lindgren, Waldemar, 1900, Colfax folio: U. S. Geol. Survey Geol. Atlas of the U. S., folio 66, 10 pp.

Lindgren, Waldemar, 1911, Tertiary gravels of the Sierra Nevada: U. S. Geol. Survey Prof. Paper 73, pp. 156–157.

Logan, C. A., 1936, Gold mines of Placer County, placer mines: California Div. Mines Rept. 32, pp. 49–96.

West Walker

This is a small district on the east flank of the Sierra Nevada in northern Mono County a few miles southwest of Coleville. The principal sources of gold have been the Al Mono and Golden Gate mines, which were active in the late 1890s and early 1900s. The deposits consist of quartz stringers containing native gold and sulfides or massive sulfide bodies containing disseminated gold. Some narrow kaolinized gold-bearing seams are found. Country rock consists of schist, slate, greenstone, and quartzite. Copper, lead, zinc, and cadmium also occur here.

Bibliography

Eakle, A. S., and McLaughlin, R. P., 1919, Mono County, gold: California Min. Bur. Rept. 15, pp. 139–142 and 165.

Wheatland

This is a small placer-mining district in the vicinity of the town of Wheatland on the lower Bear River in western Placer and southern Yuba Counties. During the gold rush placer gold was recovered from the creeks and streams. During the 1930s gold was recovered from the gossan by cyanidation at the Dairy Farm copper mine a few miles to the east and by 'dragline dredging in some of the ravines.

Bibliography

Lindgren, Waldemar, and Turner, H. W., 1895, Smartsville folio: U. S. Geol. Survey Geol. Atlas of the U. S., folio 18, 6 pp.

White Oak Flat

White Oak Flat is in north-central Amador County about 10 miles northeast of Volcano. Several moderate-sized deposits of Tertiary channel gravels were mined years ago. There are several gold-quartz deposits including the Marklee mine, which was prospected recently. Bedrock consists of slate, chert, and quartzite. In places the gravels are overlain by andesite and rhyolite.

Bibliography

Turner, H. W., 1894, Jackson folio: U. S. Geol. Survey Geol. Atlas of the U. S., folio 11, 6 pp.

White River

Location and History. White River is in southern Tulare County and northern Kern County approximately 25 miles southeast of Porterville. Gold was discovered here in 1853. The town was originally known as Tailholt, but the name was changed to White River in 1870. Mining continued until around 1906, and there has been minor activity since. The district was estimated to have yielded a total of $750,000 worth of gold by 1914.

Geology. The area is underlain by granodiorite and smaller outcrops of more basic intrusives. Small amounts of schist and slate and a few limestone lenses lie to the west. A series of west-northwest-trending parallel quartz veins occur in shear zones in the granodiorite. The ore contains free gold and small amounts of pyrite.

Mines. Bald Mountain (several hundred thousand dollars), Eclipse No. 2, Josephine, Last Chance, Stencil.

Bibliography

Laizure, C. McK., 1923, Tulare County, White River mining district: California Min. Bur. Rept. 18, pp. 524–527.

Tucker, W. B., 1919, Tulare County, White River mining district: California Min. Bur. Rept. 15, pp. 912–915.

Whitlock

Location. The Whitlock district is in west-central Mariposa County five miles north of the town of Mariposa. The district is east of the Mother Lode gold belt and includes the Colorado, Sherlock Creek, and Whiskey Flat areas. The area was placer-mined soon after the beginning of the gold rush, and lode mining began shortly afterward. A number of mines were active here during the 1930s, and a few, such as the Diltz and Schroeder mines, have been intermittently prospected in recent years.

Geology. Greenstone and green schist underlie much of the district, with some slate, phyllite, and mica schist in the north portion. Granitic intrusives and serpentine are to the south. There is an appreciable number of diorite, quartz-diorite, and aplite dikes that commonly are associated with the gold-quartz veins. A northwest-trending fault extends along the west side of the district (see fig. 18).

Ore Deposits. Numerous north- and northwest-striking quartz veins contain small but rich ore shoots. The veins usually are one to five feet thick, and a number dip at low angles. The veins have a tendency to roll or bend, and it is in these bends or rolls that the high-grade pockets often occur. Much specimen ore has been produced in the district; in 1932 the Diltz mine yielded 52- and 40-pound masses of gold and quartz. The greatest depth of development is about 900 feet.

Mines. Buffalo, Champion, Colorado $50,000, Diltz $750,000 to $1 million, Geary, Golden Key $154,000+, King Solomon, Landrum, Nutmeg $180,000+, Our Chance, Permit, Schroeder $200,000 to $300,000, Spread Eagle $425,000, Whitlock $500,000.

Bibliography

Bowen, O. E., Jr., 1957, Mariposa County, lode mines: California Jour. Mines and Geology, vol. 53, pp. 69–187.

Castello, W. O., 1921, Colorado district—Colorado, Diltz, Schroeder, and Whitlock mines: California Min. Bur. Rept. 17, pp. 93, 111, 113, 137 and 142.

Lowell, F. L., 1916, Mariposa County, Colorado, Diltz, and Whitlock mines: California Min. Bur. Rept. 14, pp. 579, 581, and 599–600.

Ransome, G. L., 1900, Mother Lode district folio: U. S. Geol. Survey Geol. Atlas of the U. S., folio 63, 11 pp.

Yankee Hill

Location. This district is in east-central Butte County about 15 miles northeast of Oroville. It is fairly extensive and includes the Concow and Big Bend areas.

History. The streams and surface placers were first worked during the gold rush. For a time the locality was known as Rich Gulch and Spanishtown. In those days much gold was recovered from the North Fork of the Feather River, and a diversion tunnel was driven through Big Bend. Numerous Chinese miners reworked the old placer tailings later on. Lode mining began in the 1850s, and there was much activity during the 1890s and early 1900s. The Surcease mine was worked on a major scale from 1933 to 1942, and copper was mined at the nearby Big Bend mine during World War II. The estimated output of the district is slightly more than 100,000 ounces of gold.

Geology. A northwest-trending belt of slate and quartzite four to five miles wide, with some limestone that is part of the Calaveras Formation (Carboniferous to Permian), crops out in the central part of the district. Interbeds of amphibolite and serpentine lie to the north. Granodiorite stocks are to the east and southeast.

Ore Deposits. A number of quartz veins contain some free gold and often abundant sulfides, especially chalcopyrite. The veins are in the metamorphic rocks. Milling ore commonly averages ½ ounce of gold per ton, much of the values being in the sulfides. The Surcease vein has been mined to a depth of more than 1000 feet. A gold-bearing barite vein occurs at the Pinkston mine.

Mines. Berry Creek, Bunker Hill, Evening Star, Hearst, Madre de Oro, Pinkston, Porter, Rainbow, Sunbeam, Surcease $1 million+, Treasure Hill.

Bibliography

Logan, C. A., 1930, Butte County, gold quartz mines: California Div. Mines Rept. 26, pp. 369–383.

Miner, J. A., 1890, Butte County, quartz mines and mills: California Min. Bur. Rept. 10, pp. 125–133.

O'Brien, J. C., 1949, Butte County, gold: California Jour. Mines and Geology, vol. 45, pp. 426–433.

Turner, H. W., 1898, Bidwell Bar folio: U. S. Geol. Survey Geol. Atlas of the U. S., folio 43, 6 pp.

You Bet

Location and History. You Bet and Red Dog are in south-central Nevada County, eight miles southeast of Nevada City. This district also includes the "diggings" at Little York. You Bet sometimes is known as Chalk Bluffs. The region was first placer-mined in 1848 or 1849. The name "You Bet" is supposed to have originated in 1857 from saloon keeper Lazarus Beard's favorite expression. Red Dog was named by Charley Wilson after his former home, Red Dog Hill, Illinois. The district was hydraulicked on a large scale from 1855 until the 1880s. There was some drift mining. Later, the area was mined on a moderate scale, chiefly by Chinese. It was intermittently active until about 1935. The total ouput is valued at more than $3 million. Lindgren (1911) estimated 47 million yards were removed and 100 million remained. Jarman (1927) estimated 20 million yards of 10- to 15-cent gravel remained at Red Dog.

Geology. The gravels were deposited by a Tertiary channel of the Yuba River that extended north and northwest to Hunts Hill and Scotts Flat. The lower gravels are 30 to 40 feet thick, well-cemented, and contain a high percentage of quartz including a number of large boulders. It is capped by as much as 350 feet of fine gravel with some interstratified clay and sand. Bedrock consists of slate and some chert.

Bibliography

Averill, C. A., 1946, Placer mining for gold in California, You Bet mines: California Div. Mines Bull. 135, pp. 269–270.

Hobson, J. B., and Wiltsee, E. A., 1892, You Bet mining district: California Min. Bur. Rept. 11, pp. 317–318.

Jarman, Arthur, 1927, You Bet district: California Min. Bur. Rept. 23, pp. 99–100.

Lindgren, Waldemar, 1900, Colfax folio, California: U. S. Geol. Survey Geol. Atlas of the U. S., folio 66, 10 pp.

Lindgren, Waldemar, 1911, Tertiary gravels of the Sierra Nevada: U. S. Geol. Survey Prof. Paper 73, p. 144.

MacBoyle, Errol, 1919, Nevada County, You Bet mining district: California Min. Bur. Rept. 16, pp. 63–66.

KLAMATH MOUNTAINS PROVINCE

The Klamath Mountains region in northwestern California is the second-most gold-productive province in California. The principal gold districts are in Shasta, Siskiyou, and Trinity Counties. Although there are several important lode-gold districts, the placer deposits have been the largest sources of gold. The Klamath Mountains consist of a number of complex and rugged ranges that continue north into Oregon. The entire mountain mass is essentially an irregular and deeply dissected uplifted plateau. It is underlain by a series of complexly folded and faulted metamorphic rocks of Paleozoic and Mesozoic age that have been invaded by batholiths of Late Jurassic and possibly Early Cretaceous age. In some respects the Klamath Mountains resemble the Sierra Nevada, and sometimes the two mountain ranges are classified as a single metallogenetic province.

The major rock units of the Klamath Mountains include the Abrams Schist and Salmon Schist (pre-Silurian?); the Copley Greenstone, Balaklala Rhyolite and Kennett Shale (Devonian); slate of the Bragdon Formation (Mississippian), and the younger granitic rocks of the Shasta Bally batholith. On the west side of the province are extensive beds of sandstone, shale, and conglomerate of Jurassic age, and ultramafic rocks that are in part serpentinized. Between these two rock sequences lie beds of phyllite, chert, limestone and metavolcanic rocks of Paleozoic and Triassic age. The batholiths are composed chiefly of granodiorite or quartz diorite and are either round or elongated in a northerly direction. The largest ones are the Wooley Creek, Ironside Mountain, and Shasta Bally batholiths.

The most productive placer deposits in the Klamath Mountains have been those associated with the Klamath and Trinity Rivers and their tributaries. Gold is found not only in the gravels in the present stream channels, but also in older terrace and bench deposits adjacent to the channels. The terrace and bench deposits often were mined by hydraulicking.

Rising in southern Oregon, the Klamath River flows west across the Klamath Mountains and empties into the Pacific Ocean. The most important tributary streams of the Klamath River are the Shasta, Scott, and Salmon Rivers, and Cottonwood, Horse, Seiad, Thompson, Indian, Clear, Dillon, and Camp Creeks. Important centers of placer mining in the Klamath River system have been at Hornbrook, Yreka, Scott Bar, Hamburg, Somesbar, Orleans, Sawyers, Forks of Salmon, Callahan, and Cecilville.

The Trinity River, which flows into the Klamath River at Weitchpec, drains the southern portion of the Klamath Mountains. The most productive placer deposits of the Trinity River are those located along its main channel. These include the deposits at Carrville, Trinity Center, Minersville, Lewiston, Weaverville, Junction City, and Salyer. The principal tributaries of the Trinity River are Coffee Creek, Stewart's Fork, East Fork, New River, Indian Creek and Hayfork Creek. The La Grange mine, a few miles west of Weaverville, was one of the largest hydraulic mines in California. Other sources of placer gold in the Klamath Mountains have been the Smith River region in Del Norte County and the upper Sacramento River and its tributaries, which include Backbone, Clear, Cottonwood, and Beegum Creeks.

Lode-gold deposits are found throughout the Klamath Mountains. The most productive district has been the French Gulch-Deadwood district of Shasta and Trinity Counties, in the southern portion of the province. Other important sources of lode gold have been the Deadwood district of Siskiyou County (there are several Deadwood districts in California), Dillon Creek, Callahan, Oro Fino, Liberty, Sawyers Bar, Harrison Gulch, Whiskeytown, and Buckeye-Old Diggings districts. Considerable amounts of gold have been produced in the Shasta copper-zinc belt and lesser amounts in other copper deposits, such as the Copper Bluff mine at Hoopa.

Photo 66. Bully Choop Mine, Bully Choop District. This view, in about 1900, shows the 30-stamp mill and tramway at the mine, in Trinity County. The stacked cordwood (foreground and right) fueled the steam-driven machinery. *Photo courtesy of Adele Kiessling.*

The gold nearly always occurs in native form in quartz veins, usually associated with pyrite and smaller amounts of other sulfides. The veins occur in all metamorphic rocks of Jurassic and older ages, the Bragdon Formation (Mississippian) containing the most numerous and productive veins. A few lode-gold deposits are found in granitic rocks. Undoubtedly the introduction of the veins is related to the granitic intrusions. Often the gold-quartz veins and the ore shoots in the veins are associated with fine- to medium-grained diorite, quartz diorite, and aplite dikes. In several districts these dikes are known locally as "birdseye porphry" dikes.

Backbone

This district is about 10 miles north-northwest of Redding. The French Gulch district adjoins on the west. The Backbone district includes the Squaw Creek area. Although this district is in the Shasta copper-zinc belt, it has been mainly a source of gold. The principal gold mine has been the Uncle Sam, which was worked from 1886 to 1913 and later prospected. It has yielded more than $1 million. Several gold-quartz veins contain free gold and often abundant sulfides. Country rock is greenstone.

Bibliography

Averill, C. V., 1933, Shasta County, Uncle Sam mine: California Div. Mines Rept. 29, pp. 54–55.

McGregor, Alex, 1890, Squaw Creek mines—Backbone mining district: California Min. Bur. Rept. 10, pp. 639–641.

Bully Choop

The Bully Choop district is in the vicinity of Bully Choop Mountain in southeastern Trinity County about 15 miles southeast of Weaverville. The Bully Choop and Cleveland mines, the principal gold sources, were active from the late 1880s through the early 1900s. There was some prospecting here again in the 1930s. The deposits consist of zones of quartz stringers containing free gold and often abundant sulfides, which include chalcopyrite and arsenopyrite. The ore shoots had stoping lengths of more than 200 feet. Country rock consists of gneiss, hornblende schist, mica schist and some limestone.

Bibliography

Averill, C. V., 1933, Trinity County, Bully Choop and Cleveland mines: California Div. Mines Rept. 29, pp. 15–19.

Photo 67. Princess Hydraulic Mine, Shasta County. The monitors are undercutting a bank of auriferous gravel. The photo was taken in about 1900. *Photo courtesy of Adele Kiessling.*

Callahan

Callahan is in south-central Siskiyou County in the upper Scott River region. During the early days there was considerable gold production from old bench gravels and gulches tributary to the river in the south portion of the district. The town was named for M. B. Callahan, who established a store here in 1851. In the early 1900s and again in the 1930s, substantial amounts of gold were recovered by bucket-line dredges that worked the Scott River north of the town of Callahan for a distance of five miles. There are a number of gold-quartz deposits in the district, the most productive having been the Cummings or McKeen mine, which had a total output valued at $500,000. The veins usually are in or near a granitic body and the ore bodies are small but often are rich.

Bibliography

Brown, G. C., 1916, Siskiyou County, Callahan district: California Min. Bur. Rept. 14, p. 825.
Dunn, R. L., 1893, Callahan's Ranch: California Min. Bur. Rept. 11, pp. 433–434.

Cecilville

Cecilville is in southwestern Siskiyou County near the junction of the East and South forks of the Salmon River. Gold was discovered here in 1849 by James Abrams, and the district soon became an important mining center with a population of several thousand persons. Later, from 3000 to 5000 Chinese were reported to have worked the Salmon River by means of flumes and wing dams. Substantial amounts of lode gold have been recovered in the district, the most notable source having been the King Solomon mine, which was active in the 1930s. The region is underlain by slate, greenstone, limestone, and serpentine, with schist to the east. The lode deposits consist of either massive gold-quartz veins or zones of quartz seams and stringers that in places contain high-grade pockets.

Bibliography

Brown, G. C., 1916, Siskiyou County, King Solomon mine: California Min. Bur. Rept. 14, p. 836.
Irwin, W. P., 1960, Geologic reconnaissance of the northern Coast Ranges and Klamath Mountains: California Div. Mines Bull. 179, 80 pp.
Siskiyou County Historical Society, 1957, Guidebook to Siskiyou's gold fields: Siskiyou Pioneer, vol. 2, no. 10, pp. 14–17.

Cottonwood

Cottonwood Creek, which forms the southwest border of Shasta County and is a tributary to the upper Sacramento River, has yielded substantial amounts of gold. Its principal tributaries, Crow, Antelope, Dry, and Roaring River Creeks, also have been productive. During the 1930s, the area was worked by both dragline and bucket-line dredges. Digging depths were mostly 10 feet or less.

Bibliography

Averill, C. V., 1938, Gold dredging in Shasta, Siskiyou and Trinity Counties: California Div. Mines Rept. 34, pp. 96–126.

Deadwood

Deadwood is in central Siskiyou County about 10 miles north of Fort Jones. It was an important town from 1851 to 1861 and a stop on the California-Oregon stage line from 1851 until 1886. Hooperville, a few miles to the west, also was an important early-day settlement. Deadwood, Cherry, Indian, French, and McAdam Creeks all yielded large amounts of placer gold during the gold rush and were later dredged. Cherry Creek is believed to have been worked six different times. Snipers and part-time prospectors are still active in the district.

The principal lode-gold mines here are the Franklin, Cherry Hill, Golden Eagle, New York, Mt. Vernon, and Schroeder mines. The Golden Eagle has a total production of about $1 million. Some of these mines have been intermittently worked in recent years. The veins occur in greenstone with some slate and contain free gold and varying amounts of sulfides.

Bibliography

Brown, G. C., 1916, Siskiyou County, gold mines: California Min. Bur. Rept. 14, pp. 825–865.

O'Brien, J. C., 1947, Siskiyou County, gold: California Jour. Mines and Geol., vol. 43, pp. 428–453.

Siskiyou County Historical Society, 1957, Guidebook to Siskiyou's gold fields: Siskiyou Pioneer, vol. 2, no. 10, pp. 80–82.

Dedrick-Canyon Creek

These districts are in north-central Trinity County about 15 miles northwest of Weaverville. The bench gravels at Canyon Creek are very extensive and apparently were quite rich; hydraulic mines are almost continuous throughout its length of more than 12 miles. A few of the hydraulic mines have been worked in recent years.

Dedrick is a lode-gold district near the head of Canyon Creek. The mines were developed during the 1880s and 1890s and were active until the 1930s. The Alaska, with an output of $600,000, and the Globe Consolidated, with a total output of more than $700,-000, have been the principal lode mines. Others include the Ralston, Maple, Silver Gray, and Mason and Thayer mines. The lode deposits consist of parallel quartz veins containing fine free gold with some sulfides. Country rock consists of hornblende schist with granitic stocks lying just to the north and east.

Bibliography

Dunn, R. L., 1893, Canon Creek district: California Min. Bur. Rept. 11, pp. 482–483.

Photo 68. Placer Mine, Siskiyou County. Water wheel and flume appear in this view of a mine operated by Chinese on the Klamath River in 1933. At one time these Chinese water wheels were widely used in river mining. *Photo by Olaf P. Jenkins.*

Ferguson, H. G., 1914, Gold lodes of the Weaverville quadrangle: U. S. Geol. Survey Bull. 540, pp. 22–79.

Logan, C. A., 1926, Trinity County, Globe Consolidated mine: California Min. Bur. Rept. 22, pp. 20–21.

Dillon Creek

Dillon Creek, a tributary of the lower Klamath River, is in western Siskiyou County. It was originally placer-mined during the gold rush, and the general area was prospected again in the 1930s. The Siskon mine was discovered in 1951 and operated on a large scale from 1953 until 1960, one of the last significant gold discoveries and operations in the state. Although the value of its production is unrecorded, it has been estimated at several million dollars. The deposit consisted of an extensive mass of gold-bearing gossan that overlay a body of pyrite-bearing schist and quartz stringers. It was worked chiefly in benched cuts, although some of the stringers were mined underground. The ore was concentrated and the concentrates treated in a cyanidation plant.

Bibliography

Symons, H. H., and Davis, F. F., 1958, Gold: California Jour. Mines and Geology, vol. 54, p. 102.

Dog Creek

The Dog Creek or Delta district is in northwestern Shasta County, about 25 miles north of Redding and just west of the town of Delta. Some placer gold was recovered from Dog Creek and other nearby streams during the gold rush. The Delta mine, the principal source of lode gold, was active in the 1890s and early 1900s. During this period it was connected to the Southern Pacific Railroad by a 6½-mile-long narrow-gauge railroad. A number of narrow quartz veins in greenstone contain free gold and small amounts of sulfides. The ore bodies usually are low in grade, but were reported to have had stoping lengths of as much as 800 feet.

Bibliography

Brown, G. C., 1916, Shasta County, Delta Consolidated mine: California Min. Bur. Rept. 14, p. 784.

Dorleska

This district is in the Salmon Mountains on both sides of the Trinity-Siskiyou County line and on the divide between the Salmon and Trinity Rivers. It is near the headwaters of Coffee Creek about 12 miles southwest of Coffee Creek Ranch. The name derives from the Dorleska mine, discovered in 1897. The Dorleska, with a total output of $200,000, and the Yellow Rose mine, which has yielded more than $100,000, have been the chief sources of gold. Other properties include the Upper Nash, LeRoy, and Keating mines.

The ore deposits occur in a north-northwest-trending zone of mineralization that is at least five miles long. The deposits consist of narrow gold-quartz veins and mineralized shear zones in and along the contacts of lamprophyre dikes, especially where these dikes cut serpentine. Serpentine lies to the west and schist to the

east. The ore contains free gold, pyrite, and smaller amounts of tellurides and galena. A number of high-grade pockets have been found here.

Bibliography

Averill, C. V., 1931, Trinity County, Yellow Rose mine: California Div. Mines Rept. 27, p. 55.

Averill, C. V., 1941, Trinity County, Dorleska mine: California Div. Mines Rept. 37, p. 33.

MacDonald, D. F., 1913, Gold lodes of the Carrville district—Dorleska and Yellow Rose mines: U. S. Geol. Survey Bull. 530, part I, pp. 38–39.

French Gulch

Location. This district lies astride the Shasta-Trinity County line in the general vicinity of the town of French Gulch and includes the Deadwood area to the west. It is the most important lode-gold district in the Klamath Mountains.

History. French Gulch was originally prospected in 1849 by French miners, from whom the town received its name in 1856. Clear Creek, which drains the area, yielded large amounts of placer gold at this time. The Washington mine, discovered in 1852, was the first quartz mine worked in Shasta County. From around 1900 to about 1914 the output for the district averaged between $300,000 and $500,000 worth of gold per year. There was some activity during the 1920s and 1930s, and there has been minor prospecting and development work since. The value of the total output of the district is estimated at more than $30 million.

Geology. The district is underlain predominantly by slate, shale, and siltstone of the Bragdon Formation (Mississippian). Copley Greenstone (Devonian) lies to the northeast and south, and, to the southwest, there is quartz diorite of the Shasta Bally batholith. In addition, numerous porphyritic quartz diorite and diorite dikes, locally known as "birdseye porphyry", occur.

Ore Deposits. The quartz veins usually strike west, with a few northwest exceptions, and range from a few inches to several feet thick. They are predominantly in the rocks of the Bragdon Formation and often occur near or adjacent to the dikes, which apparently have had some effect on the localization of the ore bodies. The latter consist of numerous parallel stringers rather than a single massive vein. Calcite is commonly present in the veins. The ore contains coarse, free gold usually associated with considerable pyrite and smaller amounts of galena, sphalerite, arsenopyrite, chalcopyrite, and occasionally scheelite. Numerous high-grade pockets have been recovered here. A number of large ore bodies occur in the district, several of which were more than 1000 feet long.

Mines. Accident, American $300,000, Army Batch, Blue Jay, Bright Star, Brown Bear $15 million, Brunswick $100,000, Calmich, Centennial, El Dorado, Fairview $200,000, Gambrinus $125,000, Gladstone $6.9 million, Henry Clay $100,000 to $300,000, Highland $300,000, Honeycomb, Jacoby, J.I.C., Larry, Mad Mule $1 million, Mad Ox $500,000, Milkmaid and Franklin $2.5 million, Montezuma 7,150+ ounces, Mt. Shasta 8,500 ounces, Niagara $1 million, Niagara Sum-

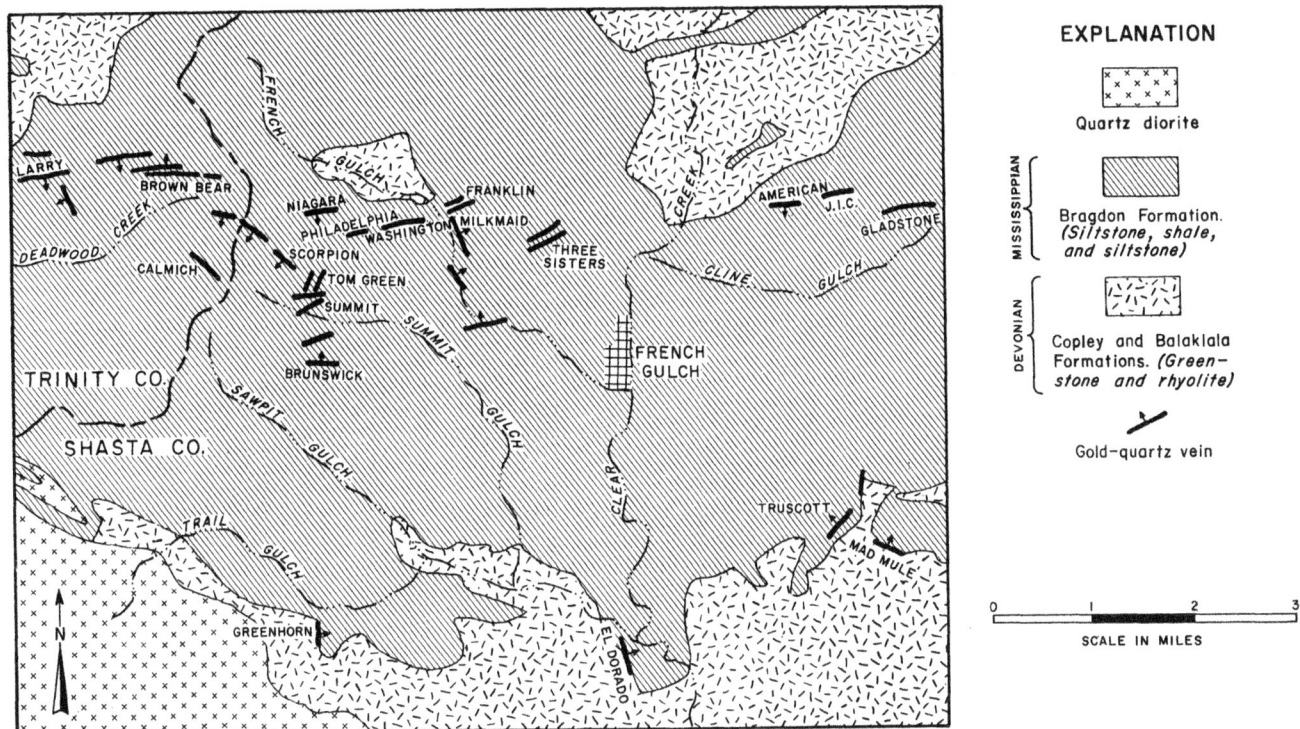

Figure 26. Geologic Map of French Gulch District, Shasta County. The principal gold-quartz veins are shown. *Modified from Albers, 1961, p. C-2.*

mit, Philadelphia, St. Jude $280,000+, Scorpion 7,140 ounces, Summit $200,000, Sybel $600,000, Three Sisters $100,000, Tom Green, Truscott, Venecia $500,000, Vermont and Montezuma, Washington $2.5 million.

Bibliography

Albers, J. P., 1961, Gold deposits in the French Gulch-Deadwood district, Shasta and Trinity Counties, California: U. S. Geol. Survey Prof. Paper 424-C pp. 1–4.

Albers, J. P., 1964, Geology of the French Gulch quadrangle, Shasta and Trinity Counties: U. S. Geol. Survey Bull. 1141-J, 70 pp.

Albers, J. P., 1965, Economic geology of the French Gulch quadrangle: California Div. Mines and Geology Spec. Rept. 85, 41 pp.

Averill, C. V., 1933, Gold deposits of the Redding and Weaverville quadrangles: California Div. Mines Rept. 29, pp. 2–72.

Brown, G. Chester, 1916, Shasta County, French Gulch district: California Min. Bur. Rept. 14, p. 775.

Crawford, J. J., 1894, Gladstone and Green mines: California Min. Bur. Rept. 12, pp. 248–249.

Ferguson, H. G., 1913, Gold lodes of the Weaverville quadrangle: U. S. Geol. Survey Bull. 540-A, pp. 16–73.

Logan, C. A., 1926, Shasta County, French Gulch district: California Min. Bur. Rept. 22, pp. 167–168.

MacGregor, Alex., 1890, French Gulch mining district: California Min. Bur. Rept. 10, pp. 635–638.

Gazelle

This district is in south-central Siskiyou County west of the town of Gazelle. The Dewey mine, which was worked from the 1880s until about 1907, has been the chief source of gold, its total output valued at about $900,000. The region is underlain by metasedimentary rocks, which are of Silurian age, and small stocks of granodiorite. The gold-quartz veins occur in granodiorite and contain abundant sulfides.

Bibliography

Logan, C. A., 1925, Siskiyou County, Dewey mine: California Min. Bur. Rept. 21, p. 438.

Gilta

Gilta is in the southwestern corner of Siskiyou County near the head of Knownothing Creek and about five miles south of Forks of Salmon. Most of the gold output here has come from the Gilta or Gold Hill mine, which has a reported production of $1 million, and from the extremely rich placers of nearby Knownothing Creek. The Gilta mine was worked on a large scale, mostly prior to 1900. The gold-quartz veins occur in slate and schist and are associated with diorite dikes. One ore shoot at this mine was 250 feet long.

Bibliography

Brown, G. C., 1916, Siskiyou County, Gold Hill mine: California Min. Bur. Rept. 14, p. 833.

Harrison Gulch

Harrison Gulch is in the extreme southwest corner of Shasta County about six miles west of Platina and 40 miles southwest of Redding. It was named for Judge W. H. Harrison, who settled here in 1852. The Midas mine, the principal source of gold in the district and one of the major lode mines in the Klamath Mountains, has a total output of nearly $4 million. It was discovered in 1894 and was worked on a major scale from 1896 to 1914. Placer gold was recovered here and in the Platina and Beegum areas to the east.

The lode deposits consist of lenticular ore bodies in quartz veins that range from one to three feet in thickness. They contain free gold and some sulfides. Much of the ore produced at the Midas mine yielded more than one ounce of gold per ton. The vein was mined to a depth of 1500 feet. Country rock consists of greenstone and schist. Granite lies to the west, and sandstone and shale are to the east.

Bibliography

Brown, G. C., 1916, Shasta County, Midas mine: California Min. Bur. Rept. 14, pp. 792–793.

Logan, C. A., 1926, Shasta County, Harrison Gulch mines: California Min. Bur. Rept. 22, pp. 173–174.

Helena-East Fork

The Helena and East Fork districts are in north-central Trinity County about 20 miles west of Weaverville and adjoin the Dedrick and Canyon Creek districts on the west. The old town of Helena, which is quite well-preserved, was an important mining center in the early days. The once-important nearby towns of Bagdad and Coleridge have long since disappeared. The bench gravels along the East Fork were quite productive during the early days but are not as extensive as those at Canyon Creek to the east. There are several lode-gold mines, the most productive having been the Enterprise mine, which has been intermittently worked since 1884. Its estimated total production is valued at $500,000. Others include the Yellowstone, Lone Jack, and Ozark mines. The deposits consist of parallel quartz veins containing free gold and varying amounts of sulfides. Some of the veins are fairly flat, and in places these contain high-grade pockets. Tellurides have been found. Country rock is hornblende schist and granitic rock.

Bibliography

Ferguson, H. G., 1914, Gold lodes of the Weaverville quadrangle: U. S. Geol. Survey Bull. 540, pp. 22–79.

Logan, C. A., 1926, Trinity County, Enterprise mine: California Min. Bur. Rept. 22, pp. 18–19.

Hoopa

This is a copper-gold district in the Hoopa Indian Reservation in northeastern Humboldt County. Placer gold has been recovered from the Trinity River and by-product gold from the Copper Bluff copper mine, active in 1965. The placer deposits occur both in the present river channel and in older terrace deposits along the bank. The ore deposits at the Copper Bluff mine consist of mineralized schist and quartz veins containing gold associated with chalcopyrite, sphalerite, galena, and pyrite.

Bibliography

Symons, H. H., and Davis, F. F., 1959, Copper: 55th Report of the State Mineralogist, p. 122.

Humbug

Humbug is in north-central Siskiyou County about 10 miles northwest of Yreka. Humbug Creek, which flows into the upper Klamath River, was extremely rich during the early days; it is credited with an output of nearly $15 million. The town of Humbug City,

which was founded in 1851, has largely disappeared. Some dragline dredging has continued until the present time. The gold-quartz veins were fairly productive, the Eliza, Spencer, Hegler, McKinley, and Mono mines all having yielded substantial amounts of gold. The veins range from one to five feet in thickness and contain free gold, pyrite, galena, and smaller amounts of other sulfides. Several high-grade pockets have been mined. Country rock consists of greenstone and granite with smaller amounts of schist and slate.

Bibliography

Brown, G. C., 1916, Siskiyou County, Humbug Creek district: California Min. Bur. Rept. 14, p. 824.

Irwin, W. P., 1960, Geologic reconnaissance of the northern Coast Ranges and Klamath Mountains: California Div. Mines Bull. 179, 80 pp.

Igo—Ono

Location and History. These two adjacent placer-mining districts are in southwestern Shasta County about 15 miles southwest of Redding. The South Fork silver-mining district lies just to the north. The region was first mined soon after the beginning of the gold rush. From the 1860s through the 1880s the hydraulic and drift mines were highly productive, especially the Hardscrabble and Russell mines near Igo. Many Chinese placer miners were here during this period. The origin of the two names is reputed to have been from the expressions "I go?" and "Oh no!", derived either from the pidgin English spoken by the Chinese miners when they were told to move on or from statements made by a young son of the superintendent of the Hardscrabble mine. There was appreciable activity in these districts in the 1930s, much of the gold output having come from the use of power shovels and dragline dredges. From 1933 to 1959 the districts were credited with an output of 115,000 ounces of gold.

Geology. The gold production has come from Recent stream gravels in South Fork, Eagle, Dry, North Cottonwood, and Clear Creeks, and older terrace deposits. Some of the older terrace deposits are quite extensive, the gravels at the Hardscrabble mine being as deep as 50 feet. Bedrock consists of slate, schist, greenstone and granite. There are some small gold-quartz veins that have yielded high-grade pockets.

Bibliography

Averill, C. V., 1939, Shasta County, gold: California Div. Mines Rept. 35, pp. 129–159.

Brown, G. C., 1916, Shasta County, Igo district: California Min. Bur. Rept. 14, p. 775.

Jelly Ferry

Jelly or Jelly Ferry is on the upper Sacramento River about 10 miles north of Red Bluff in Tehama County. At one time Andrew Jelly operated a ferry across the river. Later the state operated the ferry; it was finally replaced by a bridge in 1950. During the early days, Chinese mined gold-bearing gravels in the area by ground sluicing, which was followed by an unsuccessful early dredge. There was some dragline dredging here in the 1930s. The gold occurs in both the river gravels and terraces adjacent to the river.

Bibliography
O'Brien, J. C., 1946, Tehama County, gold: California Jour. Mines and Geol., vol. 42, p. 190.

Klamath River

The Klamath River flows across the northern portion of the Klamath Mountain province. It enters the Klamath Mountains in the vicinity of Hornbrook, flowing southwest and then generally west for more than 50 miles, crossing a number of well-known mining districts. The placer-gold production has come from the present channel and a succession of terraces and benches ranging from less than 50 to more than 200 feet above the present channel and its tributaries. These older benches often are miles in extent and in places are cut by younger and deeper channels. The present streams were mined by hand methods during the early days, later, by wingdams, flumes, and tunnels, and, more recently, by bucket-line or dragline dredges. The benches were worked by hydraulicking, ground sluicing and some by drift mining.

At Hornbrook the river is joined by Cottonwood Creek, which was noted for extremely rich but shallow deposits. A number of lode mines are found here and in the Paradise or Fool's Paradise district, which lies to the southeast. The Shasta River enters the Klamath about five miles south of Hornbrook and both the river and two of its tributaries, Yreka and Greenhorn Creeks, were extremely rich. Between 1850 and 1900 Greenhorn Creek was reported to have yielded $11 million. The Yreka or Hawkinsville district also was nearly as productive. Farther west, the Klamath River was extensively placer-mined between Humbug Creek and the Scott River, especially at Masonic, Skeahan, and Kanaka Bars and at Gottville. Humbug Creek also was very rich (see separate section on the Humbug district). Lumgrey, Empire, and Dutch Creeks, all of which have been mined, enter the river here. The Hazel lode mine, which has yielded more than $800,000, is a few miles north of Gottville. The gold-quartz veins in this mine occur in slate.

Farther downstream are Oak Bar, Beaver and Horse Creeks and Hamburg, where the Scott River flows into the Klamath (see separate sections on the Scott Bar and Callahan districts). At Seiad Valley some 10 miles to the west, substantial mining was done by both dredging and hydraulicking. In the Happy Camp district the river flows around several sharp bends and then turns south. Here the China Creek, Davis, Reeves, Woods Bar, Richardson Bedrock and Muck-a-Muck hydraulic mines were important, as was Indian Creek, which flows into the river from the northwest.

From Happy Camp the river flows in a general south-southwest direction for approximately 50 miles. It runs through the Clear Creek area, where the Siskiyou and Bunker Hill mines are located, Cottage Grove, the Dillon Creek areas (see section on the Dillon Creek district), Rattlesnake Bar, Ti Bar, and Somesbar, where the Salmon River flows into the Klamath (see section on Salmon River). The Klamath River then flows through the highly productive Orleans district in Humboldt County and on to Weitch-pec, where it is joined by the Trinity River from the south. At Weitchpec the Klamath River turns west and then northwest for about 45 miles and empties into the Pacific Ocean near the town of Requa in Del Norte County.

Liberty

Location and History. The Liberty or Black Bear district is in the Salmon Mountains in southwestern Siskiyou County. The Sawyer's Bar district lies immediately to the north and the Cecilville district is to the south. The area was placer-mined in the 1850s. Lode gold was discovered in 1860, both at White's Gulch, a few miles to the northeast, and at the Black Bear mine. From 1865 until about 1910, the lode mines were highly productive, especially the Black Bear, Klamath, and California Consolidated mines. There was some activity in the district again in the 1930s, and there has been some prospecting since.

Geology and Ore Deposits. The district is underlain by slate, phyllite, greenstone, and chert of Paleozoic and Mesozoic age, with a few lenses of serpentine and small granitic bodies. Diorite dikes are often associated with the veins. The ore deposits consist of lenticular gold-quartz veins, usually five feet or less in thickness. Milling-grade ore usually averages only a few dollars per ton, but a considerable number of high-grade pockets have been found in the district.

Mines. Advance $250,000+, Ball, Black Bear $3.1 million, California Cons. $473,000+, Cleaver, Hanson, Hickey, Jumbo, Klamath $600,000, Lanky Bob, Mountain Laurel $600,000, White Bear.

Bibliography
Brown, G. C., 1916, Siskiyou County, gold mines—quartz: California Min. Bur. Rept. 14, pp. 825–842.
Irwin, W. P., 1960, Geologic reconnaissance of the northern Coast Ranges and the Klamath Mountains: California Div. Mines, Bull. 179, 80 pp.
Logan, C. A., 1925, Siskiyou County, Salmon River district: California Min. Bur. Rept. 21, pp. 419–420.

Monumental

This small lode-gold district is in northern Del Norte County about 45 miles northeast of Crescent City. Most of the work here has been done at the Monumental Consolidated mine, which has been intermittently developed since about 1900. Several quartz veins in greenstone and slate contain some gold, pyrite, chalcopyrite, and hematite. The greatest depth attained here is about 250 feet.

Bibliography
Lowell, F. L., 1916, Del Norte County, Monumental Consolidated quartz mine: California Min. Bur. Rept. 14, pp. 389–390.
Maxon, J. H., 1933, Economic geology of portions of Del Norte and Siskiyou Counties, northeastern California: California Div. Mines Rept. 29, pp. 123–160.

New River-Denny

New River is an important tributary of the Trinity River. It rises in the Salmon Mountains in northwestern Trinity County and flows in a southwesterly direction through the Denny area and eventually joins the Trinity River at Burnt Ranch. Gold was discovered here in 1849, and placer mining followed for some years. Lode mining began later, the chief producer having been the Mountain Boomer mine, which has a total output of more than $350,000. The area was prospected again in the 1930s. A number of gold-quartz veins occur in slate and greenstone. Other mines in the district include the Uncle Sam, Modoc, Sherwood, Mary Blaine, Hunter, Live Oak, and Gun Barrel mines.

Bibliography

Brown, G. C., 1916, Trinity County, Modoc and Mountain Boomer mines: California Min. Bur. Rept. 14, p. 895.

Min. and Sci. Press, Jan. 24, 1885, New River district, p. 53.

Old Diggings

The Old Diggings or Buckeye district is about five miles due north of Redding in the vicinity of the towns of Buckeye and Summit City. The area was settled by miners from Ohio, the "Buckeye State". It was extremely productive during the gold rush, when large amounts of placer gold were recovered by hydraulicking. Later, considerable amounts of lode gold were produced, particularly from 1904 until 1919, when large tonnages of siliceous gold-bearing copper ore were recovered from the Reid mine and used as flux in the Mammoth smelter at Kennett. The area was prospected again in the 1930s.

The district is underlain largely by greenstone of Devonian age. The veins consist of white sugary quartz that contain free gold, pyrite, and small amounts of chalcopyrite. Tellurides have been reported to occur in the deposits. The veins range from a few feet to as much as 25 feet in thickness and were mined to depths up to 1000 feet. Milling ore contained from less than 1/6 to one or more ounces of gold per ton. Some high-grade pockets were found. Most of the placer gold was recovered from older bench gravels.

Mines. Calumet, Central $500,000, Evening Star, Mammoth, National $200,000, Texas $750,000, Reid $2.5 million+, Walker.

Bibliography

Averill, C. V., 1933, Gold deposits of the Redding-Weaverville districts: California Div. Mines Rept. 29, pp. 2–73.

Logan, C. A., 1926, Shasta County, gold: California Min. Bur. Rept. 22, pp. 167–186.

McGregor, Alex, 1890, Old Diggings district: California Min. Bur. Rept. 10, pp. 629–632.

Orleans

Location. This district is on the Klamath River in the extreme northeast corner of Humboldt County in the vicinity of the town of Orleans. It is mainly a placer-mining district. There was mining here during and after the gold rush that continued through the early 1900s. The Pearch hydraulic mine was active during the 1930s.

Geology. The gold-bearing deposits are stream gravels in the Klamath River and extensive older bench gravels about 50 to 80 feet above the level of the present river. Bedrock is slate and schist. The bench gravels were mined by hydraulicking. The gold is fine to medium, and some platinum is present.

Mines. Allen, Bondo, Orcutt, Orleans Bar, Pearch, Rocky Point, Rough and Ready, Salstrom.

Bibliography

Averill, C. V., 1941, Humboldt County, Pearch mine: California Div. Mines Rept. 37, pp. 512–513.

Irelan, William, 1888, Orleans Bar, Orleans, and Pearch mines: California Min. Bur. Rept. 8, pp. 219–221.

Lowell, F. L., 1916, Humboldt Co., placer gold, California Min. Bur. Rept. 14, pp. 401–407.

Oro Fino

Location. The Oro Fino district, in central Siskiyou County about five miles west of Fort Jones, includes the Quartz Valley and Mugginsville areas. The placer deposits were first worked during the gold rush and the lode mines began operating soon afterward. There was appreciable activity here again in the 1930s and 1940s, when the lode mines were active.

Geology and Ore Deposits. Many of the lode deposits are in Quartz Hill, a steep resistant peak in the central portion of the district. It consists largely of hard dark massive pyritic greenstone. Schist, limestone, small amounts of serpentine, and valley alluvium are present. Diorite dikes often are associated with the veins. Numerous quartz-calcite veins contain free gold and often abundant pyrite. The quartz is white to smoky in color. The placer deposits occur in the various creeks, and the gold generally is fine but rough and angular. Some of the placers were extremely rich.

Mines. Fino, Gibralter, Gold Reef, Golden Eagle, Morrison and Carlock $500,000+, Providence, Quartz Hill, Star, Umpah.

Bibliography

Brown, G. C., 1916, Siskiyou County, gold: California Min. Bur. Rept. 14, pp. 825–865.

Irwin, W. P., 1960, Geologic reconnaissance of the northern Coast Ranges and Klamath Mountains: California Div. Mines Bull. 179, 80 pp.

Siskiyou County historical Society, 1957, Guidebook to Siskiyou's gold fields: Siskiyou Pioneer, vol. 2, no. 10, pp. 83–88.

Redding

Location and History. Redding is in south-central Shasta County. Originally named Reading for Major Pierson B. Reading, who discovered gold in the Trinity River, the district was renamed for Benjamin Redding, land agent for the Central Pacific Railroad Company. During the gold rush appreciable amounts of placer gold were recovered in the area, from the upper Sacramento River and from Oregon, Flat, and Clear Creeks, which are to the southwest. Also, high-grade surface pockets were mined. During the 1930s a number of dragline and bucket-line dredges were active in the area. The Yankee John lode mine has been intermittently worked in recent years.

Ore Deposits. The lode deposits consist of narrow quartz veins and seams containing native gold, small amounts of sulfides and some silver. The deposits occur in a belt that extends southwest to Centerville. Much of the output has been from small but rich pockets. The Yankee John, the chief lode mine in the district, has a total output of slightly more than $200,000. Country rock consists of greenstone, slate, and granitic rocks. There are numerous dioritic dikes.

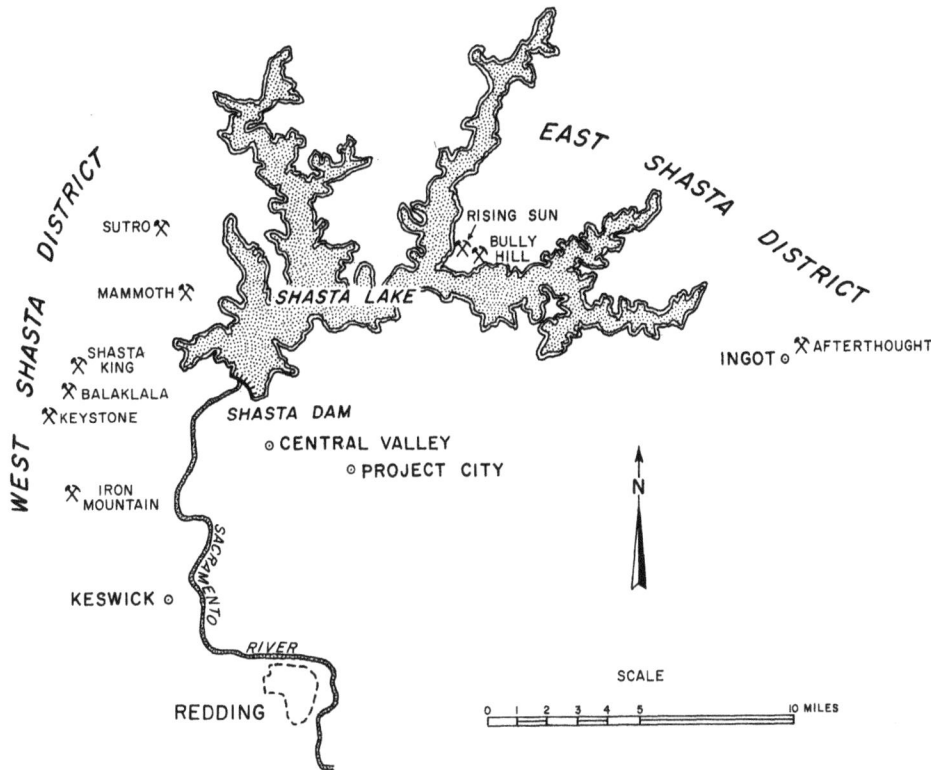

Figure 27. Sketch Map of Shasta Copper-Zinc Belt. The major mines are shown.

Bibliography
Averill, C. V., 1933. Gold deposits of the Redding and Weaverville quadrangles: California Div. Mines Rept. 29, pp. 3–73.

Salmon River

The Salmon River drains the Salmon Mountains, which are in the central portion of the Klamath Mountains province in Siskiyou County. This river is not as long as the Klamath or Trinity Rivers, but it flows through several rich and famous placer-mining districts. The most productive districts have been at Snowden and Sawyer's Bar on the North Fork, Cecilville and Knownothing on the South Fork, and Forks of Salmon at the junction of the North and South Forks. Probably the richest portion of the river was the 17-mile stretch of the North Fork between Sawyer's Bar and Forks of Salmon, a segment that had an estimated gold production of $25 million. Eddy's Gulch, just south of Sawyer's Bar, yielded $4 million. As in the other streams in this province, the river bars were first worked by hand methods and later by wing-dams and flumes. The bench gravels were hydraulicked or worked by drift mining. Placer gold was discovered on the Salmon River in 1849 at Cecilville and in 1850 near Sawyer's Bar. Most of the other important placer "diggings" were developed soon afterward. Among the important lode-gold mines are those in the Liberty and Gilta districts (see separate sections on these two districts).

Scott Bar

Scott Bar is on the lower Scott River in Siskiyou County a few miles south of Hamburg, where the Scott joints the Klamath River. John Scott discovered gold here in 1850. The district was an important center during the early days when the river and older bench gravels were extensively mined. The Quartz Valley mine just east of town has been worked on a large scale both by hydraulicking and as a lode mine. At this property the gold occurs in thin stringers or lenses of quartz and calcite in micaceous schist. A number of rich pockets have been found in this district.

Bibliography
Brown, G. C., 1916, Siskiyou County, Scott Bar district: California Min. Bur. Rept. 14, pp. 823–824.
O'Brien, J. C., 1947, Siskiyou County, Quartz Hill mine: California Jour. Mines and Geology, vol. 43, p. 447.

Shasta Copper-Zinc Belt

The Shasta copper-zinc belt is in west-central Shasta County in the foothills of the Klamath Mountains and a few miles north of Redding. The two main areas of mineralization are known as the West and East Shasta districts (fig. 27). Part of the East Shasta district has been inundated by Shasta Lake.

Gold- and silver-bearing gossans were originally mined in these districts during the 1860s. Later, from the 1890s to about 1920, copper and zinc ores were mined in large quantities and treated in several nearby smelters. Substantial amounts of by-product gold were recovered in these operations, especially in the West Shasta district. For example, the Mammoth mine, during the period of 1905 to 1925, yielded 132,510 ounces of gold from copper-zinc ore at an average of .039 ounces of gold per ton. During the 1930s substantial amounts of gossan were mined for gold near the Iron Mountain mine. The total gold output of the West Shasta district is estimated at 520,000 ounces. The total production of the East Shasta district is unknown, but during the period 1900–52, the district was credited with an output of 44,000 ounces of gold.

The ore deposits consist of either bodies of massive pyrite with varying amounts of chalcopyrite and sphalerite in rhyolite or sulfide minerals disseminated in schist. Veins or replacement deposits exist in schist and limestone. Most of the massive sulfide bodies are lenticular and range from a few tens of feet long to one at the Iron Mountain mine that is 4500 feet long. The gold content of these deposits usually ranges from .01 to .1 ounce per ton.

Shasta-Whiskeytown

Location and History. Shasta and Whiskeytown are in western Shasta County about 10 miles west-northwest of Redding. The Iron Mountain copper-zinc district is to the north, and the French Gulch gold district is to the northwest. Gold was discovered in Clear Creek, which flows through the area in 1849, and many mining camps were soon established. The largest and best known were Horsetown and Whiskeytown, which no longer exist, and Shasta, which was the first seat of government of Shasta County.

The Shasta camp is now a state historical monument, and many of the old buildings have been restored. There was some dragline dredging in the district in the 1930s.

Geology and Ore Deposits. Much of the gold production was from placer deposits in Clear Creek and its tributaries. Lode gold was recovered from pocket mines. Narrow and shallow quartz veins contain free gold and abundant sulfides in places. The largest source of lode gold apparently was the Mt. Shasta mine, which has yielded about $180,000. The gold-bearing veins occur either in granite or in greenstone and schist near granitic contacts.

Bibliography

Averill, C. V., 1933, Gold deposits of the Redding and Weaverville quadrangles: California Div. Mines Rept. 29, pp. 2–73.

Logan, C. A., 1926, Shasta County, Whiskeytown and Shasta districts: California Min. Bur. Rept. 22, pp. 168–169.

Smith River

Most of the gold produced in Del Norte County has come from placer-mining operations along the Smith River and its tributaries. These operations include the placer mines of Hurdy Gurdy, Monkey, Myrtle, and Craigs Creeks and the French Hill area. Gold has been obtained by mining the present stream gravels, terrace gravels adjacent to the present streams, and patches of the so-called Klamath "oldland cycle" gravels at such places as French Hill and Haines Flat. The terrace and "oldland" gravels were mined by hydraulicking. The principal period of mining was from the 1850s through the 1870s, but there has been small-scale intermittent work ever since. The estimated

Photo 69.   Carrville Gold Company Dredge, Trinity River District. This photo was taken on the upper Trinity River in Trinity County in 1940.
Photo courtesy of Yuba Consolidated Industries.

Photo 70. Steam Dragline Operation, Trinity River District. The photo was taken at Coffee Creek, Trinity County, in the 1920s. *Photo by C. V. Averill.*

total production is 40,000 ounces of gold. Chrome ore also was mined at French Hill during World Wars I and II.

Bibliography

Lowell, F. L., 1916, Del Norte County, placer gold: California Min. Bur. Rept. 14, pp. 386–389.

Maxon, J. H., 1933, Economic geology of portions of Del Norte and Siskiyou Counties: California Div. Mines Rept. 29, pp. 123–160.

O'Brien, J. C., 1952, Del Norte County, gold: California Jour. Mines and Geology, vol. 48, pp. 277–279.

Trinity River

The Trinity River, the southern part of the Klamath-Trinity River system, drains the southern part of the Klamath Mountain province. As this stream and its tributaries flow across auriferous rocks for much of their total lengths, they have been the sources of vast amounts of placer gold. The estimated output of the Trinity River placers is $35 million worth of gold. The Trinity River rises in the northeast corner of Trinity County in an area known as the Dodge district and flows in a general south-southwesterly direction for about 60 miles.

In the upper Trinity River area, the principal sources of gold have been the Carrville and Trinity Center districts. At Carrville the river was dredged until comparatively recently. Among several impor-

tant lode-gold mines here, the most productive have been the Trinity Bonanza King with a total output valued at $1.25 million and the Headlight, which has yielded $500,000. The gold-quartz veins occur in slate and greenstone, but granitic bodies are nearby. Coffee Creek, an important tributary, enters the area from the west. This creek was placer-mined for many years, and several high-grade quartz mines near its headwaters produced gold.

Trinity Center was settled in 1851. Several older bench gravel deposits were extremely rich in the early days and bucket-line dredges were active recently. From Trinity Center south through the Minersville district, including part of the Stuart Fork, the region is covered by the Trinity Reservoir. The well-known Fairview lode mine was in the Minersville district.

The Trinity River and adjoining terraces were extensively mined in the Eastman Gulch and Lewiston districts. One of the better-known lode mines was the Venecia mine. The gold-quartz veins at this mine have yielded more than $500,000. In the Lewiston and Douglas City districts, the river makes a number of extremely sharp bends that have formed several wide river bars, particularly at Starvation Flat near the town of Lewiston and at Gold Bar. The Douglas City area was extremely rich in the early days where the

Photo 71.   Hydraulic Mining of Bench Gravels, Trinity County. This is a 1933 scene at the Salyer mine. *Photo by Olaf P. Jenkins.*

highly productive Weaver, Indian, and Reading Creeks empty into the Trinity River.

Downstream, in the Junction City district, the river swings to the north and then west. The river has been dredged for a distance of at least eight miles here. The bench gravels are extensive and thick, and some of the hydraulicked banks are several hundred feet high. The largest bench deposits are at Coopers Bar, Hocker Flat, Benjamin Flat, and Chapman Ranch. Canyon Creek and Oregon Gulch flow into the river at Junction City, and the North and East Forks, at Helena. Moderate amounts of older gravels exist at Big Bar to the northwest.

Hayfork, which is about 15 miles southwest of Junction City, is a branch of the South Fork of the Trinity River. Most of the gold recovered in Hayfork Valley has been by dragline dredging, upstream in Hayfork Gulch, by hydraulicking. A number of narrow but often rich gold-quartz veins are found in slate and related rocks in the mountains just south of Hayfork. These include the recently active Kelly mine.

From Big Bar the main channel of the Trinity River flows in a northwesterly direction through the Burnt Ranch and Salyer districts. The highly productive New River, which drains the Denny district to the northeast, enters the Trinity River here (see section on New River-Denny district). At both Hawkins Bar and Salyer are bench gravels high above the present streams. Beyond Salyer the Trinity River is joined by the South Fork, and from this junction the river flows northerly through Willow Creek and Hoopa, site of

the recently active Copper Bluff copper-gold mine. The Trinity River then continues on to Weitchpec, where it empties into the Klamath River.

Weaverville

Location and History. Weaverville, the seat of government of Trinity County, is about 50 miles west-northwest of Redding. For years it was one of the major centers of gold mining in the Klamath Mountains. The area was settled son after Major Pearson B. Reading's gold discovery on the Trinity River in 1848. The stream and bench gravels were highly productive during the gold rush. The town was named for John Weaver, a prosperous Forty-Niner. By the middle 1850s many persons lived here, including several thousand Chinese, some of whom became involved in a tong war in 1854. The old Chinese joss house still stands and is now a state historical monument.

The La Grange mine, a few miles west of town and one of the major hydraulic mines in California, was opened in 1851. Large-scale hydraulicking began in 1862 and continued until 1918. From 1932 until 1942, further excavation at the mine, for the state highway, brought some gold production. The total output of the mine has been estimated by the author to be at least $8 million, although the commemorative plaque states that it is $3.5 million. More than 100 million cubic yards of material were excavated. Water was delivered from Stuart's Fork of the Trinity River via a 29-mile system of canals, flumes, and tunnels. During the 1930s

Photo 72. La Grange Hydraulic Mine, Weaverville District. This northward view of the mine, in Trinity County, was taken in about 1914. La Grange was one of the largest hydraulic mines in the state.

several other hydraulic mines and a considerable number of dragline dredges were active in the district.

Geology. The central portion of the district is underlain by an extensive series of continental sedimentary rocks known as the Weaverville Formation. This formation includes the auriferous channel gravels, as well as shale, sandstone, and tuff. In places the gravels are partly cemented and as much as 400 feet thick. The Weaverville Formation is underlain by schist, limestone, slate, and shale and to the east by granite of the Shasta Bally batholith. The gravel deposits at the La Grange mine lie in a trough bounded by a fault. Schist lies on the northwest and slate and limestone on the southeast. The base of the fault plane is a soft gouge. The richest zone in this mine was a 15-foot layer of blue gravel that yielded up to $2 in gold per yard. Some gold-quartz veins in the district contain free gold and varying amounts of sulfides. The ore bodies commonly are associated with diorite or "birdseye porphyry" dikes.

Bibliography

Averill, C. V., 1933, Gold deposits of the Redding and Weaverville quadrangles: California Div. Mines Rept. 29, pp. 2–73.
Averill, C. V., 1941, Trinity County, La Grange placer mines, Ltd.: California Div. Mines Rept. 37, pp. 43–44.
Diller, J. S., 1914, Auriferous gravels in the Weaverville quadrangle: U. S. Geol. Survey Bull. 540, pp. 11–21.
Diller, J. S., 1911, The auriferous gravels of the Trinity River basin: U. S. Geol. Survey Bull. 470, pp. 11–29.
Ferguson, H. G., 1914, Gold lodes of the Weaverville quadrangle: U. S. Geol. Survey Bull. 540, pp. 22–79.
Hinds, N. E. A., 1933, Geologic formations of the Redding-Weaverville districts: California Div. Mines Rept. 29, pp. 77–122.
Irwin, W. P., 1963, Preliminary geologic map of the Weaverville quadrangle: U. S. Geol. Survey Mineral Investigations Field Studies Map MF-275.
Logan, C. A., 1926, Trinity County, La Grange and Lorenz hydraulic mines: California Min. Bur. Rept. 22, pp. 39–43.
MacDonald, D. F., 1910, The Weaverville-Trinity Center gold gravels: U. S. Geol. Survey Bull. 430, pp. 48–58.

BASIN RANGES PROVINCE

The Basin Ranges occupy most of Mono and Inyo Counties and small portions of several other counties, including Modoc County in the northeast corner of the state (see fig. 3). These mountain ranges lie east of the Sierra Nevada and north of the Garlock fault, which separates them from the Mojave Desert. The Basin Ranges province is a region of roughly parallel mountain ranges alternating with basins or troughs and controlled by fault block structure. The region is underlain by granitic, sedimentary and metamorphic rocks of Precambrian, Paleozoic and Mesozoic ages, which in places are overlain by Cenozoic sedimentary and volcanic rocks. As in the Mojave Desert province, the gold occurs either as epithermal deposits in silicified and brecciated zones in volcanic rocks or as mesothermal gold-quartz veins in older metamorphic or granitic rocks. The largest source of gold in the Basin Ranges province has been the Bodie district in Mono County. Appreciable amounts of gold also have been mined in the Argus, Chloride Cliff, Russ, Skiddoo, and Ballarat districts.

Argus

Location and History. This district, in southern Inyo County in the Argus Range about 10 miles north of Trona, has also been known as the Kelley or Sherman mining district. The mines here apparently were first worked in the 1890s, although gold may have been discovered earlier. Considerable mining activity during the early 1900s and again in the 1930s was followed by intermittent prospecting and development work until the present time.

Geology. The rocks that underlie the district range from quartz monzonite to gabbro in composition, but the acidic intrusives predominate. The ore deposits occur either in quartz veins or in zones consisting of cemented, silicified breccia containing jasper, quartz veinlets, calcite, and abundant iron oxide. The gold is usually in a very fine state; sulfides are present only in some of the deposits.

Mines. Arondo $200,000, Davenport, Mohawk, Ruth $700,000+, Star of the West, Stockwell.

Bibliography

Norman, L. A., Jr., and Stewart, R. M., 1951, Inyo County, Arondo, Mohawk, and Ruth mines: California Jour. Mines and Geology, vol. 47, pp. 38–39, 46, and 49–50.
Tucker, W. B., 1938, Inyo County, gold: California Div. Mines Rept. 34, pp. 379–424.

Ballarat

Location and History. Ballarat is in south-central Inyo County on the west flank of the Panamint Range and just west of Death Valley National Monument. It was named for the Ballarat mining district in Australia. It includes the South Park area to the south. The old silver-mining camp of Panamint City is just to the north. The Ratcliff mine, the largest gold source in the district, was discovered in 1897, and considerable mining activity lasted until about 1915. The mines were active again from 1927 until 1942, and there has been intermittent prospecting and development work since.

Geology. The district is underlain by schist, dolomitic limestone, and gneiss of Precambrian age, which in places have been cut by granitic dikes. The ore deposits consist of quartz veins containing free gold and occasionally abundant sulfides.

Mines. Cecil R., Knob, Lestro Mountain, Lotus, Porter, Ratcliff $1.3 million+, Thorndyke, World Beater.

Bibliography

Jennings, C. W., 1958, Death Valley sheet: California Div. Mines geologic map of California, Olaf P. Jenkins edition.
Norman, L. A., Jr., and Stewart, R. M., 1951, Inyo County, Lotus and Ratcliff mines: California Jour. Mines and Geology, vol. 47 pp. 45–48.

Photo 73. Standard Consolidated Mine, Bodie District. This northeast view shows the mill buildings and mine dumps at the mine, in Mono County. The Standard Consolidated has produced more than $18 million in gold and silver. Photo courtesy of Calif. Division of Beaches and Parks.

Beveridge

Location and History. The Beveridge district is in west-central Inyo County in the Inyo Mountains. The area was first worked in the late 1870s, when the Big Horn and Keynote mines were discovered. Mining operations continued fairly steadily until the early 1900s. There was activity in the district again in the 1930s, and there has been intermittent prospecting since.

Geology and Ore Deposits. The region is underlain by a series of northwest-trending beds of limestone, quartzite, and schist that have been intruded by quartz monzonite and other granitic rocks. Quartz veins occur both in the granitic and metamorphic rocks. The veins strike north and usually range from two to eight feet in thickness. The ore contains some free gold, but much of the value is in sulfides, which are abundant in places. Some copper, lead, silver, and zinc have been produced in the district. The greatest depth of development is about 500 feet.

Mines. Big Horn, Burgess, Cinnamon, Golden Eagle, Gold Standard, Keynote $500,000, Mountain View, Tom Casey.

Bibliography

Jennings, C. W., 1958, Death Valley sheet: California Div. Mines geologic map of California, Olaf P. Jenkins edition.

Tucker, W. B., and Sampson, R. J., 1938, Inyo County—gold mines: California Div. Mines Rept. 34, pp. 379–424.

Waring, C. A., 1919, Inyo County, gold mines: California Min. Bur. Rept. 15, pp. 75–85.

Big Pine

Several gold mines and prospects exist on the west slope of the White Mountains in northern Inyo County a few miles northeast of Big Pine and southeast of Bishop. The deposits consist of narrow quartz veins in Paleozoic metasediments cut by granitic dikes. The ore contains free gold, pyrite, and varying amounts of other sulfides.

Bodie

Location. The Bodie district is in eastern Mono County about 18 miles southeast of Bridgeport.

History. Gold was discovered here in 1859 by William S. Bodey, and the district was organized in 1860, but activity was minor until 1872, when rich ore was discovered. From 1876 to about 1884, a rush was on with much production from rich but shallow deposits. By 1888, the district had yielded more than $18 million, but, thereafter until World War II, mining was confined to lower-grade deposits and reworking of old tailings.

From 1881 until 1914 timber was delivered to the mines by a narrow-gauge railroad from the east side of Mono Lake. Bodie was one of the first mining camps to use electricity (1893). Most of the important mines came under the control of J. S. Cain in 1915. The town became a noted tourist attraction in the 1940s, although many of the buildings had been destroyed by fire in 1932. The remaining portion be-

Photo 74.   Red Cloud Mine, Bodie District. This 1931 view of the mine, in Mono County, looks north. The Red Cloud was a part of the Standard Consolidated group. *Photo by Olaf P. Jenkins.*

came a state park in 1961. Studies have been made to determine the feasibility of working the entire Standard Hill area as a large open-pit operation. Bodie is the most productive district in the Basin Ranges, with a total production estimated to be valued at slightly more than $30 million. The district also has yielded more than 1 million ounces of silver.

Geology and Ore Deposits.   A number of steep, north-trending silicified and brecciated zones and narrow quartz veins occur in Tertiary andesite. They are especially common in the Standard mine area, where the mineralized zone is as much as 1000 feet wide. Most of these veins and brecciated zones pinch out at depths of around 1000 feet, but some in the central portion of the district are reported to be deeper. The deepest shaft is 1200 feet. Most of the values have been recovered from above 500 feet. The ore contains finely disseminated free gold in both the quartz and silicified breccia with little or no sulfides. In only one vein, the Addonda Oro of the Southern Consolidated group, is pyrite abundant. The high-grade ore recovered in the early days from shallow workings yielded from $100 to $300 of gold and silver per ton.

Mines.   Bechtel Cons. $200,000+, Bodie Tunnel $200,000+, Bulwar $428,000+, Mono $122,000+, Southern Cons. $1 million+, Standard Cons. $18 million+, Syndicate $1 million+.

Bibliography

Brown, R. A., 1908, The vein system of the Standard mine, Bodie, California: Trans. Am. Inst. Min. Engrs., vol. 38, pp. 343–357.
Cain, Ella M., 1956, The Story of Bodie, Fearon Publishers, San Francisco, 196 pp.
Crawford, J. J., 1896, Standard mine: California Min. Bur. Rept. 13, p. 231.
Eakle, A. S., and McLaughlin, R. D., 1919, Mono County, Bodie district: California Min. Bur. Rept. 15, pp. 149–160.
Tucker, W. B., and Sampson, R. J., 1940, Mono County, Southern Cons. and Standard Cons. mines: California Division of Mines Rept. 36, pp. 136–138.
Whiting, H. A., 1888, Bodie district: California Min. Bur. Rept. 8, pp. 382–401.

Chloride Cliff

Location and History.   Chloride Cliff is in the Funeral Range in the eastern part of Death Valley National Monument, about 20 miles north of Furnace Creek. It is sometimes known as the South Bullfrog district. Gold was probably discovered here at an early date, but the chief period of activity was from around 1900 to 1916 when the Keane Wonder and Chloride Cliff mines were active. There has been minor work since. The Keane Wonder mine is credited with a total output of more than $1 million.

Geology.   The district is underlain predominantly by Precambrian schist, quartzite and gneiss, which in places have been cut by dioritic dikes. The ore bodies occur in lenticular quartz veins as much as 30 feet thick. The ore contains fine free gold, pyrite, and galena. Most of the ore contained ½ ounce of gold or less per ton, but the ore shoots were as long as 300 feet.

Bibliography

Tucker, W. B., 1938, Inyo County, Keane Wonder mine: California Div. Mines Rept. 34, pp. 402–403.
Waring, C. A., and Huguenin, Emile, 1919, Inyo County, Chloride Cliff and Keane Wonder mines: California Min. Bur. Rept. 15, pp. 76–77 and 79–81.

Clover Patch

Gold has been recovered from a number of mines in southern Mono County, in the Clover Patch mining district. It includes the areas known as the Chidago and Indian districts. The Blind Spring Hill silver mining district is a few miles to the northeast, and Casa Diablo is to the south. The area was first worked prior to 1900, and a number of mines were worked or prospected again in the 1930s.

The quartz veins, which are as much as 10 feet thick and often brecciated, contain free gold, pyrite, and smaller amounts of other sulfides. In places the ore contained ½ to one ounce of gold per ton. Country rock consists of granite, rhyolite, quartzite and limestone.

Mines.   Casa Diablo, Clover Patch, El Dorado, Evening Star, Gold Crown, Last Chance, Mary B, Sierra Vista, Sour Dough.

Bibliography

Sampson, R. J., and Tucker, W. B., 1940, Mono County, gold: California Div. Mines Rept. 36, pp. 120–140.

El Paso Mountains

Location. The El Paso Mountains are in northeastern Kern County, some 10 miles northwest and north of Randsburg. A series of dry placer "diggings" lies between Redrock Canyon on the southwest and the Summit "diggings" to the northeast. The district includes the areas known as the Goler, Garlock and Searles districts.

History. Gold was discovered in Goler Canyon in 1893, and dry washing camps soon sprang up at Last Chance, Red Rock, Jawbone Canyon and Summit Diggings. Mining activity declined by 1900, but a number of operations were reactivated during the 1930s, and since World War II, there has been minor prospecting. In these dry placer districts, the easily recoverable gold was mined at one locality in a few months to a year or two, and the miners moved on to other areas.

Ore Deposits. Auriferous sands and gravels occur in benches above the present canyons and on bedrock in the washes and canyons themselves. Much of the gold is believed by Hulin (1934) to have been derived from the erosion and reworking of the basal conglomerate of the Ricardo Formation (lower Pliocene), which is extensive in this region. The gold particles are round and show evidence of considerable abrasion. The gold is mostly fine, although nuggets of up to several ounces have been recovered. Some narrow gold-quartz veins occur in granite and schist.

Bibliography

Dibblee, T. W., Jr., and Gay, T. E., Jr., 1952, Mineral deposits of the Saltdale quadrangle: California Div. Mines Bull. 160, pp. 47–49.

Hess, F. L., 1909, Gold mining in the Randsburg quadrangle: U. S. Geol. Survey Bull. 430, pp. 23–47.

Hulin, C. D., 1934, Geologic features of the dry placers of the northern Mojave Desert: California Div. Mines Rept. 30, pp. 417–426.

Troxel, B. W., and Morton, P. K., 1962, Kern County, El Paso Mountains district: California Div. Mines and Geology, County Rept. 1, pp. 29–31.

Tucker, W. B., and Sampson, R. J., 1933, Goler Canyon placer district: California Div. Mines Rept. 29, p. 281.

Tucker, W. B., Sampson, R. J., and Oakeshott, G. B., 1949, Kern County, Goler Canyon placer and Janney group of placers: California Jour. Mines and Geology, vol. 45, pp. 223 and 225.

Fish Springs

The Fish Springs or Tinemaha district, in northwestern Inyo County about eight miles south of Big Pine, has several small mines and prospects. The New Era mine was active in the 1940s. The deposits consist of gold-quartz veins in granitic rocks that are commonly associated with diorite dikes. The deposits usually consist of a series of narrow parallel veins.

Bibliography

Norman, L. A., Jr., and Stewart, R. M., 1951, Inyo County, New Era mine: California Jour. Mines and Geology, vol. 47, p. 47.

Grapevine

Some gold was recovered at one time from Grapevine Canyon, which is in the north end of the Grapevine Mountains in eastern Inyo County and is part of Death Valley National Monument. Several gold-quartz veins occur in metamorphosed sedimentary rocks of Paleozoic age. Scotty's Castle, for many years the home of Death Valley Scotty and now a well-known tourist attraction, is located in Grapevine Canyon.

Harrisburg

Location and History. Harrisburg or Harrisburg Flat is in east-central Inyo County in Dealth Valley National Monument. It is about five miles north of Wildrose Canyon and nine miles south of Skidoo. Gold was discovered here in 1905 by Shorty Harris, one of the most colorful and best-known "single-blanket jackass prospectors" of the Death Valley region. He was also the first settler in Rhyolite, Nevada. Harrisburg, which was mainly a tent city, lasted for only a few years. The chief source of gold in the district was the Independent or Cashier mine, which had an output valued at about $300,000.

Geology. There are several lenticular north- to northwest-striking gold-bearing quartz veins in dolomitic limestone. Granitic rocks also crop out in the area. The ore contained free gold and some sulfides. Much of the ore averaged about one ounce of gold per ton, but the values do not extend to depths of more than 150 feet.

Bibliography

Norman, L. A., Jr., and Stewart, R. M., 1951, Inyo County, Independent mine: California Jour. Mines and Geology, vol. 47, p. 44.

Waring, C. A., 1919, Inyo County, Cashier mine: California Min. Bur. Rept. 15, pp. 75–76.

High Grade

Location. The High Grade district is in northeastern Modoc County near the California-Oregon border. The district is on the crest of the Warner Mountains about 10 miles northwest of Fort Bidwell and 50 miles northeast of Alturas. It was known as the Hoag district until 1912.

History. This region was first settled in the 1850s. Several Indian wars were fought in the area, and Fort Bidwell was a U.S. Cavalry post from 1865 to 1892. According to local legend a prospector named Hoag found gold in the Warner Mountains northwest of the fort. However, he was killed by Indians soon after his discovery. Modoc County has a recorded gold production from 1880 to 1885, which may have come from this district.

Gold was rediscovered by a sheepherder in the Warner Mountains in 1905. A boom lasted for a few years, and as many as several hundred men were employed in the mines. Intermittent prospecting and development work continued through the 1920s and 1930s, and there has been minor work since. It is doubtful if the district has yielded more than several hundred thousand dollars worth of gold.

Geology and Ore Deposits. The rocks in the main part of the district consist of white to yellow rhyolite of Tertiary age. To the west and south is andesite, and to the east is basalt, both of Miocene age. The ore deposits consist of narrow, north- and northwest-trending epithermal veins and replacements in rhyolite. The vein material consists of quartz, silicified and brecciated country rock, and fault gouge. The ore contains free gold, finely divided pyrite, and manganese-stained material. A few veins in andesite contain minor amounts of copper. None of the deposits extends to a depth of more than a few hundred feet.

Mines. Blue Bell, Fort Bidwell Cons., Klondike, Modoc, Northern Star, Sunset, Sunshine.

Bibliography

Averill, C. V., 1929, Modoc County, gold: California Div. Mines Rept. 25, pp. 10–19.

Averill, C. V., 1936, Modoc County, High-Grade district: California Div. Mines Rept. 32, pp. 448–451.

Gay, T. E., Jr., and Aune, Q. A., 1958, Alturas sheet: California Div. Mines, geologic map of California.

Hill, James M., 1915, High-Grade district: U. S. Geol. Survey Bull. 594.

Tucker, W. B., 1919, Modoc County, High Grade mining district: California Min. Bur. Rept. 15, pp. 241–250.

Lees Camp-Echo Canyon

These two small adjoining districts are in eastern Death Valley National Monument in the Funeral Mountains, near the Nevada-California border. Most of small mines and prospects here have been idle for many years. The deposits consist mainly of narrow gold-quartz veins in metamorphic rocks of Precambrian age.

Bibliography

Jennings, C. W., 1958, Death Valley sheet: California Div. Mines geologic map of California, Olaf P. Jenkins edition.

Masonic

Location and History. The Masonic district, in northeastern Mono County near the Nevada line, extends into Nevada. It is 12 miles northeast of Bridgeport and 16 miles northwest of Bodie. The region was prospected for some years during and after the Comstock rush of the early 1860s, but valuable ore was not discovered until 1902. The chief period of production was 1907–10, although some activity continued through the 1930s, and the Chemung mine has been intermittently worked in recent years.

Geology and Ore Deposits. The district is underlain by coarse-grained porphyritic granite and small amounts of schist. Tertiary basalt and andesite surround the area. The ore deposits are thick silicified zones or veins in the granite that strike north, northwest, or northeast. The ore consists of brecciated and recemented chert, quartz, and chalcedony that contains fine free gold. Pyrite and chalcopyrite are present in places. The ore has an open porous appearance and is often managanese-stained, and the values commonly appear in thin seams near the openings. Milling ore usually contains less than one ounce of gold per ton, but appreciable high-grade has been recovered. None of the deposits has been developed to depths of more than a few hundred feet.

Mines. Chemung $60,000, Home View, Lakeview, Maybell, Perini, Pittsburg-Liberty $700,000, Rough-and-Ready, Serita $500,000.

Bibliography

Boalich, E. S., 1923, Mono County, Masonic district: California Min. Bur. Rept. 18, pp. 415–416.

Eakle, A. S., and McLaughlin, R. P., 1919, Mono County, Masonic district: California Min. Bur. Rept. 15, pp. 160–165.

Modoc

Gold has been produced in the Modoc lead-silver district, which is in southern Inyo County at the north end of the Argus Range and about 10 miles east of Darwin. The small mines and prospects here have been idle for some time. The deposits consist of narrow quartz veins containing free gold, pyrite, galena, and chalcopyrite. They occur in granitic rocks and schist.

Patterson

Location and History. This is a gold-silver district in the Sweetwater Mountains, in northern Mono County about 15 miles north of Bridgeport. It includes the Silverado and Fryingpan Canyon areas. The area was probably first prospected in the early 1860s, but the principal period of mining activity was 1880 to 1884, when more than $500,000 was produced. Settlements that once existed in the district were Belfort, Monte Cristo, and Star City. Some mining was done in the district again in the early 1900s and 1930s, and there has been some prospecting since.

Geology and Ore Deposits. The district is underlain by various types of granitic rocks and andesite and rhyolite. The gold- and silver-bearing deposits occur in north-trending veins of quartz and silicified breccia that are up to 10 feet thick. The ore contains pyrite, argentite, cerargyrite, and often abundant iron oxides. In places, the ore was high in grade. Ore shoots, with stoping lengths of up to several hundred feet, were mined.

Mines. Anglo Mission, Frederick, Kentuck, Longstreet, Montague, Silverado, Star and Great Western, Summers, Tiger.

Bibliography

Eakle, A. S., and McLaughlin, R. P., 1919, Mono County, Patterson district: California Min. Bur. Rept. 15, pp. 165–166.

Sampson, R. J., 1940, Mono County, Silverado and Kentuck mines: California Div. Mines Rept. 36, pp. 145–146.

Rademacher

Location. This district is in northeastern Kern County about 15 miles north of Randsburg and five miles south of Ridgecrest. It was organized in the 1890s, and the most active period was in the early 1900s.

Geology. The area is underlain by Mesozoic granitic rocks containing small pendants of metamorphic rocks and cut by many dikes. Acidic dikes are most common to the east but become more basic to the west. Numerous faults are present. A number of narrow, north-trending quartz veins often cut the dikes. The ore contains free gold with varying amounts of

Photo 75. Town of Skidoo, Skidoo District. This 1907 view of the town, in Inyo County, looks east. Photo courtesy of Calif. State Library.

sulfides and manganese oxide. Milling-ore usually averages ½ ounce or less of gold per ton, and the ore shoots usually are narrow with short stoping lengths.

Mines. Apple Green, Bellflower, Broken Axle, Butte, Crown Cons., Gold Bug, Gold Pass, Hillside, Huntington, Indian Wells Valley, Jerry, Lehigh Valley, Lost Keys, Northern View, Prize, Rademacher, Red Wing, Stardust, Star Lode, Stellar Group, Townsend, Vera Queens, White Star, Wildcat, Yellow Treasure.

Bibliography

Troxel, B. W., and Morton, P. K., 1962, Kern County, Rademacher mining district: California Div. Mines and Geology, County Rept. 1, pp. 46–47.

Tucker, W. B., and Sampson, R. J., 1933, Kern County, Rademacher mining district: California Div. Mines Rept. 29, pp. 284–285.

Russ

This district is in west-central Inyo County on the west slope of the Inyo Mountains about nine miles southeast of Independence. The largest source of gold has been the Reward or Brown Monster group of mines. The veins were discovered in 1878 and worked steadily until 1914. The Reward mine was active again during the 1930s and 1940s, and there has been some work in the area since. The ore deposits are northwest-striking quartz veins up to 12 feet thick, and the ore contains free gold and often abundant sulfides. Moderate amounts of lead, silver, and copper also have been produced here. The district is underlain by schist, slate, and limestone with granitic rocks to the east.

Bibliography

Norman, L. A., Jr., and Stewart, R. M., 1951, Inyo County, Reward mine: California Jour. Mines and Geology, vol. 47, pp. 48–49.

Tucker, W. B., and Sampson, R. J., 1938, Inyo County, Reward-Brown Monster mines: California Div. Mines Rept. 34, pp. 386–388.

Skidoo

Location and History. Skidoo is in Death Valley National Monument in east-central Inyo County. The district is in the Panamint Mountains and includes the Tucki Mountain area to the north. The town of Skidoo was an important mining center from 1905 until around 1917, with water reaching the area through 23 miles of pipe from Telescope Peak to the south. Mining was done in the district again during the 1930s, and there has been intermittent prospecting since. The total production of the district has been estimated at between $3 million and $6 million.

Geology and Ore Deposits. The area is underlain by quartz monzonite and other granitic rocks. Schist, dolomitic limestone, and gneiss of Precambrian age are just to the east. The ore deposits consist of a number of north- and northwest-striking quartz veins that contain free gold and small amounts of pyrite. The veins are chiefly in the quartz monzonite and most are only a few feet thick. A number of high-grade pockets have been found here.

Mines. Del Norte, Emigrant Springs, McBride, Napoleon, Skidoo $1.5 million+, Sunset, Treasure Hill, Tucki.

Bibliography

Jennings, C. W., 1958, Death Valley sheet: California Div. Mines geologic map of California, Olaf P. Jenkins edition.

Norman, L. A., Jr., and Stewart, R. M., 1951, Inyo County, Skidoo mine: California Jour. Mines and Geology, vol. 47, p. 51.

Waring, C. A., 1919, Inyo County. Skidoo mine: California Min. Bur. Rept. 15, pp. 83–84.

Slate Range

The Slate Range district, in northwestern San Bernardino and southern Inyo Counties, has sometimes been known as the Arondo district. Gold occurs in several places in the Slate Mountains, the principal source apparently having been the Hafford mine. The area is underlain by granite and schist. The deposits consist of narrow quartz veins that contain small but rich gold- and silver-bearing pockets. In places sulfides are quite abundant.

Bibliography

DeGroot, Henry, 1890, Slate Range district: California Min. Bur. Rept. 10, p. 533.

Tucker, W. B., and Sampson, R. J., 1943, San Bernardino County, Hafford mine: California Div. Mines Rept. 39, pp. 453–454.

Spangler

Spangler is in northwestern San Bernardino County about 10 miles northeast of Johannesburg. Most of the gold production has been from the Spangler mine, which has been intermittently prospected and developed since the 1890s. A number of narrow west-striking gold-quartz veins traverse granitic rock. Some of the ore contained more than one ounce of gold per ton.

Bibliography

Wright, L. A., et al, 1953, San Bernardino County, Spangler mine: California Jour. Mines and Geology, vol. 49, p. 58.

Tibbetts

This district is in the Inyo Range in northern Inyo County about 10 miles northeast of Independence. A number of narrow quartz veins in Peleozoic metasediments and granitic rocks bear free gold and often abundant sulfides. Placer deposits, including those of Mazourka Canyon, were worked by dry placer methods from 1894 until 1906. The area was prospected again in the 1930s.

Bibliography

Tucker, W. B., and Sampson, R. J., 1948, Inyo County, Mazourka Canyon placers: California Div. Mines Rept. 34, p. 411.

Ubehebe

Some gold has been produced in the Ubehebe copper-lead-silver district. The district is in central Inyo County about 75 miles east of Lone Pine. The chief source of gold has been the Lost Burro mine, which was worked from 1906 to 1917 and 1934 to 1942. The value of its total output is about $100,000. The gold occurs in a flat four-foot-thick vein of quartz, jasper, and calcite that is near a contact between limestone and granitic rocks. The gold occurs in the native state with pyrite and chalcopyrite.

Bibliography

McAllister, J. F., 1955, Geology of mineral deposits in the Ubehebe Peak quadrangle, California: California Div. Mines Spec. Rept. 42, 63 pp.

Waring, C. A., and Huguenin, Emile, 1919, Inyo County, Lost Burro mine: California Min. Bur. Rept. 15, pp. 81–82.

White Mountains

Gold has been mined in several areas in the White Mountains in southeastern Mono County. The streams apparently were first placer-mined in the 1860s and 1870s, and lode deposits were discovered soon afterward. There has been intermittent development work and prospecting since. The principal lode mines have been the Sacramento and Twenty Grand mines. The deposits consist of gold-quartz veins with considerable sulfides in granitic rocks. Some schist and limestone crop out in the area. The placer deposits are mostly in the canyons on the west flank of the range.

Bibliography

Sampson, R. J., and Tucker, W. B., 1940, Mono County, gold: California Div. Mines Rept. 36, pp. 120–140.

Wildrose

Some gold has been recovered from placer and lode deposits in the Wildrose Canyon area in east-central Inyo County. The area, in the Panamint Mountains, is part of Death Valley National Monument. Apparently the main sources of lode gold have been the Burro, Gem, and New Discovery mines, which were active in the 1930s and 1940s. At these mines, there is a series of northwest-striking quartz veins in granitic rocks, schist, and gneiss. Most of the gold occurs in sulfides, which are extremely abundant in places. The placers were small and discontinuous.

Bibliography

Norman, L. A., Jr., and Stewart, R. M., 1951, Inyo County, Corona (New Discovery and Gem) mine: California Jour. Mines and Geology, vol. 47, pp. 40–41.

Sampson, R. J., 1932, Inyo County, Burro, New Discovery and Gem mines: California Div. Mines Rept. 28, pp. 364–366.

Willow

This district is in southern Inyo County in the Black Mountains about 15 miles west of Shoshone and just east of Death Valley. The chief source of gold has been the Ashford mine, which has a reported output of $135,000, and the Confidence mine. A number of gold-copper-quartz veins occur in gneiss and schist. The deepest workings are about 375 feet. Some high-grade ore has been recovered. The old Ashford mill in Death Valley is now an historical exhibit.

Bibliography

Norman, L. A., Jr., and Stewart, R. M., 1951, Inyo County, Ashford and Confidence mines: California Jour. Mines and Geology, vol. 47, pp. 39–40.

MOJAVE DESERT PROVINCE

Gold deposits are widely distributed throughout this vast area in southeastern California. The Mojave Desert is a broad interior region of mountain ranges separated by expanses of desert plains. The western part of the province is wedge-shaped with the Sierra Nevada to the north and the Transverse Ranges to the south. The primary deposits consist of either mesothermal gold-quartz veins that occur in metamorphic and granitic rocks of Precambrian, Paleozoic, and Mesozoic ages or epithermal deposits in zones of silicification and brecciation in volcanic rocks of Tertiary age.

The largest sources of gold in this province have been the Rand and Mojave-Rosamond districts in Kern County. Other important gold sources have been the Dale and Stedman districts, San Bernardino County, and the Cargo Muchacho-Tumco and Picacho districts in Imperial County. Placer gold has been recovered in quantity in several of the districts, considerable amounts having come from dry desert placers. The most productive dry placers have been in the Rand, Cargo Muchacho, Chocolate Mountains, Picacho, and Potholes districts. By-product gold has been recovered from a number of silver, copper, lead, and zinc mines in this province.

Alvord

Location. This district is in central San Bernardino County about 35 miles northeast of Daggett. It is named for the Alvord mine, the chief producer in the district. Gold was discovered here in 1885, and the Alvord mine has been intermittently worked ever since.

Geology. The area is underlain by crystalline limestone, granite, and in places by volcanic rocks. A siliceous vein at the Alvord mine contains jasper, calcite, hematite, pyrite, limonite, and free gold. Minor copper mineralization is present. Ore mined in the past commonly yielded ½ ounce of gold or more per ton.

Bibliography

Storms, W. B., 1893, Alvord mine: California Min. Bur. Rept. 11, pp. 359–360.
Wright, L. A., et al, 1953, San Bernardino County, Alvord mine: California Jour. Mines and Geol., vol. 49, p. 70.

Arica

The Arica district, in the Arica Mountains of northeastern Riverside County, also has been known as the Onward district. The Brown and Lum-Gray mines, the principal sources of gold, were active during the early 1900s and again in the 1920s and 1930s. A number of gold-quartz veins occur in granite and schist. Sulfides are extremely abundant, and the milling ore was reported to have yielded as much as one ounce of gold per ton. The Lum-Gray shaft is nearly 1000 feet deep.

Bibliography

Merrill, F. J. H., 1919, Riverside County, Arica Mountain district: California Min. Bur. Rept. 15, pp. 541–542.
Tucker, W. B., and Sampson, R. J., 1945, Riverside County, Brown and Lum-Gray mines: California Div. Mines Rept. 41, pp. 128 and 138.

Arrowhead

Location. The Arrowhead district is in eastern San Bernardino County, 20 miles north-northwest of Danby in the southwest end of the Providence Mountains. The Hidden Hill mine was located in 1882. For several years following, Mexican miners recovered rich surface ores concentrated in arrastras. The Big Horn mine was worked on a large scale in 1918–19, and there was some activity during the 1920s and 1930s.

Geology. The principal rock in the district is quartz monzonite, which is cut by numerous diorite dikes, and in smaller amounts, quartzite and andesite. The ore deposits occur in north-trending quartz veins that are closely associated with the dikes. The ore contains free gold, pyrite, and chalcopyrite. Among the high-grade pockets found here, one 300-pound lot of ore recovered at the Hidden Hill mine in 1915 yielded $13,000. The chief sources of gold have been the Coarse Gold, Big Horn, and Hidden Hill mines.

Bibliography

Cloudman, H. C., et al, 1919, San Bernardino County, Arrowhead district: California Min. Bur. Rept. 15, pp. 800–801.
DeGroot, Henry, 1890, Arrowhead district: California Min. Bur. Rept. 10, p. 532.
Tucker, W. B., and Sampson, R. J., 1943, San Bernardino County, Big Horn mine: California Div. Mines Rept. 39, pp. 441–442.

Bendigo

The Bendigo or Riverside Mountain district is in the northeast corner of Riverside County. Gold was discovered here at the Mountaineer mine in 1898, and mining continued until around 1920. Some work was done in the district again in the 1930s. This region is underlain by limestone and schist, which in places are cut by diorite dikes. The ore bodies, replacement deposits that occur along limestone-schist contacts, contain gold, copper, silver, and manganese. Some of the ore deposits are as much as 15 feet thick, and extend to depths of 200 feet.

Mines. Alice, Gold Dollar, Jacknife, Mountaineer (Calzona), Morning Star, Steece.

Bibliography

Merrill, F. J. H., 1919, Riverside County, Bendigo district: California Min. Bur. Rept. 15, pp. 542–544.
Tucker, W. B., and Sampson, R. J., 1945, Riverside County, Mountaineer mine: California Div. Mines Rept. 41, p. 140.

Cargo Muchacho-Tumco

Location. This district is an extensive area in the Cargo Muchacho Mountains in southeastern Imperial County, seven miles north of Ogilby and 50 miles east of El Centro. The district includes not only the area

Photo 76. Golden Cross Mine, Cargo Muchacho District. This view of the mine, at Tumco, Imperial County, looks west. The photo was taken in about 1915.

known as the Cargo Muchacho district but also the area known as the Tumco or Hedges mining district.

History. Mining was first done in this region by Spaniards as early as 1780–81, when placers in Jackson Gulch and oxidized ores in Madre Valley were worked. This is believed to have been the first gold mined in California. Later, mining was resumed under Mexican rule. The district received its name of Cargo Muchacho, or Loaded Boy, when two young Mexican boys came into camp one evening with their shirts loaded with gold. American miners became interested in this district soon after the end of the Mexican War in 1848. Mining became firmly established in 1877 with the completion of the Southern Pacific Railroad to Yuma. Large-scale mining continued from around 1890 until 1916 and again from 1932 until 1941, with intermittent activity since World War II.

Geology and Ore Deposits. The Cargo Muchacho Mountains are composed of quartzites and schists that have been intruded by granitic rocks. In places there are andesite and dioritic dikes.

The gold deposits are on the west side of the range and occur in both the metamorphic and granitic rocks. They are tabular bodies with a definite hanging wall or footwall but rarely both. The deposits consist of quartz, calcite, sericite, and chlorite, and the values are either native gold or auriferous sulfides. Appreciable amounts of silver and copper also have been recovered in the district. The deposits, usually striking west, with a few north-strike exceptions, are up to eight feet thick and have been mined to depths of as much as 1000 feet. Appreciable high-grade ore was found here.

Mines. American Boy, Amercian Girl $1 million, Big Bear, Blossom, Butterfly, Cargo Muchacho $100,-000+, Coffee Pot, Colorado, Desert King, Golden Cross $3 million+, Golden Queen, Guadaloupe, Little Bear, Madre and Padre $100,000+, Ogilby group, Pasadena, Sovereign, Vitrafax, White Cap.

Bibliography

Crawford, J. J., 1896, Cargo Muchacho district: California Min. Bur. Rept. 13, p. 333.

Henshaw, P. C., 1942, Geology and mineral deposits of the Cargo Muchacho Mountains, Imperial County, California: California Div. Mines, Rept. 38, pp. 147–196.

Merrill, F. J. H., 1916, Imperial County, Cargo Muchacho Range: California Min. Bur. Rept. 14, pp. 725–729.

Sampson, R. J., and Tucker, W. B., 1942, Gold—Imperial County: California Div. Mines Rept. 38, pp. 112–126.

Tucker, W. B., 1926, Gold—Imperial County: California Min. Bur. Rept. 22, pp. 253–261.

Chocolate Mountains

Gold has been recovered from the southeast end of the Chocolate Mountains in eastern Imperial County in an area east of Glamis, a stop on the Southern Pacific Railroad. The district includes the area known as the Mesquite mining district. The Paymaster lead-silver and manganese mining district is just to the north. This district was first prospected prior to 1900 both by quartz mining and small-scale dry placer methods. During the 1930s, several unsuccessful short-lived attempts were made to work the dry placer deposits on a large scale. Gold and silver associated with iron oxides in quartz veins occur in granitic rocks. The veins usually are narrow, but several high-grade pockets have been discovered. The placer deposits occur in the washes along the south and west flanks of the range.

Mines. Desert Gold, Gold Basin, Gold Delta, Mary Lode, Mesquite Lode, Peg Leg, Rainbow, Vista.

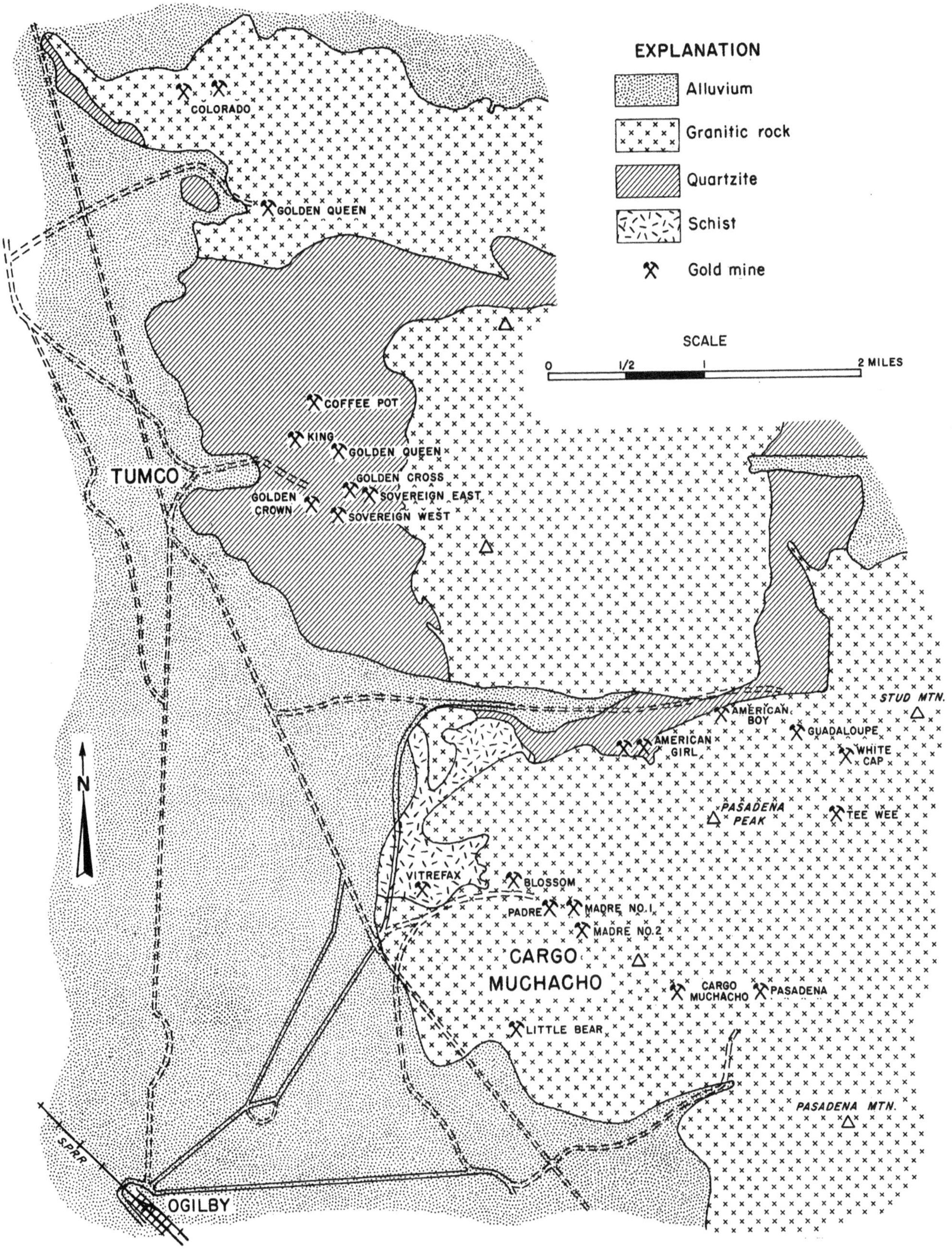

Figure 28.   Geologic Map of Cargo Muchacho-Tumco District, Imperial County. *After Henshaw, 1942, plate 2.*

Photo 77. Dry Placer Mining, Coolgardie District. The photo was taken in San Bernardino County in the early 1900s. *Photo courtesy of O. A. Russell, Yermo.*

Bibliography

Sampson, R. J., and Tucker, W. B., 1942, Imperial County, gold: California Div. Mines Rept. 38, pp. 112–126.

Chuckwalla

Location and History. The Chuckwalla district is in the Chuckwalla Mountains of southeastern Riverside County, south of Desert Center. This district at one time was also known as the Pacific mining district. It was organized some time in the 1880s, and mining continued through the early 1900s. There was activity here again in the 1930s, when the Red Cloud and other mines were worked.

Geology and Ore Deposits. The region is underlain by granitic rocks and gneiss. The gold-quartz veins often contain abundant pyrite and copper, lead, and silver minerals. A number of high-grade pockets have been recovered. Several of the veins have mined to depths of about 350 feet.

Mines. Baumonk, Bryan, Coffee, Granite, Great Western, Lost Pony, Model, Red Cloud $100,000+, Sterling, Sunnyside.

Bibliography

Merrill, F. J. H., 1919, Riverside County, Chuckwalla district: California Min. Bur. Rept. 15, pp. 538–540.

Orcutt, C. R., 1890, Pacific mining district: California Min. Bur. Rept. 10, pp. 900–901.

Tucker, W. B., and Sampson, R. J., 1945, Riverside County, Red Cloud mine: California Min. Bur. Rept. 41, pp. 141–142.

Clark

Location and History. This district is in northeastern San Bernardino County in the Clark Mountains about 35 miles northeast of Baker. Gold and other metals have been mined here since the early 1860s, and the mining district was organized in 1865. The mountains were named for Senator William A. Clark, the "copper king" of Montana. The gold mines were worked intermittently until the 1930s, and there has been prospecting since. The Mountain Pass mine located here is now an important source of rare earth minerals.

Geology and Ore Deposits. The district is underlain by a belt of limestone and dolomite in the central portion, with quartzite to the west and granite and gneiss to the east. The gold-bearing deposits consist of quartz and barite veins or mineralized breccia, the latter occurring in shear zones in gneiss that commonly are associated with rhyolitic dikes. The ore contains auriferous pyrite and chalcopyrite. Milling-grade ore usually averages 1/5 ounce of gold per ton. In this district the metal-bearing deposits are associated with major thrust zones that extend northward along the entire mountain mass.

Mines. Benson, Birthday, Colosseum, Green, Mohawk, Sulphide Queen, Taylor.

Bibliography

Hewett, D. F., 1956, Geology and mineral resources of the Ivanpah quadrangle, California and Nevada: U. S. Geol. Survey Prof. Paper 275, 172 pp.

Wright, L. A., Stewart, R. M., Gay, T. E., Jr., and Hazenbush, G. C., 1953, San Bernardino County, gold: California Jour. Mines and Geol., vol. 49, pp. 69–86.

Coolgardie

This is a dry-placer mining district in western San Bernardino County about 15 miles northwest of Barstow. The area was mined intermittently from around 1900 to 1915, with a total output valued at about $100,000. The principal operator was the Cool Gardie Mining Company, which operated a battery of gasoline-powered dry washers. Several two-man operations employed single dry washers or rockers (see photo 77). Minor prospecting was done in the district during the 1920s and 1930s. The deposits are in a broad valley; the gold apparently was derived from veins in granitic rocks that are to the east and northeast.

Bibliography

Cloudman, H. C., Huguenin, E., and Merill, F. J. H., 1919, San Bernardino County Cool Gardie: California Min. Bur. Rept. 15, p. 817.

Laizure, C. McK, 1934, San Bernardino County, Coolgardie: California Min. Bur. Rept. 30, p. 250.

Dale

Location and History. The Dale or Virginia Dale gold-mining district is in southern San Bernardino and northern Riverside County about 18 miles east of Twentynine Palms. It includes the area known as the Pinto Basin mining district. The first claims here were apparently located in the early 1880s, but the district was not too productive until the 1890s. There was moderate activity during the early 1900s and 1920s, increasing in the 1930s and early 1940s, when the Gold Crown, Supply, Virginia Dale, and Carlyle mines were active. A little work has been done since.

Geology and Ore Deposits. The district is underlain by a variety of rocks, which includes granite, quartz diorite, banded gneiss, andesite porphyry and schist. The quartz veins contain native gold, varying amounts of sulfides and iron minerals, and silver is abundant in some deposits. The veins are up to 10 feet thick and have been mined to depths of as much as 1200 feet. Several high-grade pockets have been uncovered.

Mines. San Bernardino County: Brooklyn $150,-000+, Carlyle $125,000+, Exchequer, Gypsy, Imperial, Ivanhoe, Iron Age, Supply $500,000+, Thelma, Virginia Dale. Riverside County: Cow Bell, Dalton, Duplex, Gold Crown $385,000, Gold Rose, Gold Standard, Golden Rod, Los Angeles, Louise, Mission, O.K. $200,000, Outlaw, Pinto, Zulu Queen.

Bibliography

Cloudman, H. C., 1919, San Bernardino County, Dale mining district: California Min. Bur. Rept. 15, pp. 801–803.

Tucker, W. B., and Sampson, R. J., 1930, Gold—San Bernardino County: California Div. Mines Rept. 26, pp. 221–260.

Wright, L. A., Stewart, R. M., Gay, T. E., Jr., Hazenbush, G. C., 1953, Gold—San Bernardino County: California Jour. Mines and Geology, vol. 49, pp. 69–86.

Dos Palmas

Dos Palmas is in the Orocopia Mountains northeast of the Salton Sea. Some mining was done here in the 1890s, and the area has been prospected since. Several narrow gold-quartz veins occur in gneiss, schist, and granitic rocks. There have been considerable shearing and faulting in the area; the San Andreas fault zone extends southeast along the side of the Salton Sea.

Mines. Charity, Dos Palmas, Fish, Free Coinage, Messenger.

Bibliography

Merrill, F. J. H., 1919, Riverside County, gold: California Min. Bur. Rept. 15, p. 541.

Eagle Mountains

The Eagle Mountains are in eastern Riverside County. Although this district is best known as a major source of iron ore, it has also yielded some gold, silver, lead, and copper. The principal source of gold has been the Iron Chief mine, which has an estimated total production of $150,000. The gold and base metal deposits occur either as replacements along limestone-

Photo 78. Gold Crown Mine, Dale District. This view of the mine, in Riverside County, looks east. The photo was taken in about 1936. *Photo by W. B. Tucker.*

granite contacts or in fissure veins in either granitic or metamorphic rocks.

Bibliography

Tucker, W. B., 1945, Riverside County, Iron Chief mine: California Div. Mines Rept. 41, p. 136.

Emerson Lake

Several mines and prospects lie west of Emerson Lake, which is in southern San Bernardino County about 25 miles northwest of Twentynine Palms. The principal gold sources have been the Emerson and Los Padre mines. The deposits consist of parallel veins in gneiss and granitic rocks, and the ore occurs as small but rich pockets, usually near the surface. Several high-grade pockets containing wire gold have been found here.

Bibliography

Tucker, W. B., and Sampson, R. J., 1940, San Bernardino County, Los Padre mine: California Div. Mines Rept. 36, p. 70.

Gold Reef

This district is in the Clipper Mountains in east-central San Bernardino County about five miles northwest of Danby. Gold was discovered here in 1915, and there was considerable activity for a few years following. Several wide gold-bearing quartz-calcite veins occur in fault zones in volcanic rocks of Miocene age. The ore-bearing zones are as much as 50 feet thick and 1500 feet long. The principal sources of gold have been the Clipper Mountain, Gold Reef, and Tom Reed mines.

Bibliography

Tucker, W. B., 1921, San Bernardino County, gold: California Min. Bur. Rept. 17, pp. 345–346.

Goldstone

The Goldstone district is in northwestern San Bernardino County about 35 miles north of Barstow, in what is now the U.S. Naval Ordnance Test Station, Mojave Range. The district was active in 1915 to 1918, in the 1920s, and again just before World War II. There are several shallow gold-quartz veins in limestone, siliceous shales, and associated diorite dikes. Several high-grade pockets have been discovered. Copper and silver also are present.

Bibliography

Cloudman, H. C., et al., 1919, San Bernardino County, Goldstone district: California Min. Bur. Rept. 15, pp. 804–807.

Grapevine

There are several small lode-gold mines, prospects and dry placer deposits in an area known as the Grapevine district. It is in the Paradise Mountains in western San Bernardino County about 15 miles north of Barstow. A number of quartz veins in granitic rocks contain free gold, and copper and manganese minerals. The veins are narrow and the deposits are shallow. The Olympus mine apparently in the only property

that has had much development work. Also the well-known Waterman silver mine is here.

Bibliography

Wright, L. A., Stewart, R. M., Gay, T. E., Jr., and Hazenbush, G. C., 1953, San Bernardino County, Olympus mine: California Jour. Mines and Geol., vol. 49, p. 76.

Hackberry Mountain

Gold and copper have been mined in the Hackberry Mountain-Von Trigger Spring area of eastern San Bernardino County, about 25 miles south of Ivanpah. The principal sources of production have been the Leiser Ray, True Blue, and Von Trigger mines. The mines were intermittently worked from the 1890s through the 1940s, but the most productive period was 1904 to 1915. The deposits consist either of mineralized shear zones or gold- and copper-bearing quartz veins in gneiss and schist. Yellow cuprodescloizite, a rare vanadium-bearing mineral, has been found at the Leiser Ray mine.

Bibliography

Hewett, D. F., 1956, Geology and mineral resources of the Ivanpah quadrangle, California and Nevada: U. S. Geological Survey Prof. Paper 275, 172 pp.

Ver Planck, W. E., 1961, History of mining in northeastern San Bernardino County: California Div. Mines Mineral Inf. Service, vol. 14, no. 9.

Halloran Springs

Halloran Springs is in northeastern San Bernardino County about 12 miles northeast of Baker. Indians mined turquoise here in prehistoric times. The area was probably prospected for gold during the 1890s. The Telegraph mine, the principal gold source in the district, with an output of $100,000, was discovered in 1930. The ore deposits consist of gold-quartz veins, up to eight feet thick in quartz monzonite and consisting either of massive and banded quartz or cemented silicified breccia. The ore contains native gold and often abundant chalcopyrite and bornite. Some of the ore is extremely rich.

Bibliography

Hewett, D. F., 1956, Geology and mineral resources of the Ivanpah quadrangle: U. S. Geol. Survey Prof. Paper 275, 172 pp.

Wright, L. A., Stewart, R. M., Gay, T. E., Jr., and Hazenbush, G. C., 1953, San Bernardino County, gold: California Jour. Mines and Geol., vol. 49, pp. 69–86.

Hart

The Hart or Castle Mountain mining district is in northeastern San Bernardino County about 15 miles east of Ivanpah near the Nevada border. Gold was discovered here in 1907, and the area flourished for a few years. For a time it was served by the Santa Fe Railroad's branch to Searchlight, Nevada. There was some activity again during the 1930s. The principal sources of gold were the Oro Belle, Valley View, and Hart Consolidated mines. The ore deposits consist of breccia zones along which the wall rock is silicified. The deposits contain pyrite and native gold in small grains and wires. Country rock consists of rhyolite flows, tuff, and breccia of late Tertiary age.

Photo 79. Exposed Treasure Mine, Mojave District. The photo shows the mine, in Kern County, in about 1914.

Bibliography

Hewett, D. F., 1956, Geology and mineral resources of the Ivanpah quadrangle, California and Nevada: U. S. Geol. Survey Prof. Paper 275, 172 pp.

Tucker, W. B., and Sampson, R. J., 1943, San Bernardino County, Valley View mine: California Div. Mines Rept. 39, p. 464.

Ibex

The Ibex district is in eastern San Bernardino County about 15 miles north of Needles. The Ibex or Gold Ridge mine, the principal source of gold, was active in the 1880s and 1890s. The deposits consist of massive quartz veins in gneiss and schist containing free gold and varying amounts of sulfides.

Bibliography

Miller, W. J., 1944, Geology of the Needles-Goffs region, San Bernardino County: California Div. Mines Rept. 40, pp. 113–129.

Ivanpah

Location and History. The Ivanpah mining district is in northeastern San Bernardino County about 35 miles northeast of Baker and south of the Mountain Pass-Clark Mountain area. The district includes the mines in both the Ivanpah Range and the Mescal Range, which is just to the west. Gold mining began here at least as early as 1882, when the Mollusk mine was opened. Moderate mining activity continued in the district until about 1915, and there was some work again in the 1930s.

Geology and Ore Deposits. The western part of the district is underlain predominantly by limestone and dolomite, with smaller amounts of shale, sandstone and dacite. To the east is granite and gneiss, and to the south is quartz monzonite. The gold deposits are in quartz veins or mineralized breccia, which occur chiefly in granitic rocks or gneiss, although the Mollusk vein is in dolomite. Other mineral commodities in the district are silver, copper, tungsten, tin, barite, fluorspar, and rare earths. As in the Clark mining

district to the north, the metal-bearing deposits are associated with several major thrust fault zones.

Mines. Kewanee, Mollusk $250,000, Morning Star, New Era, Teutonia.

Bibliography

Hewett, D. F., 1956, Geology and mineral resources of the Ivanpah quadrangle, California and Nevada: U. S. Geol. Survey Prof. Paper 275, 172 pp.

Tucker, W. B., and Sampson, R. J., 1943, San Bernardino County, gold: California Div. Mines Rept. 39, pp. 438–465.

Wright, L. A., Stewart, R. M., Gay, T. E., Jr., and Hazenbush, G. C., 1953, San Bernardino County, gold: California Jour. Mines and Geol., vol. 49, pp. 69–86.

Mojave-Rosamond

Location. The Mojave-Rosamond district is in southeastern Kern County. The gold deposits are associated with the five prominent buttes south of the town of Mojave and west and north of the town of Rosamond.

History. Gold was discovered in the Yellow Rover vein on Standard Hill by George Bowers in 1894, and soon afterward other discoveries were made. Activity continued until about 1910 but waned over the next 20 years. The Cactus Queen mine was discovered in 1934, and from 1931 until 1941 mining was done in the district on a major scale. The mines were shut down during World War II, but there has been some activity since. The Tropico mine is now an historical museum and a popular tourist attraction. The district is estimated to have had a total gold and silver output valued at $23 million.

Geology and Ore Deposits. The principal rocks are Tertiary rhyolite, rhyolite porphyry and quartz latite, which are underlain by Mesozoic quartz monzonite. All the ore deposits are associated with the five prominences (fig. 29), the most important of which, both in productivity and in the number of deposits, is Soledad Mountain. The ore occurs in epithermal

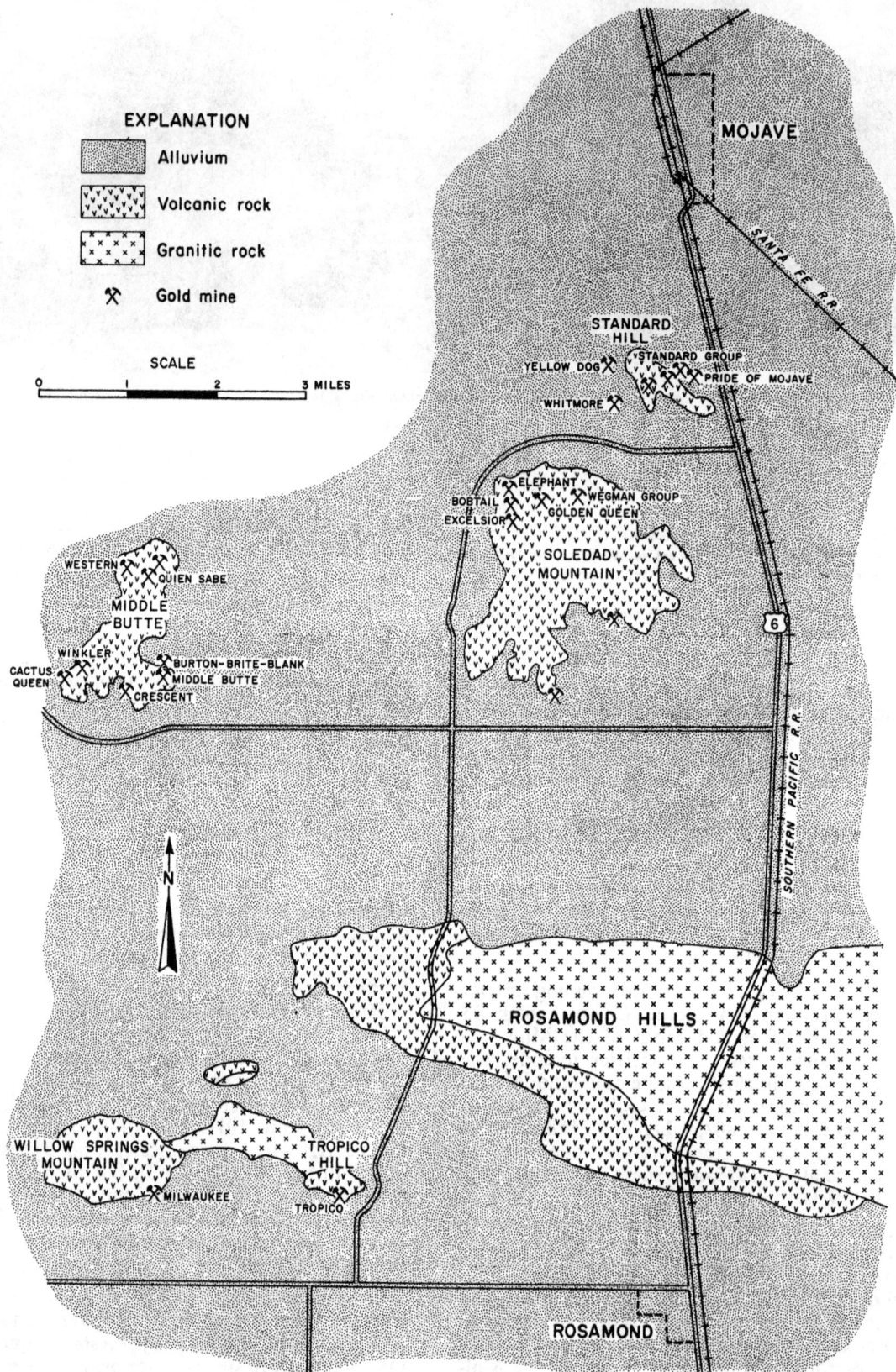

Figure 29. Geologic Map of Mojave-Rosamond District, Kern County. *After Troxel and Morton, 1962, and Dibblee, 1963.*

fissure veins that occupy brecciated and sheared zones in the rhyolitic rocks. The ore contains finely divided gold, with appreciable amounts of silver minerals, including cerargyrite, argentite, and smaller amounts of proustite, pyrargyrite, and electrum. Pyrite, arseno- pyrite, galena, and chalcopyrite also are present. The ore shoots range from a few feet to 40 feet in thick- ness, and are up to 200 feet long. The veins have been developed to depths of 1000 feet. Milling ore usually averaged about ⅓ ounce of gold per ton, but some rich ore shoots were worked in the earlier mining operations.

Mines. Burton-Brite-Blank, Cactus, Cactus Queen $5 million+, Double Eagle, Crescent, Elephant $200,000 to $400,000, Excelsior, Golden Queen (includes Echo and Gray Edge, Queen Ester and Silver Queen) $10 million+, Middle Butte $150,000+, Milwaukee, Pride of Mojave, Quien Sabe, Standard group (Desert Queen, Exposed Treasure and Yellow Rover) $3.5 million, Tropico 114,000 ounces, Wegman group (Eureka, Karma and Monarch) $100,- 000+, Western, Whitmore, Winkler, Yellow Dog 5800+ ounces.

Bibliography

Bateson, G. E. W., 1907, The Mojave mining district of California: Trans. Am. Inst. Min. Engrs., vol. 37, pp. 160–177.

Brown, G. C., 1916, Kern County, Mojave district: California Min. Bur. Rept. 14, p. 483.

Dibblee, T. W., Jr., 1963, Geology of the Willow Springs and Rosa- mond quadrangles, California: U. S. Geol. Survey Bull. 1089-C, pp. 141–253.

Simpson, E. C., 1934, Geology and mineral resources of the Elizabeth Lake quadrangle: California Div. Mines Rept. 30, pp. 371–415.

Tucker, W. B., 1923, Kern Coounty, Mojave mining district: California Min. Bur. Rept. 19, pp. 156–164.

Troxel, B. W., and Morton, P. K., 1962, Mojave mining district: California Div. Mines and Geology, County Rept. 1, pp. 43–45.

Tucker, W. B., and Sampson, R. J., 1933, Kern County, Mojave mining district: California Div. Mines Rept. 29, pp. 283–284.

Tucker, W. B., 1935, Mining activity at Soledad Mountain and Middle Buttes-Mojave mining district: California Div. Mines Rept. 31, pp. 465–485.

Tucker, W. B., Sampson, R. J., and Oakeshott, G. B., 1949, Kern County, Golden Queen Mining Company: California Jour. Mines and Geology, vol. 45, pp. 220–223.

Mule Mountains

The Mule Mountains district, which has also been known as the Hodges Mountain district, is in south- eastern Riverside County about 20 miles southwest of Blythe. Some gold and copper were recovered here years ago from several mines and prospects, the most productive of which were the Roosevelt and Rainbow group of mines. Native gold, pyrite, and chalcopyrite occur in quartz veins in granitic rock.

Bibliography

Tucker, W. B., and Sampson, R. J., 1945, Riverside County, Roosevelt and Rainbow group of mines: California Div. Mines Rept. 41, pp. 142–143.

Old Dad

This district is in the Old Dad Mountains, which are in northeastern San Bernardino County about 12 miles east of Baker. Gold was discovered here in the 1890s, and the area has been intermittently mined ever since, with considerable activity during the 1930s and early 1940s. The district is underlain by gneiss, quartz- ite, limestone, and granitic rocks. The ore deposits consist of quartz veins ranging from one to six feet in thickness that occur chiefly in granitic gneiss or quartzite. The ore bodies contain native gold, fine- grained auriferous pyrite, abundant iron oxide, and small amounts of other sulfides. Appreciable amounts of high-grade ore have been taken from this district.

Mines. Brannigan $100,000+, Lucky, Paymaster $100,000, Oro Fino.

Bibliography

Hewett, D. F., 1956, Geology and mineral resources of the Ivanpah quadrangle: U. S. Geol. Survey Prof. Paper 275, 172 pp.

Ver Planck, W. E., 1961, History of mining in northeastern San Bernardino County: California Div. Mines Mineral Inf. Service, vol. 14, no. 9.

Wright, L. A., Stewart, R. M., Gay, T. E., Jr., and Hazenbush, G. C., 1953, San Bernardino County, gold: California Jour. Mines and Geol., vol. 49, pp. 69–86.

Old Woman

The Old Woman Mountains in eastern San Ber- nardino County have yielded some gold, the main sources being the Blue Eagle and Long Shot mines, which were active during the 1930s. Native gold and often abundant sulfides occur in quartz veins in gra- nitic rocks. Most of the deposits are only a few tens of feet deep.

Bibliography

Wright, L. A., Stewart, R. M., Gay, T. E., Jr., and Hazenbush, G. C., 1953, San Bernardino County, gold tabulated list: California Jour. Mines and Geology, vol. 49, p. 259.

Ord

Location and History. The Ord district is in west- central San Bernardino County in the Ord and New- berry Mountains, about 20 miles southeast of Barstow. The mountains were named for Major General E.O.C. Ord of Civil War fame. The district was organized around 1870, and intermittent development work con- tinued for many years after. There was some work done here again in the 1930s. Although the district is reported to have been a small producer, there are many mines and prospects.

Geology and Ore Deposits. The region is underlain by granite and quartz monzonite and a variety of Ter- tiary volcanic rocks that include basalt, andesite, and rhyolite. The gold-quartz veins are confined to the granitic rocks and often are associated with dikes. The ore bodies contain abundant sulfides and iron oxide. Appreciable amounts of copper and silver minerals are present in places. There are a few placer deposits.

Mines. Alarm, Azucar, Black Butte, Camp Rock (placer), Cumberland, Elsie, Gold Banner, Gold Belt, Gold Brick, Gold Peak, Grandview, Haney and Lee, Hoover, Johnson, Lucky Strike, New Deal, Old, Ord Belt, Red Hill, Riley.

Bibliography

Cloudman, H. C., et al, 1919, San Bernardino County, Ord district: California Min. Bur. Rept. 15, pp. 808–810.

Photo 80. Picacho Mine, Picacho District. Picacho Peak rises in the right background in this 1921 view of the mine, in Imperial County. Photo by Ralph Baverstock, from collection of Dr. Horace Parker.

Gardner, D. L., 1940, Geology of the Newberry and Ord Mountains, San Bernardino County: California Div. Mines Rept. 36, pp. 257–292.

Tucker, W. B., and Sampson, R. J., 1940, Economic mineral deposits of the Newberry and Ord Mountains, San Bernardino County: California Div. Mines Rept. 36, pp. 232–240.

Weber, F. Harold, Jr., 1963, Geology and mineral deposits of the Ord Mountain district, San Bernardino County: California Div. Mines and Geology Spec. Rept. 77, 45 pp.

Oro Grande

Location. This district is in southwestern San Bernardino County, in the vicinity of the town of Oro Grande, about five miles north of Victorville and 45 miles north of San Bernardino. The gold mines were active during the 1880s, early 1900s and again in the 1930s. Large amounts of cement are produced here now.

Geology and Ore Deposits. Most of the deposits are in the hills northeast of Oro Grande. According to Bowen (1954), the area is underlain by schist, quartzite and limestone of the Oro Grande series (Carboniferous); dacite, rhyolite, and latite of the Sidewinder volcanic Series (Triassic (?)); and quartz monzonite. The quartz veins are narrow, and the ore bodies usually are small and irregular. Most of the ore has come from the oxidized zone near the surface, but a few high-grade pockets have been found in the veins. The ore contains free gold and often abundant sulfides, including pyrite, chalcopyrite, sphalerite, and bornite. The Carbonate mine has yielded appreciable amounts of gold- and silver-bearing lead carbonate.

Mines. Apex, Branch, Carbonate, Dents Grandview Lode, Gold Bullion, Gold King, Oro Grande I and II, Sidewinder, Western States.

Bibliography

Bowen, O. E., Jr., 1954, Geology and mineral deposits of the Barstow quadrangle, California: Calif. Div. Mines Bull. 165, pp. 123–134.

Picacho

Location. The Picacho district is in southeastern Imperial County about 50 miles east of El Centro and 20 miles north of Yuma, Arizona. The Colorado River is on the north side of the district, and the Chocolate Mountains extend along the southwest margin. The district was named for Picacho Peak, a prominent landmark in the area.

History. Spaniards, developing the nearby Cargo Muchacho and Potholes districts, probably mined the Picacho district as early as 1780. For many years, Mexicans and Indians mined the area by small-scale dry washing methods. Small hand-operated bellows-type washers were employed, or winnowing was done with blankets. Virtually every dry wash in the region was worked in this fashion, and many small tailings piles from these operations are still visible. Attempts were made to hydraulic the area during the 1890s.

The Picacho or Picacho Basin mine, the largest source of gold in the district, was worked on a large scale from 1904 to 1910. The reported output was $2 million. During this operation the ore was treated in a mill near the Colorado River. The mine was sampled in the 1930s, but there has been only minor prospecting in the district since, and the site of the old town of Picacho on the river was recently made into a state recreation area.

Geology The central part of the district, which is in a shallow basin, is underlain by interbedded granitic rocks and mica and hornblende schists. A number of diorite dikes may be associated with the gold mineralization. The ridges surrounding the basin and Picacho Peak are composed of andesite, rhyolite, and rhyolite tuff.

Ore Deposits. The ore deposits consist of parallel thin gold-bearing stringers in schist. Although the ore bodies at the Picacho mine contained only a few dollars of gold per ton, several were quite large. The last ore shoot worked at this mine was 250 feet long and 160 feet wide. The placer deposits in the various dry washes usually were shallow and discontinuous. The gold particles were very fine.

Bibliography

Crawford, J. J., 1894, California Picacho mine: California Min. Bur. Rept. 12, p. 238.

Crawford, J. J., 1896, Picacho Basin: California Min. Bur. Rept. 13, p. 343.

Merrill, F. J. H., 1916, Imperial County, Picacho: California Min. Bur. Rept. 14, pp. 729–731.

Potholes

This district is in the southeast corner of Imperial County about 50 miles east of El Centro and 10 miles northeast of Yuma, Arizona, near the Colorado River and west of the Laguna Dam. Nearly all of the gold produced here has come from dry desert placer deposits. The value of the total output is estimated at $2 million. The district was so named because the gold was found in small depressions or pots.

Small-scale mining began here in 1775–80, when California was under Spanish rule, and continued into the early 1800s when the state was under Mexican rule. The most productive period, apparently from the 1860s to the early 1890s, saw as many as 500 Mexicans and Indians working the dry washes. Winnowing was with blankets, and, later, hand-operated bellows-type dry washers were employed. Nearly all of these were one or two-man operations. When a deposit was worked out the miners would move on to another one, usually in the same district. These operations had mostly ceased by 1900, as the deposits were largely exhausted.

Later, several attempts were made here and in the Cargo Muchacho and Chocolate Mountain districts to the west to work the dry placers by large-scale methods. All of these attempts failed because of high equipment and operating costs, erratic distribution of gold values, rough terrain, and scant moisture, which even in desert placers, makes it difficult to separate the heavy and light particles. Also, much of the easily recoverable gold had already been removed from these deposits.

Bibliography

Crawford, J. J., 1894, Pot Holes Mines: California Min. Bur. Rept. 12, p. 242.

Haley, C. S., 1923, Dry placers—Gold placers of California: California Min. Bur. Bull. 92, pp. 154–160.

W. B. Tucker, 1926, Imperial County, Potholes placer: California Min. Bur. Rept. 22, p. 261.

Photo 81. Gold Concentrating Mill, Colorado River. Steam powered this mill on the Arizona side of the Colorado River in in the early 1900s. Picacho Peak is in the left background.

Photo 82. Yellow Aster Mine, Rand District. The Yellow Aster, in Kern County, yielded more than $12 million in gold. The mine is in the background, the town of Randsburg in the foreground.

Rand

Location. The Rand or Randsburg district lies athwart the Kern-San Bernardino County line in the vicinity of the town of Randsburg, about 40 miles northeast of Mojave and 30 miles north of Kramer. The western part of the district, in Kern County, has been chiefly a source of gold, while the eastern part, in San Bernardino County, has been largely a source of silver. The Atolia tungsten district is just to the southeast (fig. 30).

History. Although this region was prospected as early as the 1860s, it was not until placer gold was discovered in 1893 in Goler Wash, in the El Paso Mountains 15 miles to the west, that there was any mineral production. Numerous short-lived dry-washing camps soon sprang up in the entire region. The Yellow Aster mine, originally known as the Olympus, was located in 1895. Other discoveries were made, and the rich ore recovered in these early operations led to a gold rush. The district was named for the Rand district in South Africa. The ore was first shipped out for treatment, but a 100-stamp mill was erected at the Yellow Aster mine in 1901 and other mills were built soon afterward. During the early days some difficulty was encountered in concentrating the gold because of the presence of "heavy spar" or scheelite.

Large-scale gold mining continued until 1918. The famous and highly productive Kelly or California Rand silver mine was discovered in 1919 and was operated on a major scale through the 1930s. Gold production from the district was substantial in the 1930s and early 1940s, and there has been intermittent prospecting and development work since. The total gold output of the district is estimated at more than $20

million. During the two world wars and the Korean War, Atolia was the source of large amounts of tungsten ore. From 1897 until 1933 Randsburg was served by a branch of the Santa Fe Railroad, which extended north from the main line at Kramer.

Geology. The principal rocks underlying the district are the Precambrian Rand Schist and the Atolia Quartz Monzonite of Mesozoic age. The Rand Schist is chiefly biotite schist with smaller amounts of amphibolite and quartzite. To the east are poorly consolidated clays, sandstones, and conglomerates of continental origin, which are overlain by andesite at Red Mountain. Rhyolite and latite intrusives are in the east-central part of the district.

Ore Deposits. Most of the lode-gold deposits are in veins that occur along faults, except at the Yellow Aster mine, where the gold is in a series of closely spaced veinlets in small fractures. The majority of the gold deposits are in the schist, which is more widespread than the quartz monzonite, and nearly all are in an area where the rocks have been colored a pale red by iron oxides (fig. 30). The veins are unoriented but usually have a well-defined hanging wall.

The ore bodies most commonly occur in the vein footwalls, usually at or near vein intersections or in sheared and brecciated zones. The ore consists of iron oxide-stained brecciated and silicified rock containing native gold in fine grains and varying amounts of sulfides. The sulfides increase at depth, but the gold values decrease. Most mining has stopped where unoxidized sulfides were found in the veins, and the maximum depth of development is 600 feet. Milling ore contains from $\frac{1}{7}$ to $\frac{1}{4}$ ounce of gold per ton. The high-grade ore nearly always occurs in pockets near the surface. Most of the placer gold has been

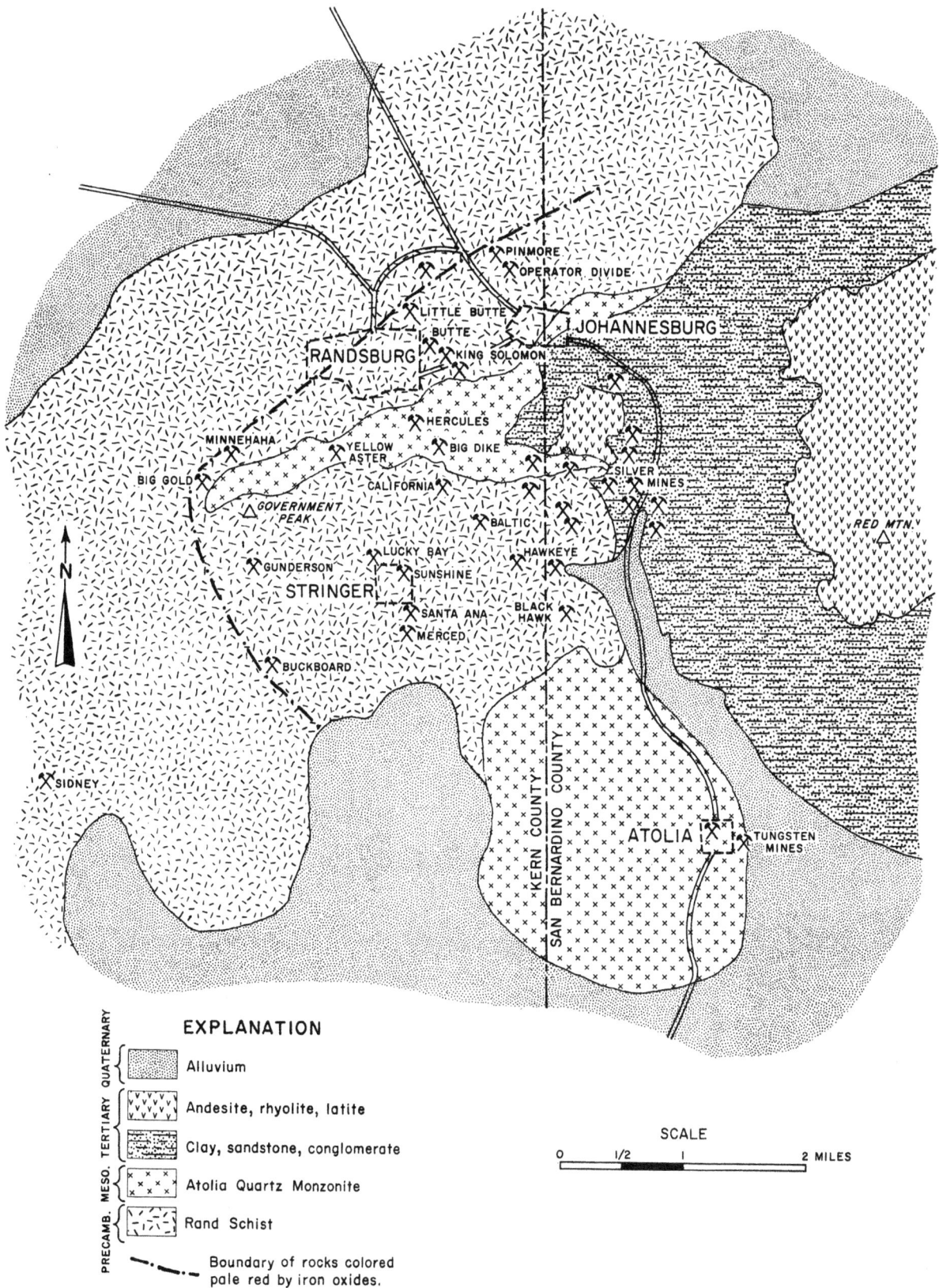

EXPLANATION

PRECAMB. MESO. TERTIARY QUATERNARY

Alluvium

Andesite, rhyolite, latite

Clay, sandstone, conglomerate

Atolia Quartz Monzonite

Rand Schist

—·—·— Boundary of rocks colored
pale red by iron oxides.

SCALE

0   1/2   1   2 MILES

Figure 30. Geologic Map of Rand District, Kern and San Bernardino Counties. *After Hulin, 1925, plate 1, and Troxel and Morton, 1962.*

Photo 83. Yellow Aster Mine Mill. Fifty stamps were "dropping" at this mill in 1912, the year the photo was taken.

Photo 84. Town of Randsburg, Kern County. This winter view, taken in the early 1900s, looks east.

recovered from dry placers at Stringer or in the Rand Mountains north of Randsburg.

Mines. Arizona, Baltic $50,000, Barnett, Beehive, Big Dike $200,000, Big Gold $500,000, Black Hawk $700,000, Buckboard $500,000, Bully Boy $120,000, Butte $2 million, California, Consolidated $50,000, Culbert, Gold Crown, Granton, Gunderson, Hawkeye, Hercules, King Solomon $500,000, Little Butte $400,-000, Lucky Boy $120,000, Merced, Minnehaha $100,-000, Mizpah Montana, Monarch Rand, New Deal, Operator Divide $600,000, Pestle, Pinmore, Red Bird, Santa Ana group $400,000, Sidney $250,000, Snowbird, Sunshine $1.06 million, Windy, Winnie, Yellow Aster $12 million.

Bibliography

Brown, G. Chester, 1916, Kern County, Randsburg district: California Min. Bur. Rept. 14, pp. 483–484.
Cooper, C. L., 1936, Mining and Milling methods and costs at the Yellow Aster mine: U. S. Bur. Mines, Inf. Circ. 6900, 21 pp.
Hess, F. L., 1909, Gold mining in the Randsburg quadrangle: U. S. Geol. Survey Bull. 430, pp. 23–47.
Hulin, Carleton D., 1925, Geology and ore deposits of the Randsburg quadrangle: California Min. Bur. Bull. 95, 152 pp.
Hulin, Carleton D., 1934, Dry placers of the Mojave Desert: California Div. Mines Rept. 30, pp. 417–426.
Newman, M. A., 1923, The Rand district: California Min. Bur. Rept. 19, pp. 61–63.
Tucker, W. B., and Sampson, R. J., 1933, Randsburg district: California Div. Mines Rept. 29, pp. 285–286.
Storms, W. H., 1909, Geology of the Yellow Aster mine: Eng. and Min. Jour., vol. 87, pp. 1277–1280.
Troxel, B. W., and Morton, P. K., 1962, Kern County, Rand district: California Div. Mines and Geology, County Rept. 1, pp. 47–51.
Tucker, W. B., 1923, Kern County, Randsburg district: California Min. Bur. Rept. 19, pp. 165–171.
Wynn, M. R., 1949, Desert Bonanza, M. W. Samuelson, publisher, Culver City, California, 263 pp.

Shadow Mountains

Small amounts of gold have been mined in the Shadow Mountains in northeastern San Bernardino County. Gold was discovered here in 1894, and there was some activity for a few years following. There are several quartz veins in granitic gneiss that are associated with aplitic dikes. The ore deposits contain nature gold and varying amounts of copper.

Bibliography

Crawford, J. J., Shadow Mountain district: California Min. Bur. Rept. 13, p. 328.

Stedman

Location and History. This district, in south-central San Bernardino County about eight miles south of Ludlow, has also been known as the Rochester or Buckeye mining district. Much of the production has been from the Bagdad-Chase or Pacific gold-copper mine. This mine has been the source of more than $6 million worth of gold.

The principal periods of mining activity were 1904 to 1910, when the Badgdad-Chase Mining Company shipped gold ore to a cyanide plant at Barstow, and 1910 to 1916, when Pacific Mines Corporation shipped gold-copper ore to Jerome, Arizona. During those years, Rochester, the principal town, and the mines were served by the Ludlow and Southern Railroad, which extended south from the Santa Fe Railroad at Ludlow. There was activity again during the 1930's and 1940's, and the area has been prospected since.

Photo 85.   Pacific Gold-Copper Mine, Stedman District. The photo of underground workings in the mine, in San Bernardino County, was taken in the early 1900s.

Geology and Ore Deposits. The ore-bearing zones are in quartz monzonite and rhyolite. The ore consists of cemented silicified breccia containing fine free gold and various copper minerals, chrysocolla being most abundant. The deposit at the Bagdad-Chase mine is eight to 20 feet thick and has been mined to a depth of 450 feet. Ore mined by Pacific Mines Corporation has an average grade of 1.82 percent copper, 0.35 ounce of gold and 1.5 ounces of silver per ton.

Bibliography

Cloudman, H C., 1919, San Bernardino County, Pacific Mine: California Min. Bur. Rept. 15, p. 790.

Tucker, W. B., and Sampson, R. J., 1930, San Bernardino County, Pacific Mine: California Div. Mines Rept. 26, pp. 218–219.

Wright, L. A., et al, 1953, San Bernardino County, Bagdad-Chase Mine: California Jour. Mines and Geology, vol. 49, pp. 71–72.

Trojan

The Trojan or Providence mining district is in eastern San Bernardino County in the Providence Mountain. Some gold, copper, silver, and lead have been mined here, the main sources having been the Barnett and Gold King mines. They were active prior to and during World War I. The deposits consist of quartz veins in quartz monzonite that contain abundant sulfides.

Bibliography

Hewett, D. F., 1956, Geology and mineral resources of the Ivanpah quadrangle: U. S. Geol. Survey Prof. Paper 275, 172 pp.

Twentynine Palms

Location and History. This is an extensive region in northern Riverside and southern San Bernardino Counties. It includes the gold mines just south of the town of Twentynine Palms and the areas to the south known as the Lost Horse, Gold Park, Hexie and Pinon districts. Gold was first mined here possibly as early as 1860, but the most productive period was during the 1890s and early 1900s. There was activity here again in the 1930s, and there has been some propecting since.

Geology and Ore Deposits. The region is underlain chiefly by quartz monzonite and gneiss, with smaller amounts of granite, diorite, and gabbro. Also there are some pegmatite and diorite dikes. The deposits consist of narrow quartz veins containing free gold, pyrite, and often abundant iron oxide. A number of small but high-grade pockets have been recovered.

Mines. Anaheim, Atlanta, Bass, Black Warrior, Desert Queen, Gold Coin, Gold Park Cons., Gold Point, Golden Bee, Golden Bell, Hexie (Hexahedran), Hornet, Lost Horse $350,000, Silver Bell.

Bibliography
Merrill, F J. H., 1919, Riverside County, Pinon Mountain district: California Min. Bur. Rept. 15, pp. 535–536.
Rogers, J. J. W., 1961, Igneous and metamorphic rocks of the western portion of Joshua Tree National Monument: California Div. Mines Spec. Rept. 68, 26 pp.
Tucker, W. B., and Sampson, R. J., 1945, Riverside County, gold: California Div. Mines Rept. 41, pp. 127–144.

Vanderbilt

Location and History. The Vanderbilt or New York mining district is in northeastern San Bernardino County, in the northeast end of the New York Mountains. Gold was first discovered here in 1861, but the principal periods of mining were 1892–98 and 1934–41. The district was so named by people who hoped it would prove to be as rich as the Vanderbilt fortune. From 1893 until 1923, the district was served by a branch of the Santa Fe Railroad that extended north from Goffs and continued northeast to Searchlight, Nevada. The chief sources of gold have been the Vanderbilt and Gold Bronze mines. Silver, copper and zinc also have been produced in this district.

Geology and Ore Deposits. The district is underlain by granitic rocks with smaller amounts of schist, gneiss, limestone, and Tertiary volcanic rocks. The ore deposits occur largely in granitic rocks and consist of quartz veins, often with abundant sulfides. One ore body in the Vanderbilt mine was mined to a depth of 400 feet and had a stoping length of 200 feet.

Bibliography
Hewett, D. F., 1956, Geology and mineral resources of the Ivanpah quadrangle, California and Nevada: U. S. Geol. Survey Prof. Paper 275, 172 pp.
Tucker, W. B., and Sampson, R. J., 1943, San Bernardino County, Vanderbilt mine: California Div. Mines Rept. 39, p. 464.

Whipple

The Whipple Mountains are in the southeast corner of San Bernardino County. Gold was probably first mined here at an early date, and there was prospecting and development work in the area again in the 1930s and early 1940s. Native gold and oxidized copper and iron minerals occur in narrow quartz veins in gneiss and metamorphic rocks of Precambrian age.

Mines. Bluff and Western, Ethel Leona, Gold Zone, Islander, Nickel Plate, Roulette, Vidal Gold.

Bibliography
Wright, L. A., Stewart, R. M., Gay, T. E., Jr., and Hazenbush, G. C., 1953, San Bernardino County, gold tabulated list: California Jour. Mines and Geol., vol. 49, p. 259.

TRANSVERSE AND PENINSULAR RANGES PROVINCES

These two provinces are in southern California. The Transverse Ranges are a complex series of nearly west-trending mountain ranges and valleys. The province includes the San Bernardino, San Gabriel, and Santa Ynez Mountains. The most productive gold-quartz mines have been in the Frazier Mountain, Acton, and Baldwin Lake districts, where the deposits occur in schist and granitic rocks. (Sometimes Frazier Mountain is considered to be in the Coast Ranges.) Placer gold has been recovered in quantity in the San Gabriel Mountains.

The Peninsular Ranges are in Orange, western Riverside, and San Diego Counties and extend southward into Lower California. These ranges are composed largely of granitic and related rocks that are part of the southern and Lower California batholith and smaller amounts of Paleozoic and Mesozoic metamorphic rocks. The principal gold sources have been the Julian-Banner, Cuyamaca, and Pinacate districts. The primary deposits consist of gold-quartz veins in schist or granitic rocks.

Acton

Location and History. This district is in central Los Angeles County in the general vicinity of the town of Acton, 20 miles north of Los Angeles. It also includes the area known as the Cedar district.

Placer gold was mined in the San Gabriel Mountains here as early as 1834. Lode mining apparently began here in the 1870s or 1880s. The district was quite productive until about 1900. A number of mines, including the Red Rover, Governor, and Monte Cristo, were active again during the 1930s and early 1940s. The district has been intermittently prospected since, but there has been very little recorded production. Acton was named for a village in Massachusetts, and the Governor mine for California's Governor Henry Gage.

Geology. The deposits consist of gold-quartz veins in quartz diorite, diorite, gabbro and schist. The veins are in faulted and fractured zones. The ore is free milling and contains varying amounts of pyrite. The ore bodies commonly consist of small parallel veins rather than a single large vein. The Governor mine has been developed to an incline depth of 1000 feet.

Mines. Buena Esperanza, Governor (New York) $1.5 million+, Helene, Hi-Grade, Red Rover $550,000, Puritan.

Bibliography
Gay, T. E., Jr., and Hoffman, S. R., 1954, Los Angeles County, Governor, Hi-Grade, and Red Rover mines: California Jour. Mines and Geology, vol. 50, pp. 497–500.
Merrill, F. J. H., 1919, Gold—Los Angeles County: California Min. Bur. Rept. 15, pp. 473–477.
Oakeshott, G. B., 1958, Geology and mineral deposits of the San Fernando quadrangle: California Div. Mines Bull. 172, 147 pp.
Sampson, R. J., 1937, Gold—Los Angeles County: California Div. Mines Rept. 33, pp. 177–196.

Azusa-Tujunga

Placer gold has been recovered in a number of canyons and washes along the south flank of the San Gabriel Mountains north and east of Los Angeles. Two of the most important sources have been San Gabriel Canyon, near Azusa, and Tujunga Canyon to the west. At the present time, some gold is recovered as a by-product in several sand and gravel plants in this area. The placer gold usually is fine. There are several narrow gold-quartz veins just to the north.

Bibliography
Gay, T. E., Jr., and Hoffman, S. R., 1954, Los Angeles County, gold: California Jour. Mines and Geology, vol. 50, pp. 493–496.

Preston, E. B., 1890, Tujunga mining district: California Min. Bur. Rept. 9, pp. 197–198.

Baldwin Lake

Location and History. This district is in the general vicinity of and east of Baldwin Lake, which is in the northern part of the San Bernardino Mountains. Gold was reported to have been mined here by Mexicans possibly as early as 1800. The Rose mine was active in 1860, and there was considerable activity in the district in the 1890s and early 1900s. The Doble mine was active again in the 1930s and 1940s. The lake was named for C. G. Baldwin, first president of Pomona College.

Geology. Rocks in this district include mica schist, quartzite, limestone, and granite. The ore deposits consist of systems of irregular quartz-calcite veins containing free gold, scheelite, and sulfides. The greatest depth of development is about 400 feet. There are also placer deposits in the district:

Mines. Lode: Christie, Doble $250,000 to $300,000, Erwin, Gem, Gold Hill, Hollie Ann, Lester, Log Cabin, Rose $450,000 to $600,000, Stewart. Placer: McClure-Bess, Parker, Rattlesnake Canyon, Vaughn, Weaver.

Bibliography
Cloudman, H. C., Huguenin, Emile, and Merrill, F. J. H., 1919, San Bernardino County gold: California Min. Bur. Rept. 15, pp. 794–797.

Wright, L. A., Stewart, R. M., Gay, T. E., Jr., and Hazenbush, G. C., 1953, San Bernardino County gold: California Jour. Mines and Geology, vol. 49, pp. 69–86.

Black Hawk

Location and History. This district is in southwestern San Bernardino County about 30 miles northeast of San Bernardino on the north side of the San Bernardino Mountains. It also has been known as the Silver Reef district. The district was organized in 1870. An English concern organized the Santa Fe group in 1890 to work the area on a large scale, but work stopped soon afterward and prospecting was minor during the early 1900s. The Santa Fe group was reopened in 1921 and operated continuously until 1940. In this last operation, the production amounted to $300,000.

Geology and Ore Deposits. The area is underlain by granitic rocks, mica schist, gneiss, and a limestone belt. A mineralized zone known as the Arlington-Santa Fe lode occurs in a thrust fault that strikes west and dips north. The ore consists of hematite-bearing gouge and a limestone breccia. Several ore bodies yielded up to one ounce of gold per ton. The ore zones are up to 75 feet thick and 1000 feet long. Much of the output in the district has come from the Santa Fe groups.

Bibliography
Cloudman, H. C., et al., 1919, San Bernardino County, the Black Hawk district: California Mining Bureau Rept. 15, pp. 797–798.

DeGroot, Henry, 1890, The Black Hawk district: California Mining Bureau Rept. 10, pp. 523–525.

Storms, W. H., 1893, Black Hawk district: California Mining Bureau Rept. 11, pp. 364–365.

Woodford, A. D., and Harris, T. F., 1928, Geology of Black Hawk Canyon: California Dept. Geol. Sci., Bull., vol. 17, pp. 265–304.

Wright, L. A., et al., 1953, San Bernardino County, Santa Fe mines: California Jour. Mines and Geol., vol. 49, p. 80.

Boulder Creek

Location. This is a small gold-mining district in central San Diego County about 50 miles east-northeast of San Diego and five miles west of Cuyamaca. Gold was discovered here about 1885, and minor work continued intermittently until the 1930s.

Geology. The district is underlain by granitic rocks and mica schist. A number of narrow north-striking quartz veins in shear zones contain free gold and varying amounts of sulfides, including pyrite, pyrrhotite, arsenopyrite, and marcasite. Some silver is present, and tellurides have been reported.

Mines. Boulder Creek, Elk, Gold Crown, Last Chance, Lucky Strike, Nona, Penny, Prosperity.

Bibliography
Everhart, D. L., 1951, Boulder Creek district: California Div. Mines Bull. 159, pp. 109–111.

Tucker, W. B., 1925, San Diego County, gold: California Min. Bur. Rept. 21, pp. 331–349.

Weber, F. Harold, Jr., 1963, San Diego County, gold: California Div. Mines and Geology County Rept. 3, pp. 115–167.

Cuyamaca

Location. This district is in central San Diego County in Cuyamaca Rancho State Park about 50 miles east-northeast of San Diego.

History. The Stonewall mine, the chief source of gold in the district, was discovered in 1870, reportedly by either William Skidmore or Charles Hensley. It operated under difficulties until 1886, when it came under the control of an ex-Governor of California, Robert W. Waterman. The mine was highly productive from then until 1895. There has been very little mining activity in the area since. The remaining equipment and surface plant of the famous old mine have been made into an outdoor museum. The mine has a total output that has been variously estimated at $2 million to $3 million.

Geology. The Stonewall mine area is underlain by mica schist, granodiorite, and gneiss. A large body of gabbro and related rocks lies immediately to the west. Ore was recovered from several north-trending and steeply dipping quartz veins. At the Stonewall mine, the vein is as much as 20 feet thick. The ore contained free gold, abundant sulfides including pyrrhotite and pyrite, and occasionally small amounts of gold tellurides. The famous ore shoot at the Stonewall mine had a pitch length of about 300 feet.

Bibliography

Creasy, S. C., 1946, Geology and nickel mineralization of the Julian-Cuyamaca area, San Diego County: California Div. Mines Rept. 42, pp. 15–29.

Everhart, D. L., 1951, Geology of the Cuyamaca Peak quadrangle, San Diego County: California Div. Mines Bull. 157, pp. 51–136.

Hanks, H. G., 1886, The Stonewall mine: California Min. Bur. Rept. 6, pt. 1, pp. 89–90.

Merrill, F. J. H., 1916, San Diego County, Stonewall mine: California Min. Bur. Rept. 14, pp. 660–662.

Weber, F. Harold, Jr., 1963, San Diego County, gold: California Div. Mines and Geology County Rept. 3, pp. 115–167.

Deer Park

This is a small district in east-central San Diego County about six miles southeast of Cuyamaca Rancho State Park and about 50 miles east of San Diego. Minor amounts of gold have been recovered from narrow north-trending quartz veins in schist, gneiss, and diorite. The area also has been prospected for tungsten.

Bibliography

Weber, F. Harold, Jr., 1963, San Diego County, gold: California Div. Mines and Geology County Rept. 3, pp. 115–167.

Dulzura

The Dulzura district is in southern San Diego County in the San Ysidro Mountains and about 25 miles southeast of San Diego. Placer gold was discovered here in 1828, but the lode deposits were not mined until 1890. The mines were worked sporadically from the 1890s until the 1930s. The principal gold source has been the Donahoe mine; others include the Johnston and Doolittle mines. The district is underlain by granitic rock, quartzite, gneiss, and schist. The deposits occur in a northwest-trending shear zone in metamorphic rock and consist of broken and crushed quartz containing native gold and sometimes abundant sulfides. The deposits are shallow and usually are discontinuous.

Bibliography

Merrill, F. J. H., 1916, Dulzura district: California Min. Bur. Rept. 14, pp. 664–665.

Weber, F. Harold, Jr., 1963, San Diego County, gold: California Div. Mines and Geology County Rept. 3, pp. 115–167.

Escondido

Location. The Escondido district is in western San Diego County about 25 miles north of San Diego and 20 miles southeast of Oceanside. Gold was first mined here many years ago by Mexicans who treated rich surface ores in arrastras. There was considerable activity here during the 1890s and early 1900s.

Geology. The area is underlain by granodiorite, diorite, and gabbro. A number of quartz veins, ranging from a few inches to several feet in thickness, contain free gold and often abundant pyrite. The greatest depth that any vein has been developed is 350 feet.

Mines. Cleveland-Pacific $250,000, Oro Fino, Cravath, Jolly Boy, Coyote.

Bibliography

Merrill, F. J. H., 1916, Escondido: California Min. Bur. Rept. 14, pp. 649–651.

Storms, W. H., 1893, Escondido mines: California Min. Bur. Rept. 11, p. 382.

Weber, F. Harold, Jr., San Diego County, gold: California Div. Mines County Rept. 3, pp. 115–170.

Frazier Mountain

Location and History. This district is in the northeast corner of Ventura County in the general vicinity of Frazier Mountain. Sometimes it is considered to be in the Coast Ranges. The Piru district is just to the south, and the towns of Gorman and Fort Tejon are to the east. The region was first placer-mined in the 1840s, and the Frazier Mountain mine was opened in 1865. This and other lode-gold mines were worked fairly steadily until around 1895. Minor prospecting and development work has been done in the district since; a small production was recorded in 1952. Fort Tejon, a U. S. Cavalry post established in 1854 to control the marauding Indians, was abandoned in 1864, but it has been restored and is now a state historical monument.

Geology and Ore Deposits. The region is underlain by granite, granodiorite, gneiss, and schist and smaller amounts of quartzite and hornfels. The gold-quartz veins strike north, range from a few inches to five feet in thickness, and occur in shear zones that are principally in gneiss and schist. The ore is free milling and contains pyrite and small amounts of other sulfides. Milling-grade ore commonly averaged ½ ounce of gold per ton. Several of the ore bodies had stoping lengths of up to 300 feet. Some placer gold was recovered in the district from the streams and older terrace gravels.

Mines. Bunker Hill, Esperanza, Fairview, Frazier $1 million, Gold Dust, Harris, Hess, Maule, Sibert, White Mule.

Bibliography

Bowers, Stephen, 1888, Gold—Ventura County: California Min. Bur. Rept. 8, pp. 680–684.

Carman, Max F., Jr., 1964, Geology of the Lockwood Valley area, Kern and Ventura Counties: California Div. Mines and Geology Spec. Rept. 81, 62 pp.

Huguenin, Emile, 1919, Gold—Ventura County: California Min. Bur. Rept. 15, pp. 759–760.

Tucker, W. B., and Sampson, R. J., 1932, Gold—Ventura County: California Div. Mines Rept. 28, pp. 253–257.

Holcomb Valley

Holcomb Valley is on the north side of the San Bernardino Mountains just north of Big Bear Lake. Placer deposits were discovered here in 1860 by W. F. Holcomb and were extensively worked for a few years following, mostly by Mexicans. The area has been intermittently prospected ever since. From 1933 to 1941 about 200,000 cubic yards were mined by power shovel, with an average recovery of 38 cents per yard. The gold-bearing material consists of Recent and older alluvium. There are a number of small lode-gold prospects. The gold occurs in thin shear and fracture zones in granitic rocks or in contacts between carbonate and intrusive rocks.

Bibliography

Cloudman, H. C., Huguenin, Emile, and Merrill, F. J. H., 1920, Holcomb Valley district: California Min. Bur. Rept. 16, pp. 798–799.

Gray, C. H., Jr., 1960, Placer gold in Geology of the San Bernardino Mountains north of Big Bear Lake: California Div. Mines Spec. Rept. 65, pp. 51–54.

Julian-Banner

Location. This district is in north-central San Diego County in the general vicinity of the towns of Julian and Banner about 50 miles northeast of San Diego. It is at the north end of a belt of gold mineralization that extends south and southwest through the Cuyamaca, Deer Park and Pine Valley districts.

History. Placer mining may have been first done here in the 1840s. The first lode claims, the George Washington and Van Wert, were located in 1870, and the Julian mining district was organized in that same year, named for Mike Julian, the recorder. Many other claims were located soon afterward. The greatest period of mining activity was from 1870 to 1880, a peak output of $500,000 having been attained in 1874. There was another period of activity from 1888, when the Gold King and Gold Queen mines were discovered, until about 1896. Since then there has been intermittent exploration and development work, particularly at the Golden Chariot and other nearby mines, but very little recorded output. The value of the total production from the district is estimated at $5 million.

Photo 86. Lytle Creek Mine, Lytle Creek District. This photo of a hydraulic mining operation in the eastern San Gabriel Mountains, San Bernardino County, was taken in about 1894. *Photo courtesy of Calif. State Library*

Photo 87. Hydraulic Mining, Lytle Creek District. The photo was taken in San Bernardino County in the 1890s. Photo courtesy of Title Insurance and Trust Company of Los Angeles.

Geology. Most of the important deposits are in or adjacent to a one- to two-mile-wide northwest-trending belt of mica schist, gneiss, and quartzite of the Julian Schist (Triassic?). On either side of this belt are quartz diorite and schist. The northwest-trending Elsinore fault extends through the area.

Ore Deposits. The ore bodies occur in lenticular quartz veins ranging from a few inches to about five feet in thickness. The veins strike northwest and dip to the northeast. The ore contains native gold and varying amounts of auriferous sulfides. Also present are small amounts of gold tellurides. Most of the ore shoots had stoping lengths of 100 feet or less, although one at the Owens mine was 400 feet. Surface ores mined during the early days contained considerable rich "jewelry" material, but at depth the ore is low in grade. The deepest working is 350 feet.

Mines. C. B. Chieftain, Cincinnati Belle, Eagle Elevada, Gardner $200,000, Gold Cross, Gold King, Golden Chariot $700,000, Gopher, Helvetia $450,000, Hidden Treasure, Kentuck group, Madden group, North Hubbard $200,000, Owens $450,000, Ranchita $150,000, Ready Relief $500,000, Redman, San Diego, Shamrock, Van Wert.

Bibliography

Crawford, J. J., 1894, Gold-San Diego County: California Min. Bur. Rept. 12, pp. 237–243.

Creasey, S. C., 1946, Geology and nickel mineralization of the Julian-Cuyamaca area, California: California Div. Mines Rept. 42, pp. 15–29.

Donnelly, Maurice, 1934, Geology and mineral deposits of the Julian district: California Div. Mines Rept. 30, pp. 331–370.

Hanks, H. G., 1886, Julian mining district: California Min. Bur. Rept. 6, pt. 1, pp. 82–89.

Merriam, Richard, and Stewart, R. M., 1958, Geology and mineral resources of the Santa Ysabel quadrangle, San Diego County, California: California Div. Mines Bull. 177, 42 pp.

Merrill, F. J. H., 1916, San Diego County, Julian district: California Min. Bur. Rept. 14, pp. 653–660.

Tucker, W. B., 1925, Gold—San Diego County: California Min. Bur. Rept. 21, pp. 331–349.

Weber, F. Harold, Jr., 1963, San Diego County, gold: California Div. Mines and Geology County Rept. 3, pp. 115–167.

Laguna Mountains

The Laguna Mountains are in east-central San Diego County east of Cuyamaca Rancho State Park and 55 miles east of San Diego. A considerable number of claims have been located here, but the principal gold source has been the Noble mine, which yielded more than $60,000 between 1888 and 1914. The district was prospected during the 1920s and 1930s. The ore deposits consist of narrow north-striking gold-quartz veins in shear zones in mica schist, gneiss, and quartz diorite. This district is at the southeast end of the same belt that includes the Julian-Banner and Cuyamaca districts.

Bibliography

Weber, F. Harold, Jr., 1963, San Diego County, gold: California Div. Mines and Geology County Rept. 2, pp. 115–167.

Lytle Creek

Lytle Creek is in southwestern San Bernardino County in the eastern San Gabriel Mountains. During the 1890s there was an appreciable amount of placer mining here. Operations extended from near the mouth of the canyon to near its headwaters on the east slope of Mt. San Antonio (Mt. Baldy). Work was done both by hydraulicking and hand methods. The stream was named for Andrew Lytle, a member of the Mormon colony that settled in San Bernardino in 1851.

Menifee

The Menifee or Auld district is in western Riverside County about eight miles south of Perris. The area was first mined many years ago, and there has been minor prospecting since. The principal gold sources have been the Menifee, Lucky Boy, and Mammoth mines, none of which has been extensively developed. A number of narrow quartz veins contain free gold and in places abundant pyrite. Country rock is quartz diorite.

Bibliography

Merrill, F. J. H., 1919, Riverside County, gold: California Min. Bur. Rept. 15, pp. 533–535.

Mesa Grande

The Mesa Grande gold-mining district is in north-central San Diego County just northeast of the town of Mesa Grande and about 50 miles northeast of San Diego. The deposits were discovered in late 1880s and were worked until about 1896. There was minor activity again in the 1930s. The principal source of gold has been the Shenandoah mine, which has yielded about $50,000. Others include the Black Eagle and Mesa Grande mines. The ore deposits consist of narrow northeast-striking gold-quartz veins in schist and gneiss. Some of the surface ore was rich.

Bibliography

Storms, W. H., 1893, Mesa Grande district: California Min. Bur. Rept. 11, p. 382.
Weber, F. Harold, Jr., 1963, San Diego County, gold: California Div. Mines and Geology County Rept. 3, pp. 115–167.

Montezuma

The Montezuma or Rice district is in northeastern San Diego County about six miles southeast of Warner's Springs and 12 miles north of Julian. Gold was first mined here about 1896, when a number of claims were located. Many of the claims were consolidated in about 1910 by the Montezuma Gold Mining Company, which worked them for a few years. The district was prospected again in the 1930s. A series of northeast-trending gold-quartz veins occur in quartzite, schist, gneiss, and quartz diorite. The veins range from one to four feet in thickness and have been developed to depths of about 230 feet.

Mines. Buckeye, Grubstake, Lucky Strike, Maid of Erin, Montezuma group.

Bibliography

Merrill, F. J. H., 1916, San Diego County, Montezuma or Rice district: California Min. Bur. Rept. 14, p. 648.
Tucker, W. B., 1925, San Diego County, Montezuma mine: California Min. Bur. Rept. 21, pp. 342–343.
Weber, F. Harold, Jr., 1963, San Diego County, gold: California Div. Mines and Geology County Rept. 3, pp. 115–167.

Morongo

This district is in the eastern San Bernardino Mountains. There are a number of small lode-gold mines and prospects, most of which have been idle for many years. The deposits are shallow and consist of narrow quartz veins containing free gold, often abundant sulfides, and scheelite. Country rock is granite, limestone, and schist.

Bibliography

Cloudman, H. C., Huguenin, Emile, and Merrill, F. J. H., 1920, San Bernardino County, Morongo district: California Min. Bur. Rept. 16, p. 800.

Mount Baldy

Location and History. This district is in the San Gabriel Mountains in eastern Los Angeles County. It is just west of Mount San Antonio, which is also known as Mount Baldy or Old Baldy. It is both a lode- and placer-mining district.

Placer mining was originally done here in the San Gabriel River in the 1840s, and for several periods following that time, this district was quite productive. Production was obtained from both the stream beds and from terrace gravels, which were mined by hydraulicking. In 1874 it was reported that more than $2 million had been produced in the previous 18 years. The principal period of lode-gold mining was 1903–1908, but there was some activity again in the 1930s. The estimated output from the lode mines is 50,000 ounces. There has been minor work in recent years, chiefly by weekend prospectors.

Geology. The gold-quartz veins occur in schist and gneiss. The ore deposits are rich in places. The ore bodies usually are three feet or less in thickness and do not extend to any great depth. The oxidized zones near the surface yielded the richest ore.

Mines. Allison $50,000, Baldora, Big Horn $40,-000+, Eagle, Gold Dollar, Holly, Heaton, Native Son, Stanley, Zanteson.

Bibliography

Gay, T. E., Jr., and Hoffman, S. R., 1954, Gold—Los Angeles County: California Jour. Mines and Geology, vol. 50, pp. 493–502.
Sampson, R. J., 1937, Gold—Los Angeles County: California Div. Mines Rept. 33, pp. 177–196.
Tucker, W. B., 1927, Gold—Los Angeles County: California Min. Bur. Rept. 23, pp. 291–295.

Mount Gleason

This district is in the central San Gabriel Mountains in Los Angeles County, in the general vicinity of Mount Gleason about 15 miles due north of Pasadena. There are a number of small lode-gold deposits, the principal ones having been the Los Padre and Mount Gleason mines. The veins are narrow and in places contain small amounts of gold and sulfides. All have been idle for many years. The country rock is granite and schist.

Bibliography

Sampson, R. J., 1937, Los Angeles County, Mount Gleason mine: California Div. Mines Rept. 33, pp. 187–188.

Neenach

This district is in northern Los Angeles County about 20 miles west-northwest of Lancaster, in the foothills on the south side of Antelope Valley. Gold was discovered here in 1899, but the bulk of the pro-

Photo 88. Lode Gold Mine, Los Angeles County. The photo was taken in the San Gabriel Mountains in the early 1900s.

duction of about $200,000 was obtained in 1935–38. There has been intermittent mining and development work here since. Most of the production has been from the Rivera or Rogers-Gentry group of mines.

The ore deposits occur in a contact zone between metasediments and quartz monzonite. The ore bodies consist of zones of narrow quartz veins and stringers containing free gold and varying amounts of pyrite. The oxidized zone yielded material valued as high as $60 of gold per ton.

Bibliography

Gay, T. E., Jr., and Hoffman, S. R., 1954, Gold—Los Angeles County: California Jour. Mines and Geology, vol. 50, pp. 497–500.

Weise, J. H., 1950, Geology and mineral resources of the Neenach quadrangle: California Div. Mines Bull. 153, 53 pp.

Pinacate

Location and History. The Pinacate district is in western Riverside County in the hills between Perris and Lake Elsinore. The area was placer-mined in the 1850s. The Good Hope vein was discovered in 1874, and there was considerable mining activity that lasted until about 1903. Some work was done in the district again in the 1930s, and there has been minor prospecting since.

Geology. The area is underlain by shales, slates, phyllites, and quartzites of the Santa Ana Formation (Triassic) and quartz diorite and granodiorite. Quartz latite lies to the west. The ore deposits consist of zones of quartz veins and seams with kaolin and gouge, which contain native gold and varying amounts of sulfides. The sulfide content increases at depth. The ore contains $\frac{1}{2}$ to one ounce of gold per ton. The Good Hope vein was mined to a depth of 575 feet.

Mines. Argonaut, Brady, Colton, Good Hope $1 million to $2 million, Hoag $140,000, Lake View, Lucky Strike, Musick, Rosalia, Santa Fe, Santa Rosa, Shay, Victor.

Bibliography

Dudley, P. H., 1935, Geology of a portion of the Perris block, southern California: California Div. Mines Rept. 31, pp. 487–515.

Engel, Rene, 1959, Geology and mineral deposits of the Lake Elsinore quadrangle: California Div. Mines Bull. 146, 154 pp.

Merrill, F. J. H., 1919, Riverside County, gold: California Min. Bur. Rept. 15, pp. 527–535.

Storms, W. H., 1893, Pinacate district: California Min. Bur. Rept. 11, pp. 384–385.

Tucker, W. B., and Sampson, R. J., 1945, Riverside County, gold: California Div. Mines Rept. 41, pp. 127–144.

Pine Valley

This district, in south-central San Diego County approximately 35 miles east of San Diego, includes the Descanso area to the west. A number of small lode-gold mines in the area were first worked around 1900, and a few have been prospected since. The deposits usually consist of a series of parallel quartz veins and stringers containing native gold and often abundant sulfides. None have been developed to depths of greater than 150 feet. The country rock is granodiorite, gneiss, diorite, and gabbro. The principal properties have been the Descanso, Free Coinage, Gold Standard, Good Luck, and Oak Canyon mines.

Bibliography

Merrill, F. J. H., 1916, San Diego County, Deer Park, Descanso, and Pine Valley Districts: California Min. Bur. Rept. 14, pp. 662–664.

Weber, F. Harold, Jr., 1963, San Diego County, gold: California Div. Mines and Geology County Rept. 3, pp. 115–167.

Piru

Location and History. The Piru district is in northeastern Ventura County in the vicinity of the creek of the same name. The Frazier Mountain district is just to the north, and the town of Gorman on the Ridge Route highway is 10 miles to the northeast. Placer mining was begun here in 1841 by Andrew Castillero, and gold from the district was shipped to the U. S. Mint in Philadelphia in 1842. Small-scale placer mining continued intermittently through the 1890s, and there was some work again in the 1920s and 1930s. Among lode-gold mines, the principal operation was the Castac mine, which has an estimated total output valued at about $160,000.

Geology and Ore Deposits. The placer deposits are in and adjacent to the upper part of Piru Creek, chiefly in the vicinity of its junction with Lockwood Creek and to the east in the Gold Hill area. The gold has been recovered both from Recent stream gravels and older terrace deposits on the hills north of the Creek. The placer gold often is coarse-grained. There are a number of north-striking gold-quartz veins that range from a few inches to about 4 feet in thickness. The veins occur in shear zones and usually in granitic gneiss or hornblende schist. The ore contains free gold and varying amounts of pyrite. Milling ore sometimes averaged ½ ounce of gold per ton.

Bibliography

Bowers, Stephan, 1888, Gold—Ventura County: California Min. Bur. Rept. 8, pp. 680–684.

Huguenin, Emile, 1919, Gold—Ventura County: California Min. Bur. Rept. 15, pp. 759–760.

Tucker, W. B., 1925, Gold—Ventura County: California Min. Bur. Rept. 21, pp. 229–232.

Saugus

This is the extensive placer-mining region in the western San Gabriel Mountains in Los Angeles County. It includes the Newhall, Soledad, Placerita Canyon, upper Santa Clara River, Sand Canyon, Pacoima Canyon, and Arrastre Canyon areas and a number of other canyons. The area sometimes is known as the San Gabriel district and also as the Newhall district. Gold was discovered in the district in the early 1800s, the exact date and place being somewhat uncertain. It is likely that the mission fathers from the San Fernando and San Buena Ventura missions worked placers in the area during the 1830s. A commemorative plaque in Placerita Canyon states that gold was discovered at that locality on March 9, 1842. Production figures are not available, but it has been estimated that $100,000 was produced during the first few years. Placer mining has been carried on intermittently ever since, mostly by small-scale methods. The gold has been recovered from the gravels in the present stream channels and from benches and terraces along the banks. Also, there are a few minor gold-quartz veins in the area.

Bibliography

Gay, T. E., Jr., and Hoffman, S. R., 1954, Mines and mineral deposits of Los Angeles County, gold: California Jour. of Mines and Geology, vol. 50, pp. 493–496.

Jamison, C. E., Santa Clara River placers, Los Angeles and Ventura Counties: Mining and Scientific Press, vol. 100, Mar. 5, 1910, pp. 360–361.

Oakeshott, G. B., 1958, Geology and mineral deposits of the San Fernando quadrangle, Los Angeles County: Calif. Div. Mines Bull. 172, pp. 108–109.

Preston, E. B., 1890, Auriferous gravels of Castaca, Palomas, and San Feliciana Canons: Calif. Min. Bur. Rept. 9, pp. 201–203.

Trabuco

Small amounts of placer gold have been recovered in some of the canyons in the Santa Ana Mountains in southeastern Orange County. The most productive have been Trabuco, Silverado, and Santiago Canyons. The town of Silverado flourished until around 1881. Prospecting was first done here many years ago, and there was some activity again during the 1930s. A number of narrow veins in the region contain varying amounts of tin, copper, zinc, and small amounts of gold.

Bibliography

Larson, E. S., 1951, Crystalline rocks of southwestern California—metals: California Div. Mines Bull. 159, pp. 46–49.

MODOC PLATEAU PROVINCE

The only sources of commercial amounts of gold in the Modoc Plateau province of northeastern California have been the Hayden Hill district in north-central Lassen County and the Winters district in southwest Modoc County. In both districts the gold-bearing veins occur in volcanic rocks of Tertiary age. The mines at Hayden Hill have yielded several million dollars worth of gold, but the Winters district has been the source of less than $200,000. A few small gold prospects occur elsewere in this region.

Hayden Hill

Location. The Hayden Hill district is in northwestern Lassen County about 20 miles southeast of Bieber and 65 miles north of Susanville. It is the only important gold-mining district in the Modoc Plateau geomorphic province.

History. Gold-bearing veins were discovered here in 1869 by J. W. Hayden and S. Lewis. The camp, established in 1871, was originally known as Providence City, renamed Hayden Hill in 1878. A rush to the district lasted until 1883. There was considerable activity again from 1903 to 1910, when the Golden Eagle mine was worked on a large scale. During the 1930s the Hayden Hill corporation operated several properties on a moderate scale, and there has been intermittent prospecting since. The district has a total output valued at about $3 million.

Geology. The district is underlain predominantly by nearly flat-lying well-bedded rhyolite tuffs of Tertiary age, some silicified and brecciated. Patches of Pliocene basalt lie to the east, and extensive beds of Miocene pyroclastic rocks lie to the west and north.

Ore Deposits. Several steeply-dipping veins and stringer zones range from one to 25 feet in thickness. These deposits consist chiefly of consolidated and cemented breccia of wall rock; only a small amount of quartz is present. The gold occurs in the free state in usually small round particles and is commonly associ-

ated with manganese. Appreciable silver is present but practically no sulfides. Nearly all of the ore has been recovered from above the 800-foot depth.

Mines. Brush Hill $400,000, Blue Bell $100,000, Evening Star $200,000, Golden Eagle $1,025,000, Hayden Gouge, Hayseed $150,000, Juniper $600,000, North Star $20,000, Providence $78,000.

Bibliography

Averill, C. B., 1936, Lassen County, Hayden Hill mining district: California Div. Mines Rept. 32, pp. 422–424.

Hanks, H. G., 1888, Lassen County: California Min. Bur. Rept. 8, pp. 329–332.

Hill, James M., 1915, Some Mining Districts in California and Nevada, Hayden Hill mining district: U. S. Geol. Survey Bull. 594 pp.

Preston, E. B., 1890, Lassen County, gold: California Min. Bur. Rept. 9, pp. 211–213.

Preston, E. B., 1893, Hayden Hill mining district: California Min. Bur. Rept. 11, pp. 241–242.

Tucker, W. B., 1919, Lassen County, Hayden Hill mining district: California Min. Bur. Rept. 15, pp. 229–235.

Winters

Location and History. This district is in southwestern Modoc County 35 miles west-southwest of Alturas and 16 miles north of Adin. The area was first prospected for gold in 1890. The vein at the Lost Cabin mine was discovered in 1904. Mining activity continued for a few years after that date, and there was prospecting here in the 1930s.

Geology. The district is underlain by andesite, andesite porphyry, and basalt of Tertiary age. There are several west- and northwest-striking veins that contain fine free gold, quartz, brecciated wall rock, calcite, and feldspar. The deposits are shallow, none of the veins having been developed to a depth of greater than 300 feet.

Mines. Dixie Queen, Lost Cabin (Hess) $150,000, Modoc.

Bibliography

Tucker, W. B., 1919, Modoc County, Winters mining district: California Min. Bur. Rept. 15, pp. 251–252.

COAST RANGES PROVINCE

Gold has been recovered in a number of places in the Coast Ranges province. The largest source of gold has been the beach placers near Orick, Humboldt County, which have yielded more than $1 million. Other producers have been the Palisade and Silverado silver-gold mines, Calistoga district, Napa County; the Island Mountain sulfide deposit, Trinity County, where gold was recovered as a by-product of copper mining; the mercury-gold mines in the Sulphur Creek district, Colusa County; the Los Burros district, Monterey County; the La Panza district, San Luis Obispo County; and the ocean beaches near Crescent City, Del Norte County.

Years ago there were small short-lived placer-mining operations at Jolon, Parkfield, and the Carmel River area, Monterey County; Panoche Valley, San Benito County; San Francisquito Creek near Palo Alto and Coyote Creek, Santa Clara County; Felton and Ben Lomond, Santa Cruz County; Mitchell Canyon north of Mount Diablo, Contra Costa County; and Putah Creek, Yolo County. Gold has been recovered from the ocean beaches at San Francisco, Half Moon Bay, Santa Cruz, Point Sal, and Surf. Small amounts of by-product gold were recovered at one time from the massive pyrite bodies at Leona Heights, Alameda County, and from a few copper prospects. Traces of gold have been noted in quicksilver ores in a few other districts besides the Sulphur Creek district.

Calistoga

Location and History. The Calistoga silver-gold district is in northwestern Napa County. Nearly all of the production has been from the Palisade mine, three miles north of Calistoga, and the Silverado mine, three miles farther north on the east flank of Mt. St. Helena. The district also has been known as the Silverado district, from the story *Silverado Squatters,* by Robert Louis Stevenson, who, with his wife, spent the summer of 1880 in a cabin at the Silverado mine.

Both mines were first worked in the 1870s. The Silverado mine was opened in 1872, and in 1874 yielded $93,000 worth of gold and silver. The Palisade mine, which was much more productive, was opened in 1876 and was worked until 1893. It has been prospected since. The total output of the Palisade mine is about $2 million worth of silver and gold, with some copper and lead. The total gold production for the district is valued at about $500,000.

Geology and Ore Deposits. Much of the district is underlain by volcanic rocks of Tertiary age. In places sandstone and shale are present. The ore deposits at the Palisade mine are in andesite, while those at the Silverado are in silicified rhyolite.

The deposits are in veins that consist of quartz and chalcedony, which often are brecciated. Some of the vein material is porous, and comb structures often are common. The gold usually is associated with silver, copper, and lead sulfides. The veins are steeply dipping, as much as 15 feet thick, and have been developed to depths of as much as 600 feet. Several high-grade pockets have been encountered.

Bibliography

Bowen, O. E., 1951, Geologic guidebook to the San Francisco Bay counties, Palisade and Silverado mines: California Div. Mines Bull. 154, pp. 361–363.

Bradley, W. W., 1916, Napa County, gold and silver: California Min. Bur. Rept. 14, pp. 269–271.

Davis, F. F., 1968, Napa County, Palisade mine: California Jour. Mines and Geology, Vol. 44, pp. 183–184.

Crescent City

Gold and minor platinum have been recovered from black sand deposits on the beaches south of Crescent City, Del Norte County, beginning in the 1850s. Most of it was recovered by small-scale methods. Several large-scale operations were attempted in the 1890s and again in 1913–14 but were unsuccessful. As in other beach deposits along the ocean, the gold-bearing black sands were deposited by shore currents and wave action. Most of the gold here was probably derived from the Smith River, which empties into the ocean a few miles to the north, and the Klamath River, a few miles to the south.

Island Mountain

The Island Mountain sulfide deposit is in the southwest corner of Trinity County about 90 miles north of Ukiah and 30 miles east of Garberville. It was discovered about 1897 but not worked until 1915, shortly after the completion of the nearby Northwestern Pacific Railroad. From 1915 until 1930, 132,000 tons of ore were mined and yielded 9 million pounds of copper, 144,000 ounces of silver, and 8,600 ounces of gold. An estimated 158,000 tons of ore remain.

The deposit is a lenticular massive sulfide body consisting predominantly of pyrite, chalcopyrite, and pyrrhotite, with smaller amounts of magnetite, arsenopyrite, galena, and bornite. The gold and silver are present in the sulfides either in solid solution or as admixtures. The ore contained an average of 1.09 ounces of silver and .065 ounces of gold per ton. Country rock consists of graywacke, shale, glaucophane schist, and chert. Greenstone and andesite are present.

Bibliography

Aubury, L. E., 1908, Island Mountain Consolidated Copper Mine: California Min. Bur. Bull. 50, pp. 148–150.

Logan, C. A., 1926, Trinity County, Island Mountain Consolidated Copper Mine: California Min. Bur. Rept. 22, pp. 14–15.

Stinson, M. C., 1957, Geology of the Island Mountain copper mine, Trinity County: California Jour. Mines and Geology, vol. 53, pp. 9-33.

Jolon

Jolon is in southern Monterey County near Mission San Antonio de Padua. Small amounts of placer gold have been recovered from several streams in the area, beginning about 1850. In 1877 and 1878 several thousand dollars worth of gold were recovered by Chinese miners and sold to the local store. There was prospecting again around 1914, but apparently nothing has been done since. The gold was recovered from Mission and Ruby Canyons and from gulches in the Santa Lucia Mountains just to the west. The gold was principally coarse nuggets, some more than ¼ ounce.

Bibliography

Hart, E. W., 1966, Monterey County, Jolon Area: California Div. Mines and Geology County Report 5, p. 45.

Waring, C. A., and Bradley, W. W., 1919, Monterey County, Jolon District: California Min. Bur. Rept. 16, p. 606.

La Panza

Location and History. This is a placer-mining district in east-central San Luis Obispo County about 40 miles east of the town of San Luis Obispo. The district is in the La Panza Mountains and includes the area around the site of the town of La Panza east of the crest of the mountain range and the Pozo area to the west. La Panza means "the paunch" in Spanish. The name was derived from the practice of the vaqueros at nearby ranchos of using the paunch or other parts of slaughtered beef as bait to trap grizzly bears, which once were common here.

Placer mining is believed to have first been done in the district in the early 1800s by Mexicans and Indians. Gold was rediscovered in 1878, and there was a rush to the area that lasted for several years. In 1888 the total output was estimated to have been valued at over $100,000. Small-scale mining continued through the early 1900s, and there was activity again in the 1930s and early 1940s. The total production of the district is estimated at $200,000. A few old buildings remain in the area.

Geology. Much of the gold apparently was obtained from San Juan Creek, which flows northward along the east flank of the mountain range, and from several of its tributaries. The richest tributaries were Navajo, McGinnis, Placer, and Hay Creeks. On the west side of the summit some gold was recovered from Pozo, Frazer, and Toro Creeks and possibly from the upper Salinas River. The placer deposits were small and discontinuous, but in places they were rich. The gold was fairly coarse and somewhat irregular. It was derived from narrow quartz veins in the granitic rocks that constitute the central part of the La Panza Mountains. The east side of the district is underlain by sandstone, shale, and conglomerate.

Bibliography

Dillon, R. H., 1961, The legends of La Panza: Westways, May 1961, pp. 10–12.

Franke, H. A., Jr., 1935, San Luis Obispo County, gold: California Div. Mines Rept. 31, pp. 420–423.

Laizure, C. McK., 1925, San Luis Obispo County, gold: California Min. Bur. Rept. 21, pp. 514–515.

Logan, C. A., 1919, San Luis Obispo County, gold: California Min. Bur. Rept. 15, pp. 687–688.

Los Burros

Location. The Los Burros district is in southwestern Monterey County in the Santa Lucia Mountains. It is about 80 miles south of Monterey and four miles east of Cape San Martin.

History. It is believed that this region was first prospected for placer gold and quicksilver in the early 1850s. Prospecting became so popular here that the Los Burros mining district was organized in 1875. In 1887 lode gold was discovered by W. D. Cruikshank at what is now the Buclimo mine. There was considerable excitement during the following few years, and a vast number of claims were located. The principal settlement was the town of Manchester or Mansfield, which burned down in 1892.

Another flurry of activity in the early 1900s followed placer gold discoveries in the various forks of Willow Creek. Intermittent small-scale prospecting and development work have continued in the district until the present time. There was a recorded production of several hundred dollars worth of gold in 1953 and again in 1963. It is believed that 2000 or more claims have been located in the district. The value of the total output is estimated to be about $150,000.

Geology and Ore Deposits. The Los Burros district is underlain by various rocks of the Franciscan Formation (Upper Jurassic). Dark sandstone is most abundant and is also the chief host rock of the gold-bearing deposits. Also present are chert, shale, serpentine, and volcanic rocks. These rocks have been strongly faulted and sheared and locally metamorphosed. Numerous narrow northeast-trending veins, composed of quartz and small amounts of calcite, occur in shear and fracture zones and commonly with fault gouge.

Most of the gold has been recovered from small lenticular ore shoots in oxidized zones near the surface. The sulfides, which consist of fine-grained pyrite and small amounts of chalcopyrite and arsenopyrite generally, are low in gold content. Most of the placer gold has come from Willow Creek, and much of it was concentrated as coarse ragged fragments. Very small amounts have been found in Alder, Plaskett, and Salmon Creeks.

Mines. Ancona, Buclimo $62,000, Bushnell, Gorda, Grizzly, Mariposa, Melville, New York, Plaskett, Plaskett (placer) $18,000, Spruce (placer) $22,000.

Bibliography

Hart, E. W., 1966, Monterey County, gold: California Div. Mines and Geology County Rept. 5, pp. 44–52.

Hill, J. M., 1923, The Los Burros district: U. S. Geol. Survey Bull. 735-J, pp. 323–329.

Irelan, William, Jr., 1888, Los Burros district: California Min. Bur. Rept. 8, pp. 405–410.

Laizure, C. McK., 1925, Monterey County, Los Burros district: California Min. Bur. Rept. 21, pp. 37–41.

Mining and Scientific Press, vol. 104, pp. 696–698, May 18, 1912.

Preston, E. B., 1892, Los Burros district: California Min. Bur. Rept. 11, pp. 259–262.

Waring, C. A., and Bradley, W. W., 1919, Monterey County, Los Burros mining district: California Min. Bur. Rept. 15, pp. 602–605.

Orick

The Orick or Gold Bluff Beach district is in northwestern Humboldt County about 50 miles north of Eureka and near the town of Orick. A series of gold-bearing black sand deposits extend along the ocean beach for a distance of about 10 miles. This area was first placer-mined about 1852, and considerable activity continued through the 1880s. There has been intermittent small-scale placer mining on the beaches since. In 1888 it was estimated that the district had yielded more than $1 million.

Gold and minor amounts of platinum occur in thin but often fairly extensive layers of black sands on the beach. Gold also is found in terrace and bench gravels in the bluffs immediately east of the beaches. The black sands were deposited by the action of shore currents and waves, which sort and distribute materials broken down from the sea cliffs or washed into the sea by streams. Some of the gold here may have come from the Klamath River, which empties into the ocean a few miles to the north. The gold is fine grained and ranges from 900 to 950 in fineness. Various types of devices have been used here to recover gold, including sluices, a special type of long tom used in surf washing, amalgamating plates, and mechanical equipment.

Bibliography

Hornor, R. R., 1918, Notes on the black sand deposits of southern Oregon and northern California: U. S. Bureau Mines Technical Paper 196, 42 pp.

Irelan, William, Jr., 1888, Gold Bluff Beach mines: California Min. Bur. Rept. 8, pp. 216–218.

Rice, Salem J., 1961, Geologic sketch of the northern Coast Ranges: California Div. Mines Mineral Information Service, vol. 14, no. 1.

Putah Creek

Placer mining was first done many years ago on lower Putah Creek in southwestern Yolo County. At one time a small mining camp existed where the creek enters the Sacramento Valley from the Coast Range near the present town of Winters. Small-scale placer mining was done here again in the 1930s, when several thousand dollars worth of gold was produced. Also at one time occasional sluicing was done on Cache Creek to the north near the town of Capay. Several narrow quartz veins containing traces of gold were prospected in the Coast Range to the west.

Bibliography

Watts, W. L., 1890, Yolo County, gold: California Min. Bur. Rept. 10, p. 790.

Red Mountain

This is a small gold- and copper-bearing district in southeastern Mendocino County in the mountains between the Russian River on the west and Clear Lake on the east. Minor amounts of placer gold were recovered from streams on the west slope of the range in the 1880s and 1890s and again in the 1920s and 1930s. Several narrow gold- and copper-bearing quartz veins occur near the summit. The region is underlain by sandstone, shale, and serpentine.

Bibliography

Crawford, J. J., 1894, Red Mountain Mining District: California Min. Bur. Rept. 12, p. 177.

San Francisco Beach

Gold occurs as fine grains in the black sands on the beach at San Francisco. From 1938 to 1950, gold was produced at the beach by people who used small washing plants. From 1938 to 1941, the recorded production was valued at about $13,000. The most productive part of the beach was south of the Fleishhacker Zoo, and the gold was most plentiful immediately after heavy winter storms. Several narrow gold-bearing quartz veins have been found in metamorphic rocks in the general area.

Santa Cruz

Some gold has been recovered in Santa Cruz County. It has been obtained from creeks in the Ben Lomond and Felton areas, small quartz veins in granitic rocks in Ben Lomond Mountain, and from the ocean beaches along Monterey Bay. Some time in the 1850s or 1860s, a large boulder was found in Gold Gulch four miles north of Santa Cruz that was reported to have contained $30,000 to $50,000, according to the various stories. Much work was done in the area following this discovery, but only small amounts of gold were found.

The black sand deposits on the beaches between Santa Cruz and Pajaro to the southeast were first placer-mined in the 1850s. Later, during the 1880s and 1890s, attempts were made to work these deposits with various mechanical devices, but none were commercially successful. During the depression years of the 1930s the beach sands were worked by small-scale hand methods. These black sand deposits are found both on the present beach and older marine terraces and low hills in back of the beaches. The black sands occur in strata that range from a few inches to several feet in thickness.

Bibliography

Huguenin, Emile, and Castello, W. O., 1921, Santa Cruz County, gold: California Min. Bur. Rept. 17, pp. 235–236.

Watts, W. L., 1890, Santa Cruz County, auriferous sand: California Min. Bur. Rept. 10, pp. 622–624.

Silver Queen

During the 1880s and early 1890s minor quantities of gold were recovered from the Silver Queen mine in western Sonoma County, five miles north of Cazedero. At this deposit there is a diabase dike up to 30 feet thick in schist that contains auriferous pyrite. Small amounts of placer gold have been recovered in the area.

Bibliography

Crawford, J. J., 1896, Silver Queen mine: California Min. Bur. Rept. 13, p. 436.

Sulphur Creek

Location and History. The Sulphur Creek mercury-gold district is in the southwest corner of Colusa County and in a small adjacent area in Lake County.

It is about 20 miles southwest of Williams and just west of Wilbur Springs. Gold was discovered here in 1865, but the chief period of production was from 1880 to 1890, with a minor output since. The total gold output of the district is valued at about $109,000. The principal source of gold has been the Manzanita mine, but some has been recovered from the Cherry Hill and Clyde mines. The Manzanita is one of the few mercury mines that also has been operated as a gold mine.

Geology and Ore Deposits. The deposits consist of narrow seams of siliceous sinter containing incrustations of free gold in the oxidized zone and auriferous pyrite at depth. Usually the gold is associated with fine-grained cinnabar but not always. Some placer gold has also been recovered here. Native sulfur and bituminous matter are present. Country rock is sandstone and shale with several bands of serpentine. Hot spring action apparently has been important in the formation of the mineral deposits in the district; ascending solfatoric waters invaded the sandstone and shale and leached out the more soluble material. Gold, cinnabar, sulfur compounds, and siliceous sinter were then deposited.

Bibliography

Becker, G. F., 1888, Quicksilver deposits of the Pacific slope: U. S. Geol. Survey Mon. 13, pp. 367–368.

Bradley, W. W., 1918, Quicksilver resources of California, Manzanita mine: California Min. Bur. Bull. 78, pp. 38–39.

Logan, C. A., 1929, Colusa County, Sulphur Creek district: California Div. Mines and Mining Rept. 25, pp. 288–290.

Surf-Point Sal

At one time appreciable quantities of gold were recovered from the ocean beaches in western Santa Barbara County. The most productive year for which there is a record was 1889, when the county's gold output was valued at $41,000, much of which may have come from these beaches. Gold and very small amounts of platinum occur as fine grains in thin layers of black sands. These deposits are more or less continuous between Point Arguello on the south and the mouth of the Santa Maria River on the north, but the most productive ones have been at Surf and Point Sal. Much of this area now is part of Vandenberg Air Force Base.

Bibliography

Irelan, William, Jr., 1888, Santa Barbara County, gold: California Min. Bur. Rept. 8, p. 537.

Tucker, W. B., 1925, Santa Barbara County, gold: California Min. Bur. Rept. 21, pp. 541–542.

GENERAL BIBLIOGRAPHY

The following is a selected general bibliography on gold in California. The total number of publications that have been written on gold in California is very large.

Allen, W. W., and Avery, R. B., 1893, California gold book: San Francisco and Chicago.

Alling, M. N., 1922, Ancient river-bed deposits in California: Engineering and Min. Jour., vol. 1, Sept. 1922, pp. 134–140, and Oct. 1922, pp. 161–166.

Aubury, L. E., Winston, W. B., and Janin, Charles, 1910, Gold dredging in California: California Min. Bur. Bull. 57, 305 pp.

Averill, C. V., et al., 1946, Placer mining for gold in California: California Div. Mines Bull. 135, 377 pp.

Bell, James E., 1956, Gold: U. S. Bur. Mines Bull. 556, pp. 315–326.

Bowie, A. J., Jr., 1905, A practical treatise on hydraulic mining in California: D. Van Nostrand Co., New York, 313 pp.

Browne, J. Ross, 1868, Mineral resources of the states and territories west of the Rocky Mountains: U. S. Government.

Browne, J. Ross, and Taylor, J. W., 1867, Mineral resources of the states and territories west of the Rocky Mountains: U. S. Government.

California Miners' Association, 1899, California's mines and minerals, San Francisco, 445 pp.

Clark, Lorin D., 1960, Foothills fault system, western Sierra Nevada: Geol. Soc. America Bull., vol. 71, pp. 483–496.

Clark, William B., 1957, Gold: in California Div. Mines Bull. 176, pp. 215–226.

Cloos, Ernst, 1935, Mother Lode and Sierra Nevada batholith: Jour. Geology, vol. 43, pp. 225–249.

Del Mar, Alexander, 1902, A history of precious metals: New York, 464 pp.

Del Mar, Alexander, 1911, Gold nuggets of California: Min. & Sci. Press, vol. 102, p. 629.

Doolittle, J. E., 1905, Gold dredging in California: California Min. Bur. Bull. 36, 120 pp.

Dunn, R. L., 1888, Drift mining in California: California Min. Bur. Rept. 8, pp. 736–770.

Dunn, R. L., 1890, River mining: California Min. Bur. Rept. 9, pp. 262–281.

Dunn, R. L., 1894, Auriferous conglomerate in California: California Min. Bur. Rept. 12, pp. 459–471.

Edman, J. A., 1907, Auriferous black sands of California: California Min. Bur. Bull. 45, 19 pp.

Emmons, W. H., 1937, Gold deposits of the world: McGraw-Hill Book Co., Inc., New York, 562 pp.

Fairbanks, H. W., 1890, Geology of the Mother Lode region: California Min. Bur. Rept. 10, pp. 22–90.

Gardner, D. L., 1954, Gold and silver mining districts in the Mojave Desert region of southern California: California Div. Mines Bulletin 170, Chap. 8, no. 6, pp. 51–58.

Gilbert, G. K., 1917, Hydraulic-mining debris in the Sierra Nevada: U. S. Geol. Survey Prof. Paper 105, 154 pp.

Goodwin, J. Grant, 1957, Lead and zinc in California: California Jour. Mines and Geology, vol. 53, pp. 353–724.

Goodyear, W. A., The auriferous gravels of California: Min. & Sci. Press, vol. 39, Sept. 20, 1879, pp. 182–183.

Haley, C. S., 1923, Gold placers of California: California Min. Bur. Bull. 92, 167 pp.

Hammond, John Hays, 1890, The auriferous gravels of California: California Min. Bur. Rept 9, pp. 105–138.

Hammond, John Hays, 1890, Mining of gold ores in California: California Min. Bur. Rept. 10, pp. 852–882.

Hanks, H. G., 1882, Placer, hydraulic, and drift mining: California Min. Bur. Rept. 2, pp. 28–192.

Henderson, C. W., 1922, The history and influence of mining in the western United States: Ore deposits of the western states, Lindgren volume, Am. Inst. Min. Engrs., New York, pp. 730–784.

Hill, J. M., 1912, The mining districts of the western United States, with a geologic introduction by Waldemar Lindgren: U. S. Geol. Survey Bull. 507, pp. 17–20, 77–133.

Hill, J. M., 1915, Some mining districts in northeastern California and northwestern Nevada: U. S. Geol. Survey Bull. 594, pp. 133–141.

Hill, J. M., 1929, Historical summary of gold, silver, copper, lead, and zinc produced in California, 1848 to 1926: U. S. Bur. Mines Econ. Paper 3, 22 pp.

Hulin, C. D., 1933, Geological relations of ore deposits in California: Ore deposits of the western states, Lindgren volume, A.I.M.E., New York, pp. 240–253.

Irwin, William P., 1960, Geologic reconnaissance of the northern Coast Ranges and Klamath Mountains, California: California Div. Mines Bull. 179, 80 pp.

Janin, Charles, 1918, Gold dredging in the United States: U. S. Bur. Mines Bull. 127, 226 pp.

Jarman, Arthur, 1927, Report of the Hydraulic Mining Commission upon the feasibility of the resumption of hydraulic mining in California: California Min. Bur. Rept. 23, pp. 44–116.

Jenkins, Olaf P., 1935, New technique applicable to the study of placers: California Div. Mines Rept. 31, pp. 193–210.

Jenkins, Olaf P., et al, 1948, Geologic guidebook along Highway 49—Sierran gold belt—the Mother Lode Country: California Div. Mines Bull. 141, 164 pp.

Joslin, G. A., 1945, Gold: California Div. Mines Bull. 130, pp. 122–151.

Julihn, C. E., and Horton, F. W., 1938, Mines of the southern Mother Lode region: U. S. Bur. Mines Bull. 413, 140 pp.

Julihn, C. E., and Horton, F. W., 1940, Mines of the Mother Lode region, II: U. S. Bur. Mines Bull. 424, 179 pp.

Knopf, Adolph, 1929, the Mother Lode system of California: U. S. Geol. Survey Prof. Paper 157, 88 pp.

Koschmann, A. H., and Bergendahl, M. H., 1962, Gold in the United States: U. S. Geological Survey Mineral Investigations Resources Map MR-24.

Koschmann, A. H., and Bergendahl, M. H., 1968, Principal gold-producing districts of the United States: U. S. Geol. Survey Prof. Paper 610, pp. 53–84.

Lindgren, Waldemar, 1894a, Sacramento folio, California: U. S. Geological Survey Geol. Atlas of the U. S., folio 5, 3 pp.

Lindgren, Waldemar, 1894b, Marysville folio, California: U. S. Geol. Survey Geol. Atlas of the U. S., folio 17, 2 pp.

Lindgren, Waldemar, 1896a, Nevada City special folio, California: U. S. Geol. Survey Geol. Atlas of the U. S., folio 29, 7 pp.

Lindgren, Waldemar, 1896b, Pyramid Peak folio, California: U. S. Geol. Survey Geol. Atlas of the U. S., folio 31, 8 pp.

Lindgren, Waldemar, 1896c, Characteristic features of the California gold-quartz veins: Bull. Geol. Soc. America, pp. 221–240.

Lindgren, Waldemar, 1897, Truckee folio, California: U. S. Geol. Survey Geol. Atlas of the U. S., folio 39, 8 pp.

Lindgren, Waldemar, 1900, Colfax folio, California: U. S. Geol. Survey Geol. Atlas of the U. S., folio 66, 10 pp.

Lindgren, Waldemar, 1909, Resources of the United States in gold, silver, copper, lead, and zinc: U. S. Geol. Survey, Bull. 394, pp. 114–156.

Lindgren, Waldemar, 1911, The Tertiary gravels of the Sierra Nevada of California: U. S. Geol. Survey Prof. Paper 73, 226 pp.

Lindgren, Waldemar, 1912, The mining districts of the western United States: U. S. Geol. Survey Bull. 507, pp. 5–13, 17–20.

Lindgren, Waldemar, 1933, Gold-quartz veins of the Sierra Nevada: Mineral Deposits, McGraw-Hill Book Company, New York, pp. 616–627.

Lindgren, Waldemar, and Turner, H. W., 1894, Placerville folio, California: U. S. Geol. Survey Geol. Atlas of the U. S., folio 3, 3 pp.

Lindgren, Waldemar, and Turner, H. W., 1895, Smartsville folio, California: U. S. Geol. Survey Geol. Atlas of the U. S., folio 18, 6 pp.

Logan, C. A., 1919, Platinum and allied metals in California: California Min. Bur. Bull. 85, 120 pp.

Logan, C. A., 1935, Mother Lode gold belt of California: California Div. Mines Bull. 108, 240 pp.

Logan, C. A., 1950, Gold: California Div. Mines Bull. 156, pp. 503–514.

Merwin, Roland W., 1968, Gold resources in the Tertiary gravels of California: U. S. Bur. Mines Technical Progress Report, 14 pp.

Mining and Scientific Press, 1860–1921, various entries, vols. 1–125.

Moore, Lyman, 1968, Gold resources of the Mother Lode belt, California: U. S. Bur. Mines Technical Progress Report 5, Heavy Metals Program, 22 pp.

Paul, Rodman W., 1947, California gold, the beginning of mining in the far west, Harvard Univ. Press, Cambridge, 380 pp.

Peterson, D. W., Yeend, W. E., Oliver, H. W., and Mattick, R. E., 1968, Tertiary gold-bearing channel gravel in northern Nevada County, California: U. S. Geol. Survey Circular 566, 22 pp.

Preston, E. B., 1895, California gold mill practices: California Min. Bur. Bull. 6, 85 pp.

Ransome, F. L., 1900, Mother Lode district folio: U. S. Geol. Survey Geol. Atlas of the U. S., folio 63, 11 pp.

Raymond, R. W., 1869, 1870, 1871, 1872, 1874, 1875, and 1876, Mineral resources of the states and territories west of the Rocky Mountains: U. S. Government.

Reid, John A., The east country of the Mother Lode: Min. and Sci. Press, Mar. 2, 1907.

Rickard, Thomas A., 1932, A history of American Mining: D. Appleton-Century Co., Inc., New York and London, 419 pp.

Ridgeway, R. H., 1929, Summarized date of gold production: U. S. Bur. Mines, Econ. Paper 6, 63 pp.

Risdon Iron Works, 1885, Gold mines and mining in California: George Spaulding and Co., San Francisco, 349 pp.

Ryan, J. P., 1960, Gold: U. S. Bur. Mines Bull. 585, pp. 347–356.

Skidmore, W. A., 1885, Gold and silver mining in California, past, present, and prospective: Report of the Director of the Mint, pp. 525–557.

Storms, W. H., 1900, Mother Lode region of California: California Min. Bur. Bull. 18, 154 pp.

Turner, H. W., 1894a, Jackson folio, California: U. S. Geol. Survey Geol. Atlas of the U. S., folio 11, 6 pp.

Turner, H. W., 1894b, Notes on the gold ores of California: Am. Jour. Sci., 3rd ser., 47.

Turner, H. W., 1897, Downieville folio, California: U. S. Geol. Survey Geol. Atlas of the U. S., folio 37, 8 pp.

Turner, H. W., 1898, Bidwell Bar folio, California: U. S. Geol. Survey Geol. Atlas of the U. S., folio 43, 6 pp.

Turner, H. W., 1899, Replacement deposits in the Sierra Nevada: Jour. Geol. pp. 389–400.

Turner, H. W., 1905, Notes on the gold ores of California: Am. Jour. Sci., 3rd ser., 49.

Turner, H. W., and Ransome, F. L., 1897, Sonora folio, California: U. S. Geol. Survey Geol. Atlas of the U. S., folio 41, 7 pp.

Turner, H. W., and Ransome, F. L., 1898, Big Trees folio, California: U. S. Geol. Survey Geol. Atlas of the U. S., folio 51, 8 pp.

Weatherbe, D'Arcy, 1907, Dredging for gold in California: Mining and Scientific Press, San Francisco, 217 pp.

Whitney, J. D., 1875, The auriferous gravels of the Sierra Nevada: Mem. Mus. Comp. Zool., Harvard College, vol. 6, 569 pp.

Yale, C. G., 1896, Total production of gold in California since 1848 by years according to different authorities: California Min. Bur. Rept. 13, pp. 64–65.

LIST OF DISTRICTS BY COUNTIES

ALPINE

Hope Valley
Mogul
Monitor
Silver King
Silver Mountain

AMADOR

Camanche (also in Calaveras and San Joaquin Counties)
Fiddletown
Forest Home
Irish Hill
Jackson-Plymouth
Lancha Plana

Pine Grove
Volcano
West Point (also in Calaveras County)
White Oak Flat

BUTTE

Bangor
Bidwell Bar
Butte Creek
Cherokee
Clipper Mills
Concow
Forbestown
Honcut
Inskip
Kimshew
Magalia

Morris Ravine
Oroville
Wyandotte
Yankee Hill

CALAVERAS

Alto
Angels Camp
Blue Mountain
Calaveritas
Camanche (also in Amador and San Joaquin Counties)
Campo Seco
Carson Hill
Collierville (also in Tuolumne County)
Esmeralda
Fourth Crossing

Glencoe
Hodson
Jenny Lind
Jesus Maria
Mokelumne Hill
Mountain Ranch
Murphys
Paloma
Railroad Flat
Rich Gulch
San Andreas
Sheep Ranch
Skull Flat
Vallecito
Valley Springs
West Point (also in Amador County)

COLUSA

Sulphur Creek

DEL NORTE

Crescent City
Monumental
Smith River

EL DORADO

Coloma
Deer Creek
Deer Valley
El Dorado
Fairplay
Garden Valley
Georgetown
Greenwood
Grizzly Flat
Hazel Valley
Indian Diggings
Kelsey
Logtown
Nashville
Newtown
Omo Ranch
Pacific
Pilot Hill
Placerville
Rattlesnake Bar
Shingle Springs
Slate Mountain
Spanish Dry Diggings
Spanish Flat
Volcanoville

FRESNO

Big Creek
Big Dry Creek
Friant
Mill Creek
Sycamore Flat
Temperance Flat

HUMBOLDT

Hoopa
Orick
Orleans
Weitchpec
Willow Creek

IMPERIAL

Cargo Muchacho
Chocolate Mountains
Mesquite
Picacho
Potholes
Tumco

INYO

Argus
Ballarat
Beveridge
Big Pine

Bishop Creek
Chloride Cliff
Echo Canyon
Fish Springs
Grapevine
Harrisburg
Kearsarge
Lee's Camp
Modoc
Russ
Skidoo
Slate Range (also in San Bernardino
    County)
South Park
Tibbetts
Tucki Mountain
Ubehebe
Wildrose
Willow

KERN

Clear Creek
Cove
El Paso Mountains
Erskine Creek
Garlock
Greenhorn Mountain
Kern River
Keyseville
Long Tom
Loraine
Mojave
Piute Mountains
Rademacher
Rand (also in San Bernardino County)
Rosamond
Tehachapi
White River (also in Tulare County)

LASSEN

Diamond Mountain
Hayden Hill
Honey Lake
Mountain Meadows

LOS ANGELES

Acton
Azusa
Mount Baldy
Mount Gleason
Neenach
Saugus

MADERA

Chowchilla
Coarsegold
Fine Gold
Fresno River
Grub Gulch
Hildreth
Potter Ridge
Quartz Mountain

MARIPOSA

Bagby
Buckeye
Cathey
Cat Town
Clearinghouse
Colorado
Coulterville
Dog Town
Granite Springs
Gravel Range (also in Tuolumne County)
Greely Hill
Hite Cove
Hornitos
Jerseydale
Kinsley

Mariposa
Mormon Bar
Mount Bullion
Peñon Blanco (also in Tuolumne County)
Whitlock

MENDOCINO

Red Mountain

MERCED

Snelling

MODOC

High Grade
Winters

MONO

Bodie
Chidago
Clover Patch
Homer
Jordan
Keith
Mammoth
Masonic
Patterson
Tioga (also in Tuolumne County)
West Walker
White Mountains

MONTEREY

Jolon
Los Burros

NAPA

Calistoga

NEVADA

Badger Hill
Blue Tent
Emigrant Gap (also in Placer County)
English Mountain
French Corral
Graniteville
Grass Valley
Lowell Hill
Meadow Lake
Moore's Flat
Nevada City
North Bloomfield
North Columbia
North San Juan
Rough-and-Ready
Scotts Flat
Smartsville (also in Yuba County)
Washington
You Bet

ORANGE

Trabuco

PLACER

Blue Canyon
Canada Hill
Colfax
Damascus
Duncan Peak
Dutch Flat
Emigrant Gap (also in Nevada County)
Forest Hill
Gold Hill
Gold Run
Iowa Hill
Last Chance
Lincoln
Michigan Bluff

Ophir
Penryn
Rocklin
Tahoe
Todd Valley
Westville
Wheatland
Yankee Jims

PLUMAS

Blue Nose Mountain
Butte Valley
Crescent Mills
Genessee
Granite Basin
Johnsville
La Porte
Lights Canyon
Meadow Valley
Mooreville Ridge
Quincy
Rich Bar
Sawpit Flat
Spring Garden
Taylorsville
Virgilia

RIVERSIDE

Arica
Bendigo
Chuckwalla
Dale (also in San Bernardino County)
Dos Palmas
Eagle Mountains
Gold Park
Hexie
Lost Horse
Menifee
Mule Mountains
Pinacate
Pinon
Twenty Nine Palms (also in San Bernardino County)

SACRAMENTO

Folsom
Michigan Bar

SAN BERNARDINO

Alvord
Arrowhead
Baldwin Lake
Black Hawk
Clark
Coolgardie
Dale (also in Riverside County)
Emerson Lake
Gold Reef
Goldstone
Grapevine
Hackberry Mountain
Halloran Springs
Hart
Holcomb
Ibex
Ivanpah
Lytle Creek
Morongo
Old Dad
Old Woman
Ord
Oro Grande
Rand (also in Kern County)
Shadow Mountains
Slate Range (also in Inyo County)
Spangler
Stedman

Trojan
Twenty Nine Palms
Vanderbilt
Whipple

SAN DIEGO

Banner
Boulder Creek
Cuyamaca
Deer Park
Dulzura
Escondido
Julian
Laguna Mountains
Mesa Grande
Montezuma
Pine Valley

SAN FRANCISCO

San Francisco Beach

SAN JOAQUIN

Camanche (also in Calaveras County)

SAN LUIS OBISPO

La Panza

SANTA BARBARA

Point Sal
Surf

SANTA CRUZ

Santa Cruz

SHASTA

Backbone
Centerville
Clear Creek
Cottonwood
Deadwood
Dog Creek
French Gulch
Gas Point
Harrison Gulch
Igo
Old Diggings
Ono
Platina
Redding
Shasta
Whiskeytown

SIERRA

Alleghany
American Hill
Brandy City
Church Meadows
Downieville
Eureka
Forest
Furnier
Gibsonville
Gold Valley
Goodyear's Bar
Pike
Poker Flat
Port Wine
Poverty Hill
Sierra City

SISKIYOU

Callahan
Cecilville
Cottage Grove
Cottonwood

Deadwood
Dillon Creek
Forks of Salmon
Gazelle
Gottville
Greenhorn
Hamburg
Happy Camp
Hornbrook
Humbug
Indian Creek
Knownothing
Liberty
Oak Bar
Oro Fino
Paradise
Sawyers Bar
Scott Bar
Seiad
Snowden
Somesbar
Yreka

SONOMA

Silver Queen

STANISLAUS

Knight's Ferry
La Grange

TEHAMA

Jelly Ferry
Polk Springs

TRINITY

Big Bar
Bully Choop
Burnt Ranch
Canyon Creek
Carrville
Coffee Creek
Dedrick
Denny
Dodge
Dorleska
Douglas City
East Fork
Eastman Gulch
Hayfork
Helena
Junction City
Lewiston
Minersville
New River
Salyer
Stuart Fork
Trinity Center
Weaverville

TULARE

Globe
Mineral King
White River (also in Kern County)

TUOLUMNE

American Camp
Big Oak Flat
Buchanan
Chinese Camp
Collierville (also in Calaveras County)
Columbia
Confidence
Garrotte
Granite Springs (also in Mariposa County)
Gravel Range (also in Mariposa County)
Hardin Flat
Jacksonville

Jamestown
Peñon Blanco (also in Mariposa County)
Sonora
Soulsbyville
Tioga (also in Mono County)
Tuolumne
Tuttletown

VENTURA

Frazier Mountain
Piru

YOLO

Putah Creek

YUBA

Browns Valley
Brownsville
Camptonville
Dobbins
Hammonton
Smartsville (also in Nevada County)

INDEX OF ALTERNATE DISTRICT NAMES

Many districts were known by more than one name. Also, the author sometimes found it convenient to include in a single district description information on areas within that district that, in the past, have been themselves loosely called "districts." The body of the text occasionally gives an alternate district name and points out those areas that have been included under a single district heading. However, the table of contents, the district headings and the illustrations refer to a single district by the same name, for the sake of consistency.

District and other place names not found in the table of contents appear below. The alternate name preferred in this report appears in *italics*, followed by the province abbreviation and the page number. The provinces and abbreviations are: Sierra Nevada, SN; Klamath Mountains, KM; Basin Ranges, BR; Mojave Desert, MD; Transverse and Peninsular Ranges, TPR; Modoc Plateau, MP, and Coast Ranges, CR.

Alta, *Dutch Flat*, SN, 45.
Altaville, *Angels Camp*, SN, 25.
Amador City, *Jackson-Plymouth*, SN, 69.
Amalie, *Loraine*, SN, 87.
Auld, *Menifee*, TPR, 174.

Baker Ranch, *Michigan Bluff*, SN, 90.
Banner, *Julian-Banner* TPR, 172.
Bath, *Forest Hill*, SN, 49.
Bear Valley, *Bagby*, SN, 29.
Ben Lomond, *Santa Cruz*, CR, 180.
Big Bar, *Trinity River*, KM 143.
Big Bend, *Yankee Hill*, SN, 131.
Birchville, *French Corral*, SN, 50.
Black Bear, *Liberty*, KM, 139.
Blue Canyon, *Emigrant Gap*, SN, 45.
Blue Nose Mountain, *Sawpit Flat*, SN, 114
Brown's Flat, *Columbia*, SN, 39.
Buchanan, *Soulsbyville*, SN, 121.
Buckeye, *Granite Basin*, SN, 52. This district is in Butte County. Another Buckeye District, in Mariposa County, is listed as such in the text.
Buckeye, *Old Diggings*, KM, 140.
Buckeye Hill, *Scott's Flat*, SN ,114.
Bull Creek, *Kinsley*, SN, 85.
Bunker Hill, *Gibsonville*, SN, 51.
Burnt Ranch, *Trinity River*, KM, 143.
Butterfly Valley, *Quincy*, SN, 111.
Byrd's Valley, *Michigan Bluff*, SN, 90.

Canyon Creek, *Dedrick-Canyon Creek*, KM, 135.
Carrville, *Trinity River*, KM, 143.
Cave City, *Mountain Ranch*, SN, 93.
Centerville, *Butte Creek*, SN, 32.
Centerville, *Redding*, KM, 140.
Chidago, *Clover Patch*, BR, 148.
Chili Gulch, *Mokelumne Hill*, SN, 91.
China Flat, *Downieville*, SN, 44.
Chip's Flat, *Alleghany*, SN, 19.
Church Meadows, *Sierra City*, SN, 117.
Clear Creek, *Redding*, KM, 140.
Clements, *Camanche-Lancha Plana*, SN, 33.
Coffee Creek, *Trinity River*, KM, 143.
Colorado, *Whitlock*, SN, 131.
Columbia Hill, *North Columbia*, SN, 101.
Concow, *Yankee Hill*, SN, 131.
Cottage Grove, *Klamath River*, KM, 139.
Cottonwood, *Klamath River*, KM, 139. This district is in Siskiyou County. Another Cottonwood district, in Shasta County, is listed as such in the text.
Craig's Flat, *Eureka*, SN, 46.
Craycroft, *Downieville*, SN, 44.

Dead Horse Flat, *Vallecito*, SN, 126.
Deadwood, *Last Chance*, SN, 86, or *Poker Flat*, SN, 108.
Deadwood, *French Gulch*, KM, 136. This district is in Trinity County. Another Deadwood district, in Siskiyou County, is listed as such in the text.
Denny, *New River-Denny*, KM, 139.
Derbec, *North Bloomfield*, SN, 101.
De Sabla, *Magalia*, SN, 88.
Descanso, *Pine Valley*, TPR, 176.
Diamond Springs, *Placerville*, SN, 107.
Dodge, *Trinity River*, KM, 143.
Dogtown, *Kinsley*, SN, 85.
Dogtown Diggings, *Jordan*, SN, 83.
Douglas City, *Trinity River*, KM, 143.
Douglas Flat, *Vallecito*, SN, 126.
Drytown, *Jackson-Plymouth*, SN, 69.
Duncan Hill, *Ophir*, SN, 102.

East Fork, *Helena-East Fork*, KM, 138.
Eastman Gulch, *Trinity River*, KM, 143.
Echo Canyon, *Lee's Camp-Echo Canyon*, BR, 150.
Edmonton, *Meadow Valley*, SN, 89.
Elizabethtown, *Quincy*, SN, 111.
Esmeralda, *Murphys*, SN, 96.
Eureka, *Graniteville*, SN, 53. This district is in Nevada County. Another Eureka district, in Sierra County, is listed as such in the text.

Feather River, *Oroville*, SN, 103.
Felix, *Hodson*, SN, 64.
Fool's Paradise, *Klamath River*, KM, 139.
Forest, *Alleghany*, SN, 19.
Forest Home, *Irish Hill*, SN, 69.
Forks of Butte, *Magalia*, SN, 88.
Forks of Salmon, *Salmon River*, KM, 141.
Fourth Crossing, *San Andreas*, SN, 114.
Furnier, *Sierra City*, SN, 117.

Garden Valley, *Kelsey*, SN, 84
Garlock, *El Paso Mountains*, BR, 149.
Garrote, *Big Oak Flat*, SN, 30.
Gas Point, *Cottonwood*, KM, 134.
Gaston, *Graniteville*, SN, 53.
Gentry Gulch, *Kinsley*, SN, 85.
Georgia Slide, *Georgetown*, SN, 51.
Glencoe, *West Point*, SN, 129.
Gold Bluff Beach, *Orick*, CR, 180.
Golden Summit, *Kimshew*, SN, 85.
Gold Hill, *Ophir*, SN, 102.
Gold Park, *Twentynine Palms*, MD, 168.

Gold Valley, *Sierra City*, SN, 117.
Goler, *El Paso Mountains*, BR, 149.
Goodyear's Bar, *Alleghany*, SN, 19.
Gottville, *Klamath River*, KM, 139.
Grass Flat, *Port Wine*, SN, 111.
Greeley Hill, *Kinsley*, SN, 85.
Greenhorn, *Klamath River*, KM, 139.
Greenville, *Crescent Mills*, SN, 42.
Groveland, *Big Oak Flat*, SN, 30.

Hamburg, *Klamath River*, KM, 139.
Hansonville, *Brownsville*, SN, 31.
Happy Camp, *Klamath River*, KM, 139.
Harrison Flat, *Sawpit Flat*, SN, 114.
Havilah, *Clear Creek*, SN, 37.
Hayfork, *Trinity River*, KM, 143.
Hazel Valley, *Grizzly Flat*, SN, 61.
Helltown, *Butte Creek*, SN, 32.
Henry Diggings, *Indian Diggings*, SN, 66.
Hexie, *Twentynine Palms*, MD, 168.
Hodges, *Mule Mountains*, MD, 161.
Hornbrook, *Klamath River*, KM, 139.
Howland Flat, *Poker Flat*, SN, 108.
Hughes Creek, *Sycamore Flat*, SN, 124.
Humbug Bar, *Damascus*, SN, 42.
Hunts Hill, *Scotts Flat*, SN, 114.
Hurleton, *Bidwell Bar*, SN, 29.

Indiana Hill, *Gold Run*, SN, 52.
Indiana Ranch, *Dobbins*, SN, 44.
Indian Creek, *Klamath River*, KM, 139, or *Trinity River*, KM, 143.
Indian Gulch, *Hornitos*, SN, 65.
Indian Valley, *Light's Canyon*, SN, 86.
Irishtown, *Pine Grove*, SN, 105.
Italian Bar, *American Camp*, SN, 24.

Jesus Maria, *Rich Gulch*, SN, 112.
Junction City, *Trinity River*, KM, 143.

Kelley, *Argus*, BR, 146.
Kentucky Flat, *Volcanoville*, SN, 127.
Kernville, *Cove*, SN, 42.
Knownothing, *Gilta and Salmon River*, KM, 137 and 141.

Lake, *Mammoth*, SN, 89.
Lancha Plana, *Camanche-Lancha Plana*, SN, 33.
Lewiston, *Trinity River*, KM, 143.
Liberty Hill, *Lowell Hill*, SN, 87.
Logtown, *El Dorado*, SN, 45.
Long Canyon, *Ralston Divide*, SN, 112.
Lost Horse, *Twentynine Palms*, MD, 168.
Lundy, *Homer*, SN, 64.

Melones, *Carson Hill*, SN, 34.
Merrimac, *Granite Basin*, SN, 52.
Mesquite, *Chocolate Mountains*, MD, 154.
Middle Bar, *Camanche-Lancha Plana*, SN, 33.
Milton, *Jenny Lind*, SN, 80.
Minersville, *Trinity River*, KM, 143.
Minnesota, *Alleghany*, SN, 19.
Mogul, *Monitor-Mogul*, SN, 92.
Monitor Flat, *Sawpit Flat*, SN, 114.
Mono Diggings, *Jordan*, SN, 83.
Monona Flat, *Iowa Hill*, SN, 67.
Moonlight Valley, *Light's Canyon*, SN, 86.
Morristown, *Eureka*, SN, 46.
Mount Fillmore, *Poker Flat*, SN, 108.
Mount Ophir, *Mount Bullion*, SN, 94.
Muletown, *Irish Hill*, SN, 69.

Natoma, *Folsom*, SN, 47.
New England Mills, *Colfax*, SN, 38.
Newhall, *Saugus*, TPR ,176.
New York, *Vanderbilt*, MD, 169.
New York Flat, *Brownsville*, SN, 31.
Nimshew, *Magalia*, SN, 88.

Oak Bar, *Klamath River*, KM, 139.
Old Gulch, *Calaveritas*, SN, 33.
Oleta, *Fiddletown*, SN, 46.
Omo Ranch, *Indian Diggings*, SN, 66.
Onion Valley, *Sawpit Flat*, SN, 114.
Ono, *Igo-Ono*, KM, 138.
Orleans Flat, *Moore's Flat*, SN, 93.

Paradise, *Klamath River*, KM, 139.
Peñón Blanco, *Coulterville*, SN, 41.
Piñón, *Twentynine Palms*, MD, 168.
Placerita Canyon, *Saugus*, TPR, 176.
Platina, *Harrison Gulch*, KM, 137.
Pleasant Valley, *Newtown*, SN, 101.

Plymouth, *Jackson-Plymouth*, SN, 69.
Point Sal, *Surf-Point Sal*, CR, 181.
Potter Ridge, *Coarsegold and Grub Gulch*, SN, 38 and 62.
Providence, *Trojan*, MD, 168.

Quaker Hill, *Scotts Flat*, SN, 114.
Quartz Mountain, *Fine Gold*, SN, 47.

Rawhide, *Jamestown*, SN, 77.
Reading Creek, *Trinity River*, KM, 143.
Red Dog, *You Bet*, SN, 131.
Relief, *North Bloomfield*, SN, 101.
Remington Hill, *Lowell Hill*, SN, 87.
Rice, *Montezuma*, TPR, 174.
Robertson Flat, *Canada Hill*, SN, 33.
Rochester, *Stedman*, MD, 167.
Rosamond, *Mojave-Rosamond*, MD, 159.

Sailor Flat, *Canada Hill*, SN, 33.
Salyer, *Trinity River*, KM, 143.
San Gabriel, *Saugus*, TPR, 176.
Sawmill Flat, *Columbia*, SN, 39.
Sawyers Bar, *Salmon River*, KM, 141.
Scales, *Poverty Hill*, SN, 111.
Seiad, *Klamath River*, KM, 139.
Shaws Flat, *Sonora*, SN, 121.
Silverado, *Calistoga*, CR, 178.
Skull Flat, *West Point*, SN, 129.
Sloughhouse, *Michigan Bar*, SN, 90.
Slug Gulch, *Fairplay*, SN, 46.
Smith Flat, *Placerville*, SN, 107.
Snowden, *Salmon River*, KM, 141.
Soledad, *Saugus*, TPR, 176.
Somesbar, *Klamath River*, KM, 139.
South Bullfrog, *Chloride Cliff*, BR, 148.
South Park, *Ballarat*, BR, 146.
Spanish Dry Diggings, *Greenwood*, SN, 60.

Spanish Ranch, *Meadow Valley*, SN, 89.
Springfield, *Columbia*, SN, 39.
Squabbletown, *Columbia*, SN, 39.
Star Town, *Last Chance*, SN, 86.
Stent, *Jamestown*, SN, 77.
St. Louis, *Port Wine*, SN, 111.
Stuart Fork, *Trinity River*, KM, 143.
Sutter Creek, *Jackson-Plymouth*, SN, 69.
Sweetland, *North San Juan*, SN, 102.

Timbuctoo, *Smartsville*, SN, 120.
Tinemaha, *Fish Springs*, BR, 149.
Todd Valley, *Forest Hill*, SN, 49.
Towle, *Dutch Flat*, SN, 45.
Trinity Center, *Trinity River*, KM, 143.
Tujunga, *Azusa-Tujunga*, TPR, 170.
Tumco, *Cargo Muchacho-Tumco*, MD, 153.
Tuolumne, *Soulsbyville*, SN, 121.
Tucki Mountain, *Skidoo*, BR, 151.

Valley Springs, *Campo Seco-Valley Springs*, SN, 33.
Virgilia, *Butt Valley*, SN, 32.
Virginia Dale, *Dale*, MD, 157.

Wallace, *Camanche-Lancha Plana*, SN, 33.
Washington, *Sheep Ranch*, SN, 115.
Weitchpec, *Klamath and Trinity Rivers*, KM, 139 and 143.
Whiskey Diggings, *Gibsonville*, SN, 51.
Whiskeytown, *Shasta-Whiskeytown*, KM, 142.
Willow Creek, *Trinity River*, KM, 143.
Willow Valley, *Nevada City*, SN, 97.
Woolsey Flat, *Moore's Flat*, SN, 93.
Wyandotte, *Bangor-Wyandotte*, SN, 29.

Yankee Jims, *Forest Hill*, SN, 49.
Yreka, *Klamath River*, KM, 139.

o

printed in CALIFORNIA OFFICE OF STATE PRINTING

ILLUSTRATIONS

Figures

Photos

Photos—Continued

Tables

www.ingramcontent.com/pod-product-compliance
Lightning Source LLC
Chambersburg PA
CBHW051213200326
41519CB00025B/7094